W0018673

Composite and Composite Coatings

Composite and Composite Coatings

Mechanical and Tribology Aspects

Edited by
Mohamed Thariq Hameed Sultan
S. Arulvel
K. Jayakrishna

CRC Press
Taylor & Francis Group
Boca Raton London New York

CRC Press is an imprint of the
Taylor & Francis Group, an informa business

First edition published 2022
by CRC Press
6000 Broken Sound Parkway NW, Suite 300, Boca Raton, FL 33487-2742

and by CRC Press
2 Park Square, Milton Park, Abingdon, Oxon, OX14 4RN

© 2022 selection and editorial matter, Mohamed Thariq Hameed Sultan, S. Arulvel and K. Jayakrishna; individual chapters, the contributors

CRC Press is an imprint of Taylor & Francis Group, LLC

Reasonable efforts have been made to publish reliable data and information, but the author and publisher cannot assume responsibility for the validity of all materials or the consequences of their use. The authors and publishers have attempted to trace the copyright holders of all material reproduced in this publication and apologize to copyright holders if permission to publish in this form has not been obtained. If any copyright material has not been acknowledged please write and let us know so we may rectify in any future reprint.

Except as permitted under U.S. Copyright Law, no part of this book may be reprinted, reproduced, transmitted, or utilized in any form by any electronic, mechanical, or other means, now known or hereafter invented, including photocopying, microfilming, and recording, or in any information storage or retrieval system, without written permission from the publishers.

For permission to photocopy or use material electronically from this work, access www.copyright.com or contact the Copyright Clearance Center, Inc. (CCC), 222 Rosewood Drive, Danvers, MA 01923, 978-750-8400. For works that are not available on CCC please contact mpkbookspermissions@tandf.co.uk.

Trademark notice: Product or corporate names may be trademarks or registered trademarks and are used only for identification and explanation without intent to infringe.

Library of Congress Cataloging-in-Publication Data
A catalog record has been requested for this book

ISBN: 978-0-367-62567-2 (hbk)
ISBN: 978-0-367-62568-9 (pbk)
ISBN: 978-1-003-10972-3 (ebk)

DOI: 10.1201/9781003109723

Typeset in Times
by Newgen Publishing UK

Contents

Preface

The application of composite materials has progressed extensively for the last few decades. This can easily be seen in structural applications such as in the construction, automotive, railway, maritime and aerospace industries. As the demands on composite materials increase, composite coatings have also become important. Composite coatings are a series of protective layers applied on any materials such as steel and concrete which are designed to meet the high standards and rigid requirements of many different industries. This book provides an insight into composite and composite coatings with a focus on the mechanical and tribological aspects. It covers in detail diverse topics such as extraction, treatment and applications of bio fiber composites; tribology properties and behaviour of fiber-reinforced polymer composites; effect of reinforcements on the tribological properties of polymer composites; mechanical and tribological behaviour of particulate-reinforced metal matrix composite; and many more topics related to mechanical and tribological aspects of composite and composite coatings.

The editors extend their sincere gratitude to all the authors who contributed to this book's chapters and provided their valuable findings, ideas and knowledge in the field of composite and composite coatings. The editors managed to gather all recognised researchers from India, the United Arab Emirates (UAE) and South Africa in the area of composite and composite coatings, mainly in the area of mechanical and tribological aspects, and finally complete this book in a most fruitful way. We greatly appreciate all the commitment and support given to compile their findings in this book. Last but not least, we are thankful to the CRC team for their generous cooperation at every stage of this book's production.

Editors:
Mohamed Thariq Hameed Sultan
S. Arulvel
K. Jayakrishna

Editors' Biographies

Prof. Ir. Ts. Dr. Mohamed Thariq Hameed Sultan completed his PhD in 2011 from the Department of Mechanical Engineering, University of Sheffield, United Kingdom, in the Field of Mechanical Engineering. He specializes in the field of Hybrid Composites, Advanced Materials, Structural Health Monitoring and Impact Studies. He is also a Professional Engineer (PEng) registered under the Board of Engineers Malaysia (BEM), and he is a Chartered Engineer (CEng) registered with the Institution of Mechanical Engineers (IMechE) United Kingdom. Recently, he was awarded recognition as a Professional Technologist (PTech) by the Malaysian Board of Technologists (MBOT). He has published more than 220 journal articles and 15 books internationally.

Dr. S. Arulvel was born in Vellore District, Tamil Nadu, India. He completed his undergraduate (UG) work (Aeronautical) in 2010 at the Hindustan College of Engineering, Tamil Nadu, and postgraduate work (PG) (Mechanical) in 2014 at Anna University, India. From 2015 to 2017, he served as a junior research fellow and senior research fellow (2017–2019) at Anna University, Chennai, India, under the University Grants Commission (UGC) scheme of the government of India. Currently he is Assistant Professor in the School of Mechanical Engineering, VIT, Vellore, India. He has five years of research experience with high impact journals of various research areas including Materials Engineering, Mechanical Behavior of Materials, Material Characterization, Metallurgical Engineering, Surface Engineering and Manufacturing Engineering. He also serves as reviewer and editor for various national and international journals and conferences. He has published various research articles on composites, surface coating and friction–wear applications. He also received a best researcher award in surface coating given by the International Journal for Research under Literal Access accredited by the World Research Council.

Dr. K. Jayakrishna is Associate Professor in the School of Mechanical Engineering at the Vellore Institute of Technology University, India. Dr. Jayakrishna's research is focused on the design and management of manufacturing systems and supply chains to enhance efficiency, productivity and sustainability performance. His more recent research is in the area of developing tools and techniques to enable value creation through sustainable manufacturing, including methods to facilitate more sustainable product design for closed-loop material flow in an industrial symbiotic setup, and developing sustainable products using hybrid bio composites. He has mentored undergraduate and graduate students (2 M.Tech Thesis and 24 B.Tech Thesis), who have so far had 40 journal publications in leading SCI/SCOPUS Indexed journals, 17 book chapters, and 85 refereed contributions to conference proceedings. Dr. Jayakrishna's team has received numerous awards in recognition of the quality of their work. He teaches undergraduate and graduate courses in the manufacturing and industrial systems area, and his initiatives to improve teaching effectiveness have been recognized through national awards. He has also been awarded the Institution of Engineers (India) Young Engineer Award in 2019 and the Distinguished Researcher Award in the field of Sustainable Systems Engineering in 2019 by the International Institute of Organized Research.

Contributors

Adel Alblawi
Department of Mechanical Engineering,
College of Engineering, Al-Dawadmi
Shaqra University, Saudi Arabia.

L. Arulmani
Department of Mechanical Engineering,
RR Institute of Technology,
Bangalore, India.

Vijay Chaudhary
Department of Mechanical Engineering,
Amity School of Engineering &
    Technology,
Amity University, Noida, India.

Kadiyala B. Drupad
Department of Mechanical Engineering,
GITAM School of Technology,
Bangalore, India

A. Elayaperumal
College of Engineering, Guindy,
Anna University, India.

P. M. Gopal
Department of Mechanical Engineering,
Karpagam Academy of Higher
    Education,
Coimbatore, India.

S. Darius Gnanaraj
Vellore Institute of Technology,
India.

Sumit Gupta
Department of Mechanical Engineering,
Amity School of Engineering &
    Technology,
Amity University, Noida, India.

M. S. Jagatheeshwaran
S. A Engineering College,
Thiruverkadu, Chennai,
Tamil Nadu, India.

R. Kamatchi
Vellore Institute of Technology,
Vellore, India.

K. Gopi Kannan
Vellore Institute of Technology,
Vellore, India.

V. Kavimani
Department of Mechanical Engineering,
Karpagam Academy of Higher Education,
Coimbatore, India.

M. Wasim Khan
College of Engineering, Guindy,
Anna University, India.

R. F. Laubscher
Department of Mechanical Engineering
    Science,
University of Johannesburg,
Auckland Park Kingsway Campus,
Johannesburg, South Africa.

R. Palanivel
Department of Mechanical Engineering,
College of Engineering, Al-Dawadmi
Shaqra University, Saudi Arabia.

Lavuluri Thrinai Pavan
Department of Mechanical Engineering,
GITAM School of Technology,
Bangalore, India.

T. Ram Prabhu
Defence R&D Organization,
India.

M. Sivanesh Prabhu
College of Engineering, Guindy,
Anna University, India.

K. Soorya Prakash
Department of Mechanical Engineering,
Anna University Regional Campus,
Coimbatore, India.

J. Ashok Raj
Department of Automobile Engineering,
Bharath Institute of Higher Education
   and Research,
Chennai, India.

Mohammad Abdur Rasheed
Department of Civil Engineering,
College of Engineering, Al-Dawadmi
Shaqra University, Saudi Arabia.

J. Jenix Rino
Associate Professor and Head,
Department of Mechanical Engineering,
Stella Mary's College of Engineering,
Kanyakumari District, Tamil Nadu, India.

D. Dsilva Winfred Rufuss
Vellore Institute of Technology,
Vellore, India.

M. Satthiyaraju
Department of Mechanical Engineering,
National Institute of Technology,
Tiruchirappalli, India.

P. Seenuvasaperumal
School of Mechanical Engineering,
Vellore Institute of Technology,
Vellore, India.

C. Chinthanai Selvan
Department of Production Engineering,
National Institute of Technology,
Tiruchirappalli, India.

A. Tajdeen
Bannari Amman Institute of Technology,
Sathy, Tamil Nadu, India.

N. Thangapandian
Department of Mechanical Engineering,
St. Joseph's Institute of Technology,
Chennai, India.

Titus Thankachan
Department of Mechanical Engineering,
Karpagam College of Engineering,
Coimbatore, India.

S. Vigneshwaran
Department of Mechanical Engineering,
National Institute of Technology,
Puducherry, Karaikal, India,

A. S. Vivekananda
Department of Mechanical Engineering,
College of Engineering, Guindy,
Chennai-25, Anna University, India.

1 Extraction, Treatment and Applications of Bio Fiber Composites
A Critical Review

S. Sathish,[1]* L. Prabhu,[1] S. Gokulkumar,[1]
N. Karthi,[1] D. Balaji,[1] N. Vigneshkumar[1]

[1]Department of Mechanical Engineering, KPR Institute of Engineering and Technology, Coimbatore, 641407, India.
*Corresponding author: sathi175@gmail.com

CONTENTS

1.1   INTRODUCTION

Cellulose-based fibers such as abaca, jute, hemp, bamboo, flax, kenaf, banana, bagasse, sisal and cotton are used as reinforcing elements for polymer matrix composites because of their lower weight, moderate stiffness, lower emissions needed for production, harmlessness, lower energy requirements for extraction, recyclability and biodegradability (Harish et al., 2009). Bio fibers are sustainable materials having excellent physical and mechanical properties compared to synthetic fibers. Greater environmental awareness among material researchers and automakers

DOI: 10.1201/9781003109723-1

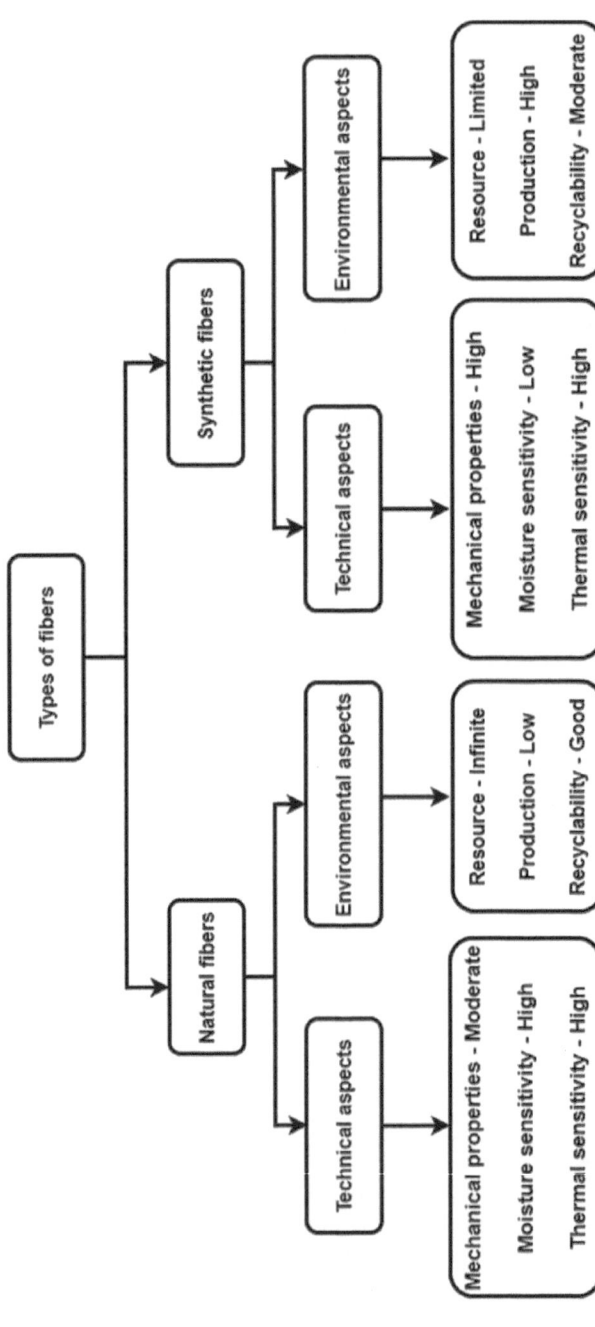

FIGURE 1.1   Technical and environmental aspects of fibers – plant and man-made.

have accelerated the development of natural fiber–reinforced composites (NFRC). Figure 1.1 shows the technical and environmental aspects of plant and man-made fibers. Government environmental guidelines have pushed many industrial sectors like aerospace, automobile, building, packaging, leisure and sports to look for sustainable and biodegradable materials (Huda et al., 2008a). Currently, bio fiber–based materials are widely used in the automobile and aircraft industries. This is the right time to investigate at the next level for novel bio fibers with required mechanical properties. Natural fibers consist of cellulose, hemicelluloses, pectin, lignin, wax and oils which cover the fiber surface (Hamidon et al., 2019, Sathish et al., 2014a, Sathish et al., 2017 and Kumaresan et al., 2015). The higher cellulose content, smaller fiber diameter, smaller spiral angle and longer fiber are advantageous features of natural fibers. Bio fiber–reinforced polymer is one of the promising replacements in reducing the ecological impact of synthetic fibers in polymers.

There is increasing awareness about the usage of cellulose-based fibers as reinforcement material. Cellulose fiber–reinforced composite companies are rising around the globe. The bio fiber–reinforced composite industry was expected to increase by up to 10% globally between 2011 and 2016 (Uddin, 2013). Bio fibers are not much denser than the petroleum-based fibers. Petroleum fiber–reinforced composites could be replaced by bio fiber–reinforced composites because of their low density, recyclability, biodegradability, availability, moderate strength and modulus, lower cost, user friendliness and lower energy consumption (Sathish et al., 2018, Prabhu et al., 2020, Gokulkumar et al., 2021, and Gokulkumar et al., 2020). Bio fiber–embedded composites are trendsetters which have greater development potential in the next few decades.

The physical and mechanical characteristics of plant fiber–embedded composites can be influenced by several factors including the fiber extraction method, moisture content and microfibrillar angle of fiber, fiber type and reinforcement form, fiber and matrix ratio, compatibility between reinforcement and polymer, manufacturing techniques, fiber direction or orientation, growing climate and cultivation method, and type of chemical agents used in the fibers (Pickering et al., 2016). The major problems in the extensive usage of these bio fibers in diverse polymer matrices are the lower fiber-matrix compatibility, poor dimensional stability and the intrinsic high water absorption which tends to lead to thick swelling in the bio fiber–reinforced composites (Sathish et al., 2014b, Gokulkumar et al., 2019, Prabhu et al., 2019). The interface between the fiber and binder is an important parameter for the composite, and effective bonding strength is a significant property of composites.

Chemical treatments can alter the surface of a fiber and its aspect ratio. Different chemical agents are available to increase the interaction between fiber and binder and to reduce the moisture assimilation by these fibers in order to improve mechanical and physical characteristics of the composites. Chemical treatment includes alkaline, silane, isocyanates, acetylation maleated coupling agents, benzoylation and acrylation. The major hitch of bio fibers is that their hydrophilic nature diminishes interaction with hydrophobic polymers while the composite is prepared. Figure 1.2 lists the different types of plant fibers and binding materials (Verma et al., 2017, Agarwal et al., 2020, Anandjiwala et al., 2007, Ramamoorthy et al., 2015, Sanjay et al., 2015, Jawaid et al., 2011, Rohit et al., 2016, L. Mohammed et al., 2015, Faruk et al., 2012, Shekar et al., 2018, Ramesh et al., 2017, Mukherjee et al., 2011, Hassan et al., 2017). Figure 1.3 shows a comparison of plant and glass

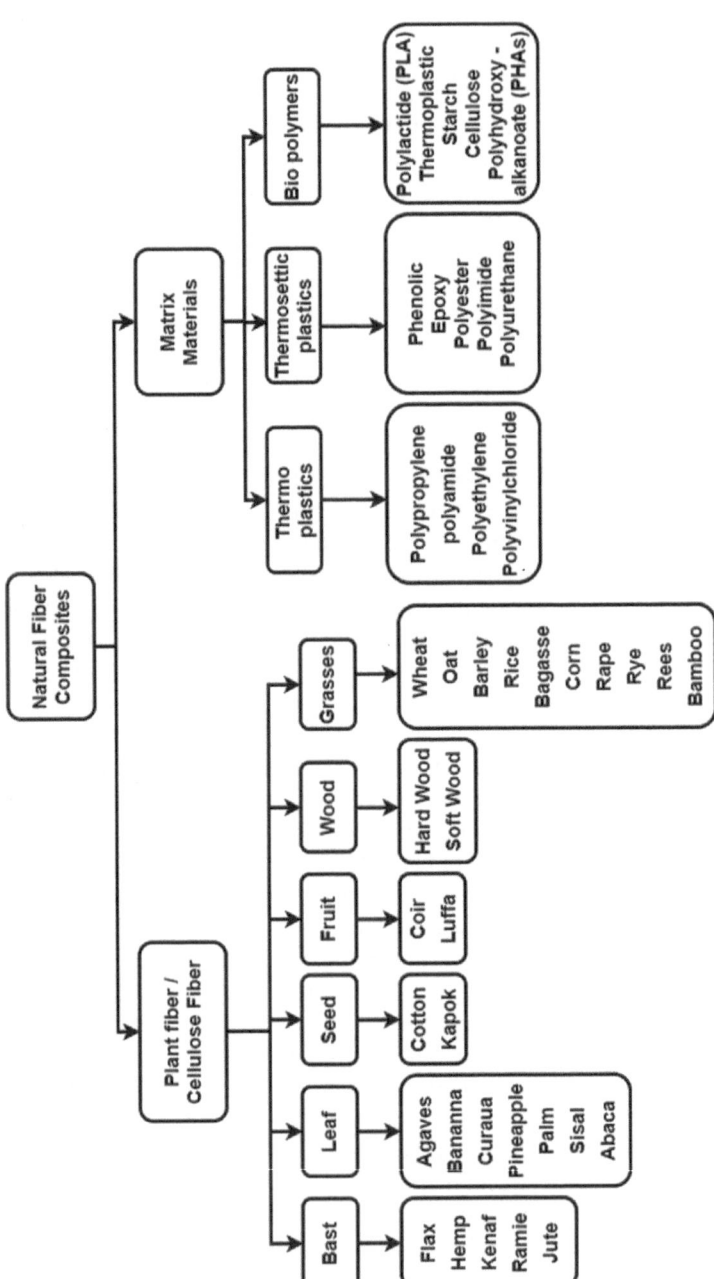

FIGURE 1.2 Diverse types of plant fibers and binding materials.

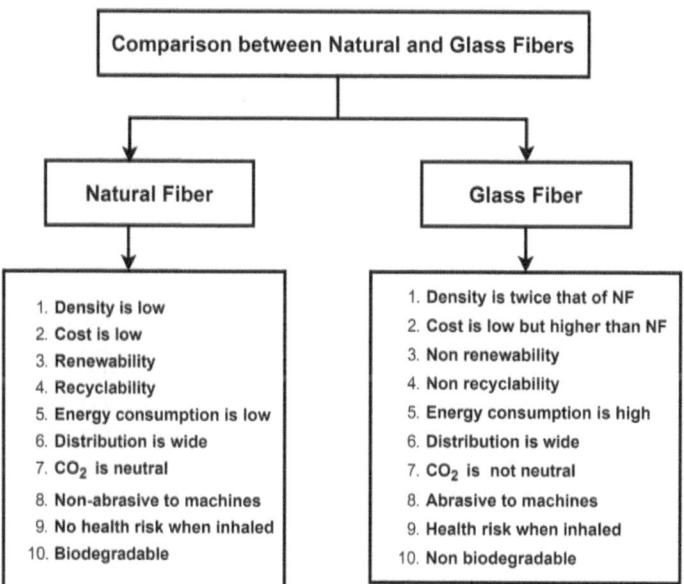

FIGURE 1.3 Plant fiber comparison with glass fiber.

fibers (Bledzki et al., 1999, Wanmbua et al., 2003). Table 1.1 represents the mechanical properties of plant fibers. Poor interaction between plant fiber and binder results in less desirable mechanical and thermal characteristics. The fiber surface modifications brought about by the chemical treatments introduce new moieties which successfully interconnect with the resin. This chapter focuses on methods of extracting fibers, different ways of treating them and their industrial applications.

1.2 RETTING PROCESS

The most familiar techniques to extract the natural fibers are the dew retting and water retting processes. Bio fibers are removed from plant leaves a few days after harvesting, which are then washed numerous times with water and allowed to dry in air to obtain clean natural fiber. Depending on the fiber type and quality, these techniques need about 14 to 28 days for the breakdown of hemicellulose, lignin, pectin and waxes (Hulle et al., 2015). The extraction time can be reduced by other techniques; for example, mechanical decortications and chemical treatments have been introduced. Yusof et al. (2015) have developed a novel mechanical decorticator system for pineapple fiber extraction. The extraction here is based on a scratching mechanism in which the decorticator blade removes some of the waxy layers on the leaves' surface. Later the leaves are pulled off the machine by hand and nearly all of the waxy part is removed. The fibers which were extracted are washed completely and allowed to dry in sunlight. Plant fibers are derived from the leaf periphery which has definite form. During the retting process the extraction of single fibers from the plant is obtained after the removal of substances like microbes, covering materials and

TABLE 1.1

Mechanical Properties of Plant Fiber

Plant Fiber	Density (g/cm³)	Tensile Strength (MPa)	Young's Modulus (GPa)	Elongation (%)
Flax	1.38–1.5	700–1500	37–55	2.5
Kenaf	1.2	295–980	41	1.7–2.7
Hemp	1.35	550–1110	45–70	1.6–3.0
Jute	1.23–1.46	325–800	10–55	1.8–2.5
Sisal	1.2–1.33	460–855	15.5–38	2–3
Pineapple	1.5	900–1600	70–82	0.8–3.0
Banana	1.3–1.35	355–900	29–33.8	2–2.5
Nettle	1.5	650–700	38	1.7
Abaca	1.5	410–980	41	3.4–3.8
Coir	1.2–1.25	140–220	6	15–27.5
Henequen	1.4	430–500	13–15	3–4.8
Ramie	1.44–1.5	500–915	23–44	2–3.7
Cotton	1.21–1.51	250–500	6–12	3–10
Bagasse	1.2	290	17–27.1	1.1

Sources: Njuguna et al., 2011, Akil et al., 2011, Alves et al., 2010.

bonding agents which are those surrounding the fiber (Gurukarthik et al., 2019). The response time ought to be closely monitored in dew or water retting because extreme retting can cause the dissociation of single fibers or might affect the quality of the fiber (Manimaran et al., 2018). Figure 1.4 shows the four natural fiber extraction techniques: dew retting, water retting, mechanical decortication, and fiber drying. Table 1.2 summarizes the diverse extraction techniques of dew retting, water retting and mechanical extraction.

1.3 CHEMICAL TREATMENTS

The natural fiber–reinforced composites show problems in several areas, including poor wettability, low fiber/polymer matrix performance, and also greater absorption of water and moisture. The hydrophilic characteristic of the plant fibers results in a less desirable fiber-matrix interface. Therefore, amending the surface of the fiber by treating it with chemicals is indispensable (Sanjay et al., 2018a, Rangappa et al., 2018, Gowda et al., 2018). The chemical treatment of bio fibers improves the interface between fiber and binder, which augments the composite's thermal and mechanical characteristics. Composite materials made with reinforcing natural fibers that are chemically treated have higher tensile, bending, impact and interlaminar shear strength as well as greater hardness compared with the untreated composites. The manufacturing of composites which possess better mechanical strength, chemical treatments of natural fiber are processed to decrease the hydrophilic property and enhances moisture absorption of fibers (John et al., 2008a and George et al., 2014). The

FIGURE 1.4 Natural fiber extraction techniques: (A) dew retting, (B) water retting, (C) mechanical decortication, (D) fiber drying.

chemical treatment of ramie fiber reveals that neither saline nor alkaline nor a combination treatment contributes to the enhancement of mechanical properties (Sanjay et al., 2018b). Chemical treatment is one of the vital processes used to diminish the hydrophilic characteristics of the fibers and enhance compatibility with the polymers. Changes in surface morphology and microstructure are observed while treating the natural fibers; these changes are due to the subtraction from the natural fibers of hemicelluose, lignin, pectin and wax elements. Notable augmentation in the mechanical and physical characteristics of the composites has been reported along with improved thermal stability for the composites reinforced by natural fibers (H. Chen et al., 2018). The chemical treatment fibers alter the structure as well as properties (Y. Chen et al., 2018). One amongst the notable chemical agents used in this process is sodium hydroxide (NaOH) (Sgriccia et al., 2008).

1.3.1 ALKALINE TREATMENT

Natural fibers are drenched for three hours in a bucket that contains 5% NaOH solution, which removes the surplus acid present in the fibers. The plant fibers are rinsed completely in distilled water. Finally, the fiber is washed with a water solution containing 1% acetic acid. Fibers are dried for 24 hours in sunlight and placed in an oven at 60°C for about 3 hours (Sathishkumar et al., 2013).

TABLE 1.2
Comparison between Dew Retting, Water Retting and Mechanical Decortication Methods (38)

Extraction Methods	Dew Retting	Water Retting	Mechanical Decortications
Description	The stems collected from the plants are uniformly distributed over the surface of grass, which allows fibers to be extracted with the aid of atmospheric air, bacteria, dew and sunlight.	Stems from plant need to be treated in various sources of water and examined periodically.	Feeding of fibers continuously between rotating rollers, which removes external layers like the gums and stem skin.
Duration	14–21 days	7–14 days	Fewer days compared to other two methods.
Advantage	Takes advantage of large spaces and warm days for efficient drying.	• Uniform diameter • Light in quality	• Can produce large quantities of fibers in short time • Good quality of fibers
Disadvantage	• Darker in colour • Poor quality • Long time for processing	• High cost, environmental concerns and inferior fibers quality • Requires high water treatment maintenance	• High cost • Required rollers and mechanical elements – rotor, blade, rope, etc.

Source: Sathishkumar et al., 2013.

1.3.2 PEROXIDE TREATMENT

Natural fibers are drenched in 6% benzoyl peroxide (BP) in acetone for 30 minutes and then continuously rinsed in distilled water and dried in an oven at 45°C for about 24 hours (Sathishkumar et al., 2013).

1.3.3 BENZOYLATION TREATMENT

Natural fibers are drenched in acetone for 30 minutes in 5% benzoyl chloride (BC). In some cases fibers are rinsed in distilled water and dried in an oven at 70°C for about 24 hours (Sathishkumar et al., 2013).

1.3.4 PERMANGANATE TREATMENT

Natural fibers are drenched for 30 minutes in acetone at 0.05% $KMnO_4$. Fibers are removed and rinsed in distilled water and dried in an oven at 70°C for about 24 hours (Sathishkumar et al., 2013).

1.3.5   STEARIC ACID TREATMENT

In ethynol, 1% of stearic acid is dissolved and the solution is slowly introduced into the plant fibers in a continuous stirring stainless steel (SS) vessel. Treated fibers are then dried at 80°C in the oven for 45 minutes.

1.3.6   CHEMICAL TREATMENT RESULTS

The alkaline treatment removes hemicellulose, pectin, impurities, lignin and fats and exposes cellulose from the natural fibers, thus elevating the natural fibers' surface roughness. Removal of lignin and hemicellulose from the plant fiber also allows the active hydroxyl groups to react better with the polymers, which augments the bonding between reinforcement and polymer matrix. The lignin and hemicellulose are also partially separated from the natural fiber by acetylation. Silane treatment involves the usage of silane as a coupling agent which enhances the fiber-matrix compatibility (Albinante et al., 2013, and John et al., 2008a). It was shown that alkaline treatment, by discarding hemicellulose and lignin, diminishes the fiber diameter with escalating surface roughness. The impact of chemical treatments on the mechanical and thermal characteristics of bio fibers was studied by many researchers. The literature shows that 2% and 5% NaOH enhance the mechanical stability of the composites (Owen et al., 2015). The simplest and cheapest approach for altering natural fiber surface is alkaline treatment or mercerization. It eliminates the non-cellulose elements in fibers such as lignin, wax and oils, causing ionization of cellulose hydroxyl groups to alcoxide and decreasing the quantity of the hydroxyl groups in the fiber surface (Al-Maharma et al., 2019). Alkaline treatment develops the fiber's surface roughness and hydrophobicity and thereby escalates the adhesion of fibers with the polymer binder. Many authors used alkali treatment to amend the plant fiber's surface, and testing showed the alkali-treated fibers possess exceptional mechanical strength, low water absorption and thermal stability (H. Chen et al., 2018 and Sayanjali Jasbi et al., 2018). During the chemical treatments, the fiber surface is cleaned to confirm that there are no impurities which escalate the roughness of the fiber. The treatments also improve the moisture resistance properties by reducing the OH groups on the fiber structure in a reaction given in the following equation (Kabir et al., 2012).

$$Fiber-OH+NaOH \rightarrow Fiber-O^-Na^+ +H_2O \qquad (1)$$

Chattopadhyay et al. (2009) reported the contribution of alkaline-treated pineapple leaf fiber and maleic anhydride–grafted polypropylene compatibilizer to a major enhancement in the tensile characteristics of polypropylene matrix. Prasad et al. (2017) reported that by introducing a variety of chemical treatments like acetylation, mercerization, peroxide, silane, and benzoylation treatment, hexadecyl pyridinium chloride and permanganate treatment, the bonding between the sisal fiber and polyesters is enhanced. Huda et al. (2008b) investigated the maleated anhydride–treated pineapple leaf fiber–PLA composite and found improved mechanical properties, thermal stability and bonding strength compared with untreated fiber composites. Kaewpirom et al. (2014) also revealed that the modified pineapple leaf fiber with maleated anhydride coupling agent exhibits enhanced adhesion with PLA polymer, which improves the tensile properties. Shinoj et al. (2011) revealed the influence of coupling agents like polyethylene maleic anhydride copolymer (MAPE) and a few chemical treatments like peroxide, benzoylation, mercerization, acetylation and silane that diminish moisture absorption by the bio fibers. Treatment techniques using benzoylation and acetylation are widely exploited by researchers to modify natural fibers. A less polar acetyl group is introduced to the fiber surface in the acetylation treatment method, whereas a benzoyl group (C_6H_5CO) is introduced in benzoylation. The acetylated flax fiber is highly tensile, has good flexural strength and has excellent thermal stability. Ferreira et al. (2019) indicated that the benzoylation treatment causes chemical changes on the sisal fiber which are responsible for its enhanced adhesion to the polymer matrix as confirmed by Fourier transform infrared (FT-IR) spectroscopy and scanning electron microscopy (SEM) analysis. Gonzalez et al. (2015) states that sisal fibers treated with *Penicillium echinulatum* cellulose show better thermal stability and crystallinity than untreated fibers. Ahmad et al. (2012) states that sisal fiber treated using peroxide has high tensile modulus and strength. Sari et al. (2019) exposed that the flexural and tensile characteristics of a composite of peroxide-treated sisal fiber with polyethylene increase compared with untreated composites. Enhancement of the flexural and tensile properties is due to polyethylene being grafted with peroxide-treated sisal fiber. Stearic acid treatment makes the fiber's surface more hydrophobic and thus enhances interaction with the binder. The increase in surface roughness of enzyme-treated fibers is a result of the elimination of impurities such as wax, hemicellulose, pectin, lignin and oil. Shinoj et al. (2011) revealed that a coupling agent like MAPE and a few chemical treatments like silane, benzoylation, acetylation, mercerization and peroxide diminish the moisture-absorbing tendency of the natural fibers. Thiruchitrambalam et al. (2009) revealed that alkali treatment enhances mechanical characteristics of banana and kenaf plant fibers reinforced with polyester hybrid to form composites. Treating the banana and kenaf fibers was carried out with 10% NaOH for 30 minutes. The effective treatment of the plant fiber–reinforced composite with an alkaline-based chemical like 5% NaOH for 1 hour in the temperature range of 30°C enhances the hydrophilic fiber–hydrophobic binder interaction, which in turn enhanced the mechanical strength (Mishra et al. 2001). Isa et al. (2014) studied the influence of chemical agents on the tensile and flexural properties of reinforced hybrid epoxy composites of okra and glass fibers. It was revealed that chemical agents enhance tensile and flexural strength by escalating bonding between reinforcement and matrix. M. H. Mohammed et al. (2014) examined the impact of alkali treatment for the mechanical ability of sisal fiber/unsaturated polyester composite materials and established that flexural, impact and tensile ability was enhanced after treatment owing to elimination of impurities. Figure 1.5 reveals the structure of alkali-treated and untreated fibers.

The critical assessment from the literature is that the mechanical characteristics are significantly improved by treating the bio fibers with various chemicals at diverse concentrations. The effect of alkaline treatment was especially significant. The concentration mostly of 5% NaOH was used to alter the fiber surface for preparing the composite. With concentration of 5% NaOH, the fiber diameter is reduced, resulting in enhanced bonding between fiber and binder owing to augmented fiber contact area

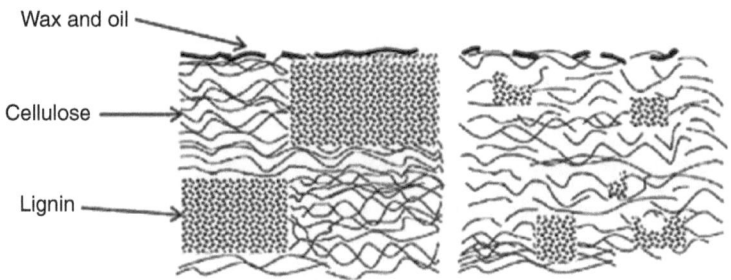

Wax and oil

Cellulose

Lignin

FIGURE 1.5 Structure of natural fiber: untreated and alkali treated.

Source: Amiandamhen et al., 2020.

as well as the aspect ratio. High NaOH concentration may cause surplus removal of covering components from the plant fiber which diminishes the mechanical capability (Vijay et al., 2019). Impacts of chemical treatments of natural fibers are tabulated in Table 1.3.

1.4 APPLICATIONS

The new trends in the improvement of plant fiber–reinforced composites have led to the substitute of composites reinforced by manmade fibers like carbon, Kevlar, boron and glass, for example, in automobile headliner panels, boot (trunk) linings, pillar cover panels, vehicle hat rack, windshields, dashboards, bumpers, bicycle frames, boat hulls, submersible pressure hulls and fishing rods (Siengchin, 2017 and Sanjay et al., 2016). Recently, these natural fibers were used in aerospace, automotive fields, sports equipment, house products, pipeline industries, packaging, paper industries and architectural structures (Puttegowda et al., 2018). Automobile, sports, marine, aircraft and musical industries are actively producing composites with different types of natural fibers for their products (Steffens et al., 2017, Ticoalu et al., 2010 and Sen et al., 2011a). The synthetic fiber–reinforced composites used in the architecture industry could be replaced by natural fiber–reinforced composites as an alternate source for walling, flooring, shielding and sunscreens (Sen et al., 2011b). Linen fibers were used as reinforcement in the Green Bente 24 yacht (Sathish et al., 2020). Flax, kenaf, rice husk, hemp, bamboo, jute and sisal are used in different applications like structural elements, worktables, skateboards, fishing rods, monobloc chairs and electric panels (Wang et al., 2009). Table 1.4 shows applications of NFRC. Figure 1.6 shows industrial applications of NFRC. Figure 1.7 shows the potential applications of NFRC (Holbery et al., 2006, Jauhari et al., 2015, John et al., 2008a, Zhao et al., 2014, Sanjay et al., 2016, Furtado et al., 2014, Norhidayah et al., 2014, Suddell, 2008).

The Ford company made floor trays for the Mondeo model using kenaf-reinforced polypropylene (PP) composites. Lotus uses natural fiber–reinforced composites to make door panels, trunk liners, interior covers and spoilers. Vauxhall utilized natural fibers to produce pillar cover casings, dashboards and headliner casings in its Corsa, Zafira and Vectra models. Toyota made spare-tire covers and door panels for the Brevis

TABLE 1.3
Impacts of Chemical Treatments of Natural Fibers

Fibers Used	Type of Chemical Treatment	Effects	Reference
Tridax procumbens fibers (TPFs)	Alkaline treatment	Enhance the thermal stability and tensile strength	Vijay et al., 2019
Coir fiber	5% NaOH	Increase the mechanical properties and reduce hydrophilicity	Rahman et al., 2007
Ramie fiber	5% NaOH	Increase the mechanical properties	Goda et al., 2006
Doum fiber	5% NaOH	Improve the tensile modulus and flexural modulus	Arrakhiz et al., 2013
Parthenium hysterophorus fibers (PHF)	Alkaline treatment	Increase the surface roughness and tensile strength	Vijay et al., 2021
Flax fiber	Alkaline treatment	Increase the mechanical properties	Sathish et al., 2018
Pineapple fiber	Alkaline treatment	Improve the thermal stability and tensile strength	Panyasart et al., 2014
Sisal fiber	Alkaline treatment	Increase the corrosion resistance and reduce water absorption	Bisanda, 2000
Vetiver fiber	5% NaOH	Increase the tensile and flexural strength	Bavan et al., 2014
Flax fiber	MAH-PP treated	Increase the tensile strength	Bledzki et al., 2004
Flax fiber	5% NaOH	Increase the tensile strength and tensile modulus	Gassan et al., 1999
Bark fiber	Alkaline treatment	Improve the mechanical properties	Donnell et al., 2004
Passiflora foetida fiber	Alkaline treatment	Enhance the mechanical properties	Akin, 2010
Sisal fiber	Benzoylation treatment	Increase the tensile strength	Joseph et al., 1995
Sisal fiber	Isocyanate treatment	Increase the tensile strength	Joseph et al., 1995
Phoenix pusilla fibers (PPFs)	Alkaline treatment	Improve the interfacial shear strength	Madhu et al., 2019
Sisal fiber	Dicumyl peroxide	Improve the mechanical properties	Sanjay et al., 2019b
Sisal fiber	Polypropylene graft	Increase the impact strength	Sanjay et al., 2019b

TABLE 1.4
Applications of NFRC

Fiber/Matrix	Components	Reference
Flax fiber/Epoxy	Aerospace: military aviation fuselage, cargo liners, rotor blades and hubs	Karthi et al., 2020
Coir fiber/Rubber/Epoxy	Domestic products: mat	Karthi et al., 2020
Coconut fiber/Polyethylene	Automotive parts	Brahmakuamr et al., 2005
Kenaf fiber/PLA	Military aviation: rotor blades	Shinji Ochi, 2008
Snake grass fiber/Isophthallic polyester	Packing applications	Sathishkumar et al., 2012
Non-woven hemp fiber/Unsaturated polyester	Domestic products: Mat	Dhakal et al., 2007
Sisal fiber/Polypropylene (PP)	Vehicle spare parts	Bakare et al., 2010
Sugar palm fiber–reinforced composites	Aerospace industries	Sahari et al., 2013
Moroccan hemp fibers/Polypropylene	Headliners and car interiors	Elkhaoulani et al., 2013
Sisal or glass/Filler or epoxy	Chairs, tables and automobile car doors	Arpitha et al., 2017

model using a natural fiber–reinforced composite. Volvo used plant fibers to make seat padding, cargo floor trays and natural foams in V70 and C70 models. Daimler Chrysler made car windshields, business tables and pillar cover panels of the A class, C class, E class and Evo bus models using plant fiber–reinforced composites. Renault, Peugeot, Saab, Mitsubishi and Rover used various plant fibers to make automobile components such as roofing sheets, decking, window frames, truck liners and parcel shelves (Pickering, 2008, Mohanty et al., 2005, Sen et al., 2011a, Mwaikambo, 2006, Kalia et al., 2011, John et al. 2008b). Sisal fiber–reinforced products in construction applications like panels, roofing sheets, fencing, door frames and shutters are light in weight for the building of less expensive houses (Chandramohan et al., 2013). Sisal and rice husk fiber–reinforced composites are used in civil industries to reduce the product weight (Taj et al., 2007). Flax, hemp and sisal fibers are being used as reinforcement in industries like transportation and sports for products such as door frames, bicycle frames, tennis rackets and snowboards, while kenaf and hemp fiber–reinforced PLA composites are used in electrical and electronics industries such as mobile phone casing and laptop casing in order to diminish the cost and weight (Peças et al., 2018, Cheung et al., 2009). Silk fiber–reinforced PLA composites were used in bone fixation. Oil palm fiber, coir, flax, sisal and ramie fibers are used in various applications like packaging materials, construction materials, household furnishings, railing systems, paper manufacture, chip boards, geotextiles and industrial sewing thread (Sen et al., 2011a). Natural fibers such as flax, cotton, coconut, wood fiber, sisal, and natural rubber are used in Mercedes-Benz parts such as trunk panels, seat backrest panels, glove boxes, door panels and instrument panels in C, S and E class models (Holbery et al., 2006).

FIGURE 1.6 Applications of natural fiber–reinforced composites: (A) guitar panel, (B) monobloc chair, (C) exterior panel of bio scooter, (D) automobile casing, (E) building material, (F) helmet.

1.5 CONCLUSIONS

Increased ecological consciousness led to exploitation of the bio fiber with polymer matrix as a remarkable reinforcing material to form composites. Natural fibers are excellent materials capable of replacing petroleum-based fibers. Mechanical decortication plays the predominant role in fiber extraction owing to the large quantity of fiber that can be extracted more quickly with uniform diameter. Critical assessments of various chemical treatments show 5% NaOH alkaline treatment is preferred for most of the surface treatment for cellulose elucidation of fibers, which enhances the bonding between fiber and binder, augmenting their mechanical properties. The future is likely to see plant fibers become important sustainable and eco-friendly engineering materials that can be substituted for petroleum-based fibers for versatile applications.

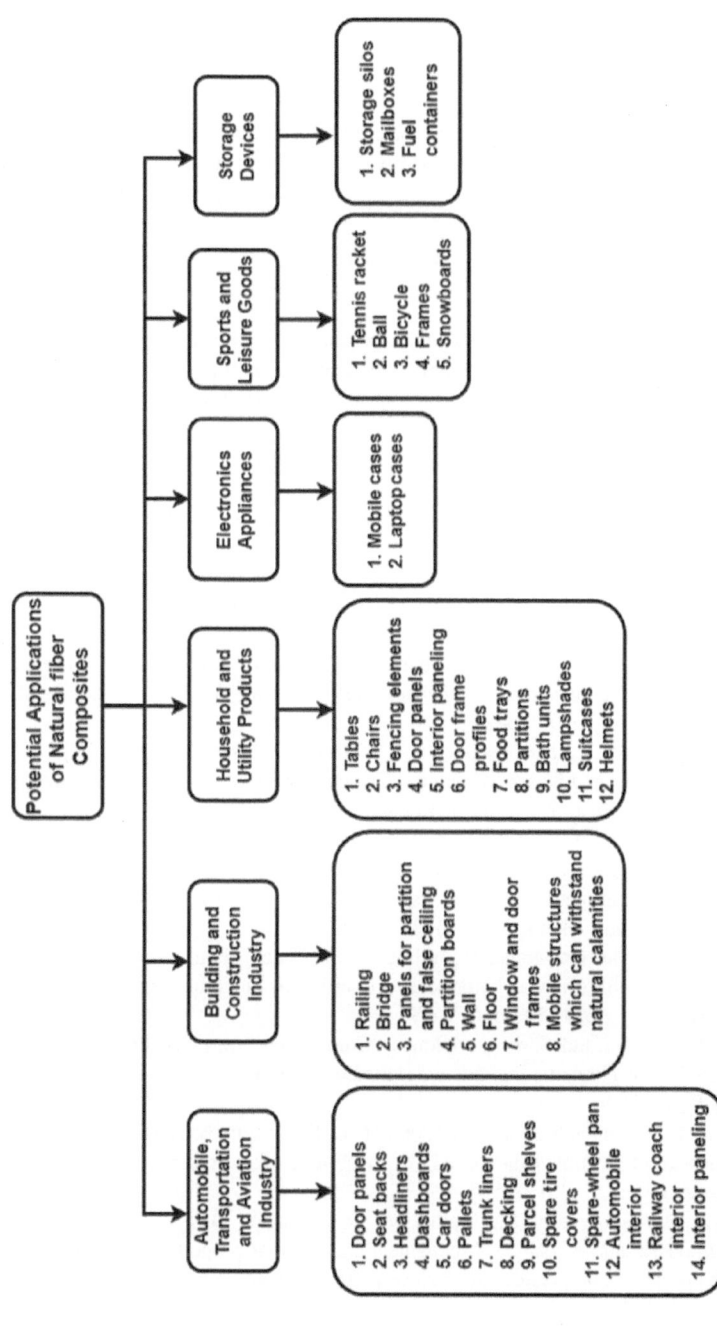

FIGURE 1.7 Potential applications of natural fiber composites.

CONFLICT OF INTEREST

None declared

ACKNOWLEDGMENTS

The authors are grateful for the support rendered by the management of KPR Institute of Engineering & Technology for providing facilities and other technical support in conducting this review.

REFERENCES

Agarwal, Jyoti, Swarnalata Sahoo, Smita Mohanty, and Sanjay K. Nayak. "Progress of novel techniques for lightweight automobile applications through innovative eco-friendly composite materials: a review." *Journal of Thermoplastic Composite Materials* 33, no. 7 (2020): 978–1013.

Ahmad, E. E. M., and A. S. Luyt. "Effects of organic peroxide and polymer chain structure on morphology and thermal properties of sisal fibre reinforced polyethylene composites." *Composites Part A: Applied Science and Manufacturing* 43, no. 4 (2012): 703–710.

Akil, H. M., M. F. Omar, A. A. M. Mazuki, S. Z. A. M. Safiee, Z. A. Mohd Ishak, and A. Abu Bakar. "Kenaf fiber reinforced composites: A review." *Materials & Design* 32, no. 8–9 (2011): 4107–4121.

Akin, D. E. 2010. "Chemistry of plant fibers." In J. Mussig, ed., *Industrial Applications of Natural Fibers, Structure, Properties and Technical Application*, ed. J. Mussig, Chichester: Wiley (2010), 13–22.

Albinante, S. R., Pacheco, E., and Visconte, L. L. Y. *New Chemistry* 36 (2013): 114.

Al-Maharma, Ahmad Y., and Naser Al-Huniti. "Critical review of the parameters affecting the effectiveness of moisture absorption treatments used for natural composites." *Journal of Composites Science* 3, no. 1 (2019): 27.

Alves, C., A. J. Silva, L. G. Reis, M. Freitas, L. B. Rodrigues, and D. E. Alves. "Ecodesign of automotive components making use of natural jute fiber composites." *Journal of Cleaner Production* 18, no. 4 (2010): 313–327.

Amiandamhen, S. O., Martina Meincken, and Luvuyo Tyhoda. "Natural fibre modification and its influence on fibre-matrix interfacial properties in biocomposite materials." *Fibers and Polymers* 21, no. 4 (2020): 677–689.

Anandjiwala, Rajesh D., and Sunshine Blouw. "Composites from bast fibres-prospects and potential in the changing market environment." *Journal of Natural Fibers* 4, no. 2 (2007): 91–109.

Arpitha, G. R., M. R. Sanjay, P. Senthamaraikannan, C. Barile, and B. Yogesha. "Hybridization effect of sisal/glass/epoxy/filler based woven fabric reinforced composites." *Experimental Techniques* 41, no. 6 (2017): 577–584.

Arrakhiz, F. Z., M. El Achaby, M. Malha, M. O. Bensalah, O. Fassi-Fehri, R. Bouhfid, K. Benmoussa, and A. Qaiss. "Mechanical and thermal properties of natural fibers reinforced polymer composites: Doum/low density polyethylene." *Materials & Design* 43 (2013): 200–205.

Bakare, I. O., F. E. Okieimen, C. Pavithran, H. P. S. Abdul Khalil, and M. Brahmakumar. "Mechanical and thermal properties of sisal fiber-reinforced rubber seed oil-based polyurethane composites." *Materials & Design* 31, no. 9 (2010): 4274–4280.

Bavan, D. Saravana, and G. C. Mohan Kumar. "Tensile and thermal degradation properties of vetiver fiber composites." *Procedia Materials Science* 5 (2014): 605–611.

Bisanda, E. T. N. "The effect of alkali treatment on the adhesion characteristics of sisal fibres." *Applied Composite Materials* 7 (2000): 331–339.

Bledzki, A. K., and Jochen Gassan. "Composites reinforced with cellulose based fibres." *Progress in Polymer Science* 24, no. 2 (1999): 221–274.

Bledzki, A. K., H-P. Fink, and K. Specht. "Unidirectional hemp and flax EP- and PP-composites: Influence of defined fiber treatments." *Journal of Applied Polymer Science* 93, no. 5 (2004): 2150–2156.

Brahmakumar, M., C. Pavithran, and R. M. Pillai. "Coconut fibre reinforced polyethylene composites: Effect of natural waxy surface layer of the fibre on fibre/matrix interfacial bonding and strength of composites." *Composites Science and Technology* 65, no. 3–4 (2005): 563–569.

Chandramohan, D., and J. Bharanichandar. "Natural fiber reinforced polymer composites for automobile accessories." *American Journal of Environmental Sciences* 9, no. 6 (2013): 494.

Chattopadhyay, Sanjay K., R. K. Khandal, Ramagopal Uppaluri, and Aloke K. Ghoshal. "Influence of varying fiber lengths on mechanical, thermal, and morphological properties of MA-*g*-PP compatibilized and chemically modified short pineapple leaf fiber reinforced polypropylene composites." *Journal of Applied Polymer Science* 113, no. 6 (2009): 3750–3756.

Chen, Hong, Wenfu Zhang, Xuehua Wang, Hankun Wang, Yan Wu, Tuhua Zhong, and Benhua Fei. "Effect of alkali treatment on wettability and thermal stability of individual bamboo fibers." *Journal of Wood Science* 64, no. 4 (2018): 398–405.

Chen, Yuxia, Na Su, Kaiting Zhang, Shiliu Zhu, Zhenzhen Zhu, Wenlian Qin, Yiwen Yang et al. "Effect of fiber surface treatment on structure, moisture absorption and mechanical properties of luffa sponge fiber bundles." *Industrial Crops and Products* 123 (2018): 341–352.

Cheung, Hoi-yan, Mei-po Ho, Kin-tak Lau, Francisco Cardona, and David Hui. "Natural fibre-reinforced composites for bioengineering and environmental engineering applications." *Composites Part B: Engineering* 40, no. 7 (2009): 655–663.

Dhakal, H. N., Z. Y. Zhang, and M. O. W. Richardson. "Effect of water absorption on the mechanical properties of hemp fibre reinforced unsaturated polyester composites." *Composites Science and Technology* 67, no. 7–8 (2007): 1674–1683.

Elkhaoulani, A., F. Z. Arrakhiz, K. Benmoussa, R. Bouhfid, and A. Qaiss. "Mechanical and thermal properties of polymer composite based on natural fibers: Moroccan hemp fibers/polypropylene." *Materials & Design* 49 (2013): 203–208.

Faruk, Omar, Andrzej K. Bledzki, Hans-Peter Fink, and Mohini Sain. "Biocomposites reinforced with natural fibers: 2000–2010." *Progress in Polymer Science* 37, no. 11 (2012): 1552–1596.

Ferreira, Diana P., Juliana Cruz, and Raul Fangueiro. "Surface modification of natural fibers in polymer composites." In *Green composites for automotive applications*, pp. 3–41. Woodhead Publishing, 2019.

Furtado, Samuel C. R., A. L. Araújo, Arlindo Silva, Cristiano Alves, and A. M. R. Ribeiro. "Natural fibre-reinforced composite parts for automotive applications." *International Journal of Automotive Composites* 1, no. 1 (2014): 18–38.

Gassan, J., I. Mildner, and A. K. Bledzki. "Influence of fiber structure modification on the mechanical properties of flax fiber-epoxy composites." *Mechanics of Composite Materials* 35, no. 5 (1999): 435–440.

George, Michael, Paolo G. Mussone, Zeinab Abboud, and David C. Bressler. "Characterization of chemically and enzymatically treated hemp fibres using atomic force microscopy and spectroscopy." *Applied Surface Science* 314 (2014): 1019–1025.

Goda, Koichi, M. S. Sreekala, Alexandre Gomes, Takeshi Kaji, and Junji Ohgi. "Improvement of plant based natural fibers for toughening green composites—Effect of load application during mercerization of ramie fibers." *Composites Part A: Applied Science and Manufacturing* 37 (2006): 2213–2220.

Gokulkumar, S., P. R. Thyla, L. Prabhu, and S. Sathish. "Measuring methods of acoustic properties and influence of physical parameters on natural fibers: A review." *Journal of Natural Fibers* (2019): 1–20.

Gokulkumar, S., P. R. Thyla, L. Prabhu, S. Sathish, and N. Karthi. "A comparative study on epoxy based composites filled with pineapple/areca/ramie hybridized with industrial tea leaf wastes/gfrp." *Materials Today: Proceedings* 27 (2020): 2474–2476.

Gokulkumar, S., P. R. Thyla, L. Prabhu, and S. Sathish. "Characterization and comparative analysis on mechanical and acoustical properties of *Camellia sinensis/Ananas comosus/glass fiber hybrid polymer composites." *Journal of Natural Fibers* 18, no. 7 (2021): –978–994.

González, Jeaneth T. Corredor, Aldo J. Pinheiro Dillon, Aly R. Pérez-Pérez, R. Fontana, and Carlos Pérez Bergmann. "Enzymatic surface modification of sisal fibers (*Agave sisalana*) by *Penicillium echinulatum* cellulases." *Fibers and Polymers* 16, no. 10 (2015): 2112–2120.

Gowda, T. G. Yashas, M. R. Sanjay, K. Subrahmanya Bhat, P. Madhu, P. Senthamaraikannan, and B. Yogesha. "Polymer matrix-natural fiber composites: An overview." *Cogent Engineering* 5, no. 1 (2018): 1446667.

Gurukarthik Babu, B., D. Prince Winston, P. Senthamarai Kannan, S. S. Saravanakumar, and M. R. Sanjay. "Study on characterization and physicochemical properties of new natural fiber from *Phaseolus vulgaris*." *Journal of Natural Fibers* 16, no. 7 (2019): 1035–1042.

Hamidon, Muhammad H., Mohamed T. H. Sultan, Ahmad H. Ariffin, and Ain U. M. Shah. "Effects of fibre treatment on mechanical properties of kenaf fibre reinforced composites: A review." *Journal of Materials Research and Technology* 8, no. 3 (2019): 3327–3337.

Harish, Sivasankaran, D. Peter Michael, A. Bensely, D. Mohan Lal, and A. Rajadurai. "Mechanical property evaluation of natural fiber coir composite." *Materials Characterization* 60, no. 1 (2009): 44–49.

Hassan, F., R. Zulkifli, M. J. Ghazali, and C. H. Azhari. "Kenaf fiber composite in automotive industry: An overview." *International Journal on Advanced Science, Engineering and Information Technology* 7, no. 1 (2017): 315–321.

Holbery, James, and Dan Houston. "Natural-fiber-reinforced polymer composites in automotive applications." *JOM* 58, no. 11 (2006): 80–86.

Huda, M. S., L. T. Drzal, D. Ray, A. K. Mohanty, and M. Mishra. "Natural-fiber composites in the automotive sector." In *Properties and performance of natural-fibre composites*, pp. 221–268. Woodhead Publishing, 2008a.

Huda, Masud S., Lawrence T. Drzal, Amar K. Mohanty, and Manjusri Misra. "Effect of chemical modifications of the pineapple leaf fiber surfaces on the interfacial and mechanical properties of laminated biocomposites." *Composite Interfaces* 15, no. 2–3 (2008b): 169–191.

Hulle, Ashish, Pradyumkumar Kadole, and Pooja Katkar. "Agave Americana leaf fibers." *Fibers* 3, no. 1 (2015): 64–75.

Isa, M. T., S. Usman, A. O. Ameh, O. A. Ajayi, O. Omorogbe, and S. U. Ameuru. "The effect of fiber treatment on the mechanical and water absorption properties of short okra/glass fibers hybridized epoxy composites." *International Journal of Materials Engineering* 4, no. 5 (2014): 180–184.

Jauhari, Nitin, Raghvendra Mishra, and Harischchandra Thakur. "Natural fibre reinforced composite laminates–a review." *Materials Today: Proceedings* 2, no. 4–5 (2015): 2868–2877.

Jawaid, M. H. P. S., and H. P. S. Abdul Khalil. "Cellulosic/synthetic fibre reinforced polymer hybrid composites: A review." *Carbohydrate Polymers* 86, no. 1 (2011): 1–18.

John, Maya Jacob, and Sabu Thomas. "Biofibres and biocomposites." *Carbohydrate Polymers* 71, no. 3 (2008a): 343–364.

John, M. J., and Anandjiwala, R. D. "Recent developments in chemical modification and characterization of natural fiber–reinforced composites." *Polymer Composites* 29 (2008b): 187–207.

Joseph, K., S. Thomas, and C. Pavithran. "Sisal fibre reinforced polyethylene composites: Effect of isocyanate treatment." *SB Academic Review, Changanacherry* 6 (1995): 85.

Kabir, M. M., Hao Wang, K. T. Lau, and Francisco Cardona. "Chemical treatments on plant-based natural fibre reinforced polymer composites: An overview." *Composites Part B: Engineering* 43, no. 7 (2012): 2883–2892.

Kaewpirom, Supranee, and Cherdthawat Worrarat. "Preparation and properties of pineapple leaf fiber reinforced poly (lactic acid) green composites." *Fibers and Polymers* 15, no. 7 (2014): 1469–1477.

Kalia, Susheel, B. S. Kaith, and Inderjeet Kaur, eds. *Cellulose Fibers: Bio-and Nano-Polymer Composites: Green Chemistry and Technology.* Springer Science & Business Media, 2011.

Karthi, N., K. Kumaresan, S. Sathish, S. Gokulkumar, L. Prabhu, and N. Vigneshkumar, N. "An overview: Natural fiber reinforced hybrid composites, chemical treatments and application areas." *Materials Today: Proceedings*, 27, no. 3 (2020): 2828–2834.

Kumaresan, M., S. Sathish, and N. Karthi. "Effect of fiber orientation on mechanical properties of sisal fiber reinforced epoxy composites." *Journal of Applied Science and Engineering* 18, no. 3 (2015): 289–294.

Madhu, P., M. R. Sanjay, S. Pradeep, K. Subrahmanya Bhat, B. Yogesha, and Suchart Siengchin. "Characterization of cellulosic fibre from *Phoenix pusilla* leaves as potential reinforcement for polymeric composites." *Journal of Materials Research and Technology* 8, no. 3 (2019): 2597–2604.

Manimaran, P., P. Senthamaraikannan, K. Murugananthan, and M. R. Sanjay. "Physicochemical properties of new cellulosic fibers from *Azadirachta indica* plant." *Journal of Natural Fibers* 15 (2018): 29–38.

Mishra, S., M. Misra, S. S. Tripathy, S. K. Nayak, and A. K. Mohanty. "Graft copolymerization of acrylonitrile on chemically modified sisal fibers." *Macromolecular Materials and Engineering* 286, no. 2 (2001): 107–113.

Mohammed, Layth, Mohamed N. M. Ansari, Grace Pua, Mohammad Jawaid, and M. Saiful Islam. "A review on natural fiber reinforced polymer composite and its applications." *International Journal of Polymer Science* 2015 (2015).

Mohammed, Mustafa Hauwa, and Benjamin Dauda. "Unsaturated polyester resin reinforced with chemically modified natural fibre." *IOSR Journal of Polymer and Textile Engineering* 1 (2014): 31–38.

Mohanty, Amar K., Manjusri Misra, and Lawrence T. Drzal, eds. *Natural Fibers, Biopolymers, and Biocomposites.* CRC Press, 2005.

Mukherjee, Tapasi, and Nhol Kao. "PLA based biopolymer reinforced with natural fibre: A review." *Journal of Polymers and the Environment* 19, no. 3 (2011): 714.

Mwaikambo, L. "Review of the history, properties and application of plant fibres." *African Journal of Science and Technology* 7, no. 2 (2006): 121.

Norhidayah, M. H., Arep Ariff Hambali, Mohd Yuhazri bin Yaakob, M. Zolkarnain, and H. Y. Saifuddin. "A review of current development in natural fiber composites in automotive applications." In *Applied Mechanics and Materials*, vol. 564, pp. 3–7. Trans Tech Publications Ltd., 2014.

Njuguna, James, Paul Wambua, Krzysztof Pielichowski, and Kambiz Kayvantash. "Natural fibre-reinforced polymer composites and nanocomposites for automotive applications." In *Cellulose Fibers: Bio- and Nano-Polymer Composites*, pp. 661–700. Berlin: Springer, 2011.

Ochi, Shinji. "Mechanical properties of kenaf fibers and kenaf/PLA composites." *Mechanics of Materials* 40, no. 4–5 (2008): 446–452.

O'Donnell, A., M. A. Dweib, and R. P. Wool. "Natural fiber composites with plant oil-based resin." *Composites Science and Technology* 64, no. 9 (2004): 1135–1145.

Owen, M. M., C. O. Ogunleye, and E. O. Achukwu. "Mechanical properties of sisal fibre-reinforced epoxy composites-effect of alkali concentrations." *Advances in Polymer Science and Technology* 5 (2015): 26–31.

Panyasart, Kloykamol, Nattawut Chaiyut, Taweechai Amornsakchai, and Onuma Santawitee. "Effect of surface treatment on the properties of pineapple leaf fibers reinforced polyamide 6 composites." *Energy Procedia* 56 (2014): 406–413.

Peças, Paulo, Hugo Carvalho, Hafiz Salman, and Marco Leite. "Natural fibre composites and their applications: A review." *Journal of Composites Science* 2, no. 4 (2018): 66.

Pickering, Kim, ed. *Properties and Performance of Natural-Fibre Composites*. Elsevier, 2008.

Pickering, Kim L., M. G. Aruan Efendy, and Tan Minh Le. "A review of recent developments in natural fibre composites and their mechanical performance." *Composites Part A: Applied Science and Manufacturing* 83 (2016): 98–112.

Prabhu, L., V. Krishnaraj, S. Gokulkumar, S. Sathish, and M. Ramesh. "Mechanical, chemical and acoustical behavior of sisal–tea waste–glass fiber reinforced epoxy based hybrid polymer composites." *Materials Today: Proceedings* 16 (2019): 653–660.

Prabhu, L., V. Krishnaraj, S. Sathish, S. Gokulkumar, and N. Karthi. "Study of mechanical and morphological properties of jute-tea leaf fiber reinforced hybrid composites: Effect of glass fiber hybridization." *Materials Today: Proceedings* 27 (2020): 2372–2375.

Prasad, G. L. Easwara, B. S. Keerthi Gowda, and R. Velmurugan. "Comparative study of impact strength characteristics of treated and untreated sisal polyester composites." *Procedia Engineering* 173 (2017): 778–785.

Puttegowda, Madhu, Sanjay Mavinakere Rangappa, Mohammad Jawaid, Pradeep Shivanna, Yogesha Basavegowda, and Naheed Saba. "Potential of natural/synthetic hybrid composites for aerospace applications." In *Sustainable composites for aerospace applications*, pp. 315–351. Woodhead Publishing, 2018.

Rahman, M. Mizanur, and Mubarak A. Khan. "Surface treatment of coir (*Cocos nucifera*) fibers and its influence on the fibers' physico-mechanical properties." *Composites Science and Technology* 67, no. 67 (2007): 2369–2376.

Ramamoorthy, Sunil Kumar, Mikael Skrifvars, and Anders Persson. "A review of natural fibers used in biocomposites: Plant, animal and regenerated cellulose fibers." *Polymer Reviews* 55, no. 1 (2015): 107–162.

Ramesh, M., K. Palanikumar, and K. Hemachandra Reddy. "Plant fibre based bio-composites: Sustainable and renewable green materials." *Renewable and Sustainable Energy Reviews* 79 (2017): 558–584.

Rangappa, Sanjay Mavinkere, and Suchart Siengchin. "Natural fibers as perspective materials." *King Mongkut's University of Technology North Bangkok International Journal of Applied Science and Technology* 11, no. 4 (2018).

Rohit, Kiran, and Savita Dixit. "A review-future aspect of natural fiber reinforced composite." *Polymers from Renewable Resources* 7, no. 2 (2016): 43–59.

Sahari, J., S. M. Sapuan, E. S. Zainudin, and M. Abdul Maleque. "Mechanical and thermal properties of environmentally friendly composites derived from sugar palm tree." *Materials & Design* 49 (2013): 285–289.

Sanjay, M. R., G. R. Arpitha, and Basavegowda Yogesha. "Study on mechanical properties of natural-glass fibre reinforced polymer hybrid composites: A review." *Materials Today: Proceedings* 2, no. 4–5 (2015): 2959–2967.

Sanjay, M. R., G. R. Arpitha, L. Laxmana Naik, K. Gopalakrishna, and B. Yogesha. "Applications of natural fibers and its composites: An overview." *Natural Resources* 7, no. 3 (2016): 108–114.

Sanjay, M. R., G. R. Arpitha, P. Senthamaraikannan, M. Kathiresan, M. A. Saibalaji, and B. Yogesha. "The hybrid effect of jute/kenaf/E-glass woven fabric epoxy composites for medium load applications: Impact, inter-laminar strength, and failure surface characterization." *Journal of Natural Fibers* (2018a).

Sanjay, M. R., P. Madhu, Mohammad Jawaid, P. Senthamaraikannan, S. Senthil, and S. Pradeep. "Characterization and properties of natural fiber polymer composites: A comprehensive review." *Journal of Cleaner Production* 172 (2018b): 566–581.

Sanjay, M. R., S. Siengchin, J. Parameswaranpillai, M. Jawaid, C. I. Pruncu, and A. Khan, A. "A comprehensive review of techniques for natural fibers as reinforcement in composites: preparation, processing and characterization." *Carbohydrate Polymers* 207 (2019): 108–121. https://doi.org/10.1016/j.carbpol.2018.11.083.

Sari, Nasmi Herlina, M. R. Sanjay, G. R. Arpitha, Catalin Iulian Pruncu, and Suchart Siengchin. "Synthesis and properties of pandanwangi fiber reinforced polyethylene composites: Evaluation of dicumyl peroxide (DCP) effect." *Composites Communications* 15 (2019): 53–57.

Sathish, S., M. Kumaresan, N. Karthi, and T. Dhilip Kumar. "Tensile and impact properties of natural fiber hybrid composite materials." *International Journal of Modern Engineering Research* 4 (2014a): 9–12.

Sathish, S., T. Ganapathy, and Thiyagarajan Bhoopathy. "Experimental testing on hybrid composite materials." In *Applied Mechanics and Materials*, vol. 592, pp. 339–343. Trans Tech Publications Ltd, 2014b.

Sathish, S., K. Kumaresan, L. Prabhu, and N. Vigneshkumar. "Experimental investigation on volume fraction of mechanical and physical properties of flax and bamboo fibers reinforced hybrid epoxy composites." *Polymers and Polymer Composites* 25, no. 3 (2017): 229–236.

Sathish, S., K. Kumaresan, L. Prabhu, S. Gokulkumar, and S. Dinesh. "Experimental testing on mechanical properties of various natural fibers reinforced epoxy hybrid composites." *International Journal of Pure and Applied Mathematics* 118, no. 16 (2018): 873–888.

Sathish, S., K. Kumaresan, L. Prabhu, S. Gokulkumar, N. Karthi, and N. Vigneshkumar. "Experimental investigation of mechanical and morphological properties of flax fiber reinforced epoxy composites incorporating SiC and Al_2O_3." *Materials Today: Proceedings* 27 (2020): 2249–2253.

Sathishkumar, T. P., P. Navaneethakrishnan, and S. Shankar. "Tensile and flexural properties of snake grass natural fiber reinforced isophthallic polyester composites." *Composites Science and Technology* 72, no. 10 (2012): 1183–1190.

Sathishkumar, T. P., P. Navaneethakrishnan, S. Shankar, and R. Rajasekar. "Investigation of chemically treated longitudinally oriented snake grass fiber-reinforced isophthallic polyester composites." *Journal of Reinforced Plastics and Composites* 32, no. 22 (2013): 1698–1714.

Sayanjali Jasbi, Moghgan, Hossein Hasani, Ali Zadhoush, and Somayeh Safi. "Effect of alkali treatment on mechanical properties of the green composites reinforced with milkweed fibers." *Journal of the Textile Institute* 109, no. 1 (2018): 24–31.

Sen, Tara, and H. N. Jagannatha Reddy. "Various industrial applications of hemp, kinaf, flax and ramie natural fibres." *International Journal of Innovation, Management and Technology* 2, no. 3 (2011a): 192.

Sen, Tara, and H. N. Jagannatha Reddy. "Application of sisal, bamboo, coir and jute natural composites in structural upgradation." *International Journal of Innovation, Management and Technology* 2, no. 3 (2011b): 186.

Sgriccia, N., M. C. Hawley, and M. Misra. "Characterization of natural fiber surfaces and natural fiber composites." *Composites Part A: Applied Science and Manufacturing* 39, no. 10 (2008): 1632–1637.

Shekar, H. S. Sharath, and M. Ramachandra. "Green composites: A review." *Materials Today: Proceedings* 5, no. 1 (2018): 2518–2526.

Shinoj, S., Rangaraju Visvanathan, S. Panigrahi, and M. Kochubabu. "Oil palm fiber (OPF) and its composites: A review." *Industrial Crops and Products* 33, no. 1 (2011): 7–22.

Siengchin, S. "Editorial corner–a personal view: Potential use of 'green' composites in automotive applications." *Express Polymer Letters* 11, no. 8 (2017): 600.

Steffens, Fernanda, Henrique Steffens, and Fernando Ribeiro Oliveira. "Applications of natural fibers on architecture." *Procedia Engineering* 200 (2017): 317–324.

Suddell, B. C. "Industrial fibres: Recent and current developments." In *Proceedings of the Symposium on Natural Fibres*, vol. 20, pp. 71–82. Rome: FAO and CFC, 2008.

Taj, Saira, Munawar Ali Munawar, and Shafiullah Khan. "Natural fiber-reinforced polymer composites." *Proceedings-Pakistan Academy of Sciences* 44, no. 2 (2007): 129.

Thiruchitrambalam, M., A. Alavudeen, A. Athijayamani, N. Venkateshwaran, and A. Elaya Perumal. "Improving mechanical properties of banana/kenaf polyester hybrid composites using sodium laulryl sulfate treatment." *Materials Physics and Mechanics* 8, no. 2 (2009): 165–173.

Ticoalu, A., T. Aravinthan, and F. Cardona. "A review of current development in natural fiber composites for structural and infrastructure applications." In *Proceedings of the Southern Region Engineering Conference (SREC 2010)*, pp. 113–117. Engineers Australia, 2010.

Uddin, Nasim, ed. *Developments in fiber-reinforced polymer (FRP) composites for civil engineering*. Elsevier, 2013.

Verma, Deepak, and Sanjay Sharma. "Green biocomposites: A prospective utilization in automobile industry." In *Green biocomposites*, pp. 167–191. Cham: Springer, 2017.

Vijay, R., D. Lenin Singaravelu, A. Vinod, M. R. Sanjay, Suchart Siengchin, Mohammad Jawaid, Anish Khan, and Jyotishkumar Parameswaranpillai. "Characterization of raw and alkali-treated new natural cellulosic fibers from *Tridax procumbens*." *International Journal of Biological Macromolecules* 125 (2019): 99–108.

Vijay, R., D. Lenin Singaravelu, A. Vinod, M. R. Sanjay, and Suchart Siengchin. "Characterization of alkali-treated and untreated natural fibers from the stem of *Parthenium hysterophorus*." *Journal of Natural Fibers* 18 (2021): 1–11.

Wang, Wei, and Gu Huang. "Characterisation and utilization of natural coconut fibres composites." *Materials & Design* 30, no. 7 (2009): 2741–2744.

Wanmbua, P., J. Ivens, and I. Verpoest. "Natural fibres: Can they replace glass in fibre reinforced plastic." *Composite Science and Technology* 63 (2003): 1259–1264.

Yusof, Yusri, Siti Asia Yahya, and Anbia Adam. "Novel technology for sustainable pineapple leaf fibers productions." *Procedia CIRP* 26 (2015): 756–760.

Zhao, Da, and Zhou Zhou. "Applications of lightweight composites in automotive industries." In *Lightweight materials from biopolymers and biofibers*, pp. 143–158. American Chemical Society, 2014.

2 Tribology Properties of Fiber-Reinforced Polymer Composites

Sumit Gupta,[1] Vijay Chaudhary[1]*

[1]Department of Mechanical Engineering, Amity School of Engineering & Technology, Amity University, Noida, PIN: 201313.

*Corresponding author: vijaychaudhary111@gmail.com

CONTENTS

2.1   INTRODUCTION

Today's material scientists have a great interest in fiber-based composites due to their numerous properties such as light weight, high specific strength, good tribological properties, and high thermal and chemical stability. These composites are replacing the existing metal and metal matrix composite materials in many applications in the automobile and aerospace industries (Agarwal et al., 2017, Mazumdar, 2001, Daniel et al., 2006).

Fiber-reinforced polymer composites are fabricated with two constituents: reinforcing fiber and polymer matrix. Reinforcing fiber strengthens the developed composites, and polymer matrix surrounds the reinforcing fiber and shapes the developed composite (Chaudhary et al., 2018a). These can be categorized as bio fiber and bio polymers, synthetic fibers and synthetic polymers. Bio fiber reinforcement of bio polymer fabricates the bio-composite materials, and synthetic fiber reinforcement of synthetic polymer fabricates the synthetic fiber–based polymer composites (Chaudhary et al., 2017, Chaudhary et al., 2018b). Bio fibers are extracted from natural resources such as plants, animals and minerals. Some bio fibers are jute, hemp, wool, cotton, flax, kenaf, ramie, asbestos, banana, pineapple, rice husk and sugarcane. Synthetic fibers are developed by polymerization, in which monomers are combined to make a polymer. Some important synthetic fibers are glass, carbon aramid, Kevlar, nylon, acrylic, rayon and polyester. Polymers consist of a large chain of molecules and macromolecules, which are fabricated by repeating monomers. Some natural bio polymers are starch, rubber, cellulose, chitin and protein. Synthetic polymers include epoxy, polycarbonate and polyurethane. (Chaudhary et al., 2020a, Chaudhary et al., 2019).

Developed bio-fiber-based polymer composites have drawn great interest from scientists and researchers. Bio-composites are eco-friendly, non-toxic and bio-degradable in nature. They can be decomposed easily after the successful completion of their life. Bio-composites are a sustainable resource (Chaudhary et al., 2020b). In the application phase of fiber-reinforced polymer composites, various governing parameters during and after fabrication influence the performance of composites. Some parameters in fabrication are percentage volume of fiber and polymer, orientation of fiber, shape of fiber, wettability of fiber with polymer matrix, surface properties of fiber and method of processing the composite. After the successful fabrication of the polymer, the performance of the composite is largely influenced by fiber matrix interfacial adhesion, alignment of fiber inside the polymer and voids due to air entrapment (Koronis et al., 2013, Wei et al., 2016, Pickering et al., 2016).

Sometimes, developed fiber-reinforced polymer (FRP) composite in a bearing application comes into contact with other objects. As a result of contact and rubbing with another object, wear and tear takes place on the rubbing surface of the composite material. Therefore, after the successful fabrication of FRP composites and before their use in real applications, these composites should be characterized on the basis of their mechanical, tribological, thermal and chemical properties. In tribological characterization, specific wear rate and friction properties are evaluated using tribometer machines. Evaluation of tribological properties helps the material scientist to decide the appropriate applications for a developed fiber-reinforced polymer composite (Chaudhary et al., 2018c, Chaudhary et al., 2020c).

In the study of tribological properties, the specific wear rate (adhesive wear, abrasive wear, erosive wear) and the coefficient of friction are experimentally calculated for the developed composite. Tribological performance of fiber-reinforced polymer composites is influenced by various parameters like orientation of fiber, shape and size of fiber (loose fiber, bi-directional mat, uni-directional mat) and interfacial adhesion between fiber and matrix interphase (El-Tayeb et al., 2008, Nirmal et al., 2012, Correa et al., 2015, Nasir et al., 2013). Various authors carried out studies on fiber-reinforced polymer composites. Chaudhary et al. (2018c) developed jute, hemp and flax-reinforced epoxy composites and their hybrids using a hand lay-up technique. They then investigated the complete mechanical properties of these developed fiber-reinforced polymer composites. Wei et al. (2016) manufactured rice husk ash and sisal fiber–reinforced cement composites and conducted the mechanical characterization of the developed composites.

In this chapter, tribological characteristics and interfacial strength of fiber-reinforced polymer composites are discussed in detail. Mechanisms of wear and friction generation during the tribological testing are also shown.

2.2 INTERFACIAL ADHESION BETWEEN FIBER AND POLYMER PHASE

Interfacial adhesion restricts the debonding of fiber into polymer phase during the sliding action in tribological analysis. During tribological analysis, some common fracture phenomena are fiber debonding, fiber pullout and fiber fracture. Good interfacial adhesion always minimizes the causes of failure of fiber-reinforced polymer composites (Chaudhary et al., 2017, Chaudhary et al., 2018a, Chaudhary et al., 2018b, Chaudhary et al., 2018c). Chemical treatments (alkali treatment, silane treatment and others) improve the surface characteristics of the reinforced-fiber phase, which enhance the wettability of fiber (Chaudhary et al., 2018a, Chaudhary et al., 2019). Figure 2.1 shows the failure mechanism of natural fiber–reinforced polymer composites using scanning electron micrographs (SEM) (Chaudhary et al., 2020c).

2.3 TRIBOLOGICAL CHARACTERIZATION OF FIBER-REINFORCED POLYMER COMPOSITES

Tribology is part of our everyday life. For example, rubbing two stones against each other produces flame, increases the temperature between the mating surfaces and wears out the surfaces. Rubbing between two mating surfaces generates friction and wear between surfaces. The study of friction, wear and lubrication is the science of tribology, a term derived from the Greek word *tribo*, "to rub." Friction plays a vital role in human life. Walking is impossible without the friction between ground and human feet, and vehicles cannot move on the roads without friction between the road and the tires. So the science of tribology is required in every field of engineering to analyze the study of friction, wear and lubrication required between two mating surfaces (Chand et al. 2008).

So analyses of three factors – friction, wear and lubrication – make up the tribological characterization of material. Tribological analysis of composite material is conducted using a tribo testing machine. Various tribo testing machines are a pin-on-disc tribometer, a block-on-disc tribometer, a cylinder-on-plate tester, a pin-on-ring tester, an abrasion resistance tester, as well as others. Input parameters in tribological testing are normal load on sample, sliding speed, temperature between counterface and test sample, sliding time, counterface roughness, and type of coolant between counterface and test sample. Normal load on a sample may be a static or a dynamic load applied by hydraulic and electromagnetic means which gives the value of load force F_N. The sliding speed of the test sample on a tribo machine counterface plays a vital role as it gives the value of friction. It is also responsible for the coefficient of friction and wear rate. The temperature between counterface and test sample gives an idea of the thermal stability of the test sample, and it should be maintained below the glass transition temperature (T_g) of the test sample. Sliding time is the time in which a test sample rubs against the counterface of the tribo testing machine and is responsible for much of the value of wear and temperature generation on the rubbing surface. Counterface roughness varies for different tests and definitely affects the wear and friction value. Coolant applied between the counterface and test sample reduces the temperature and also affects the value of friction. Figure 2.2 shows the schematic arrangement of a pin-on-disc tribometer (Chaudhary et al., 2018c, Chaudhary

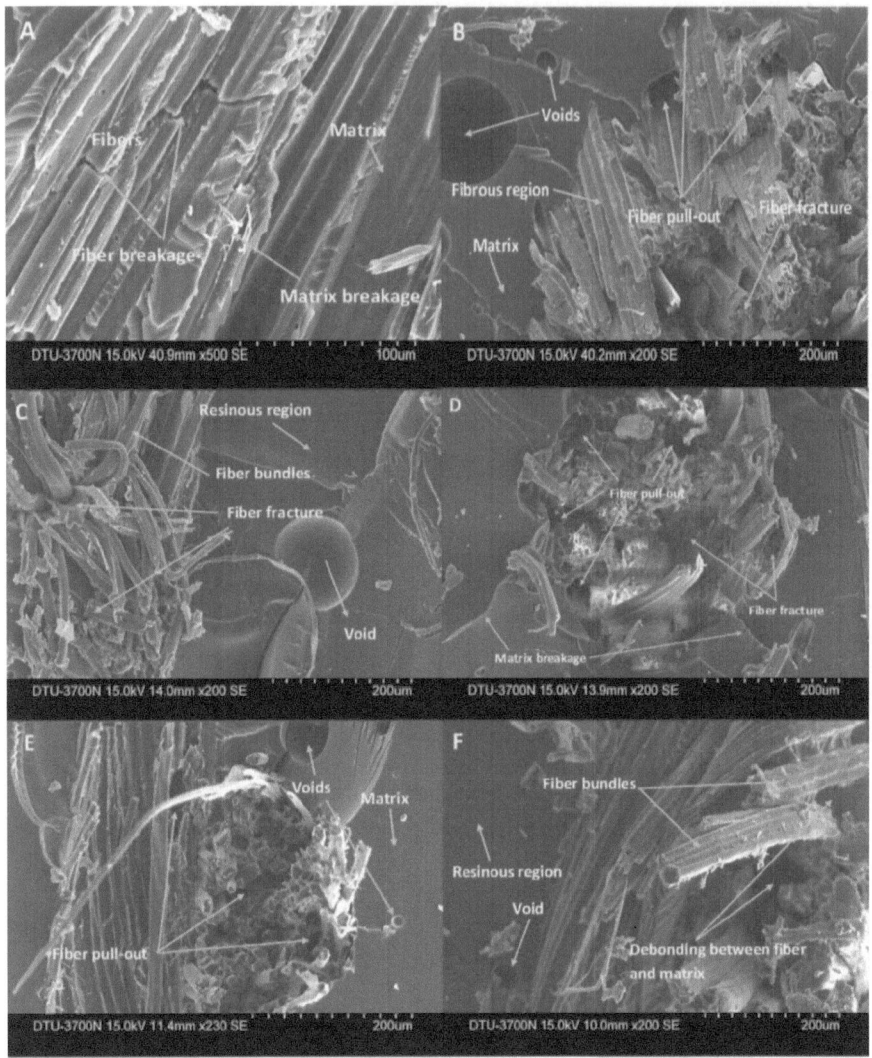

FIGURE 2.1 Fractured surfaces of fiber-reinforced polymer composites.

Source: Chaudhary et al. (2020c).

et al., 2020a). The results of various studies on tribological characterization of fiber-reinforced polymer composites are shown in Table 2.1.

2.4 MECHANISMS OF MATERIAL WEAR DURING TRIBOLOGICAL TESTING

In tribological testing, a specimen of fiber-reinforced polymer composite rubs against the sliding surface of the counterface plate of the tribo testing machine. In fiber-reinforced

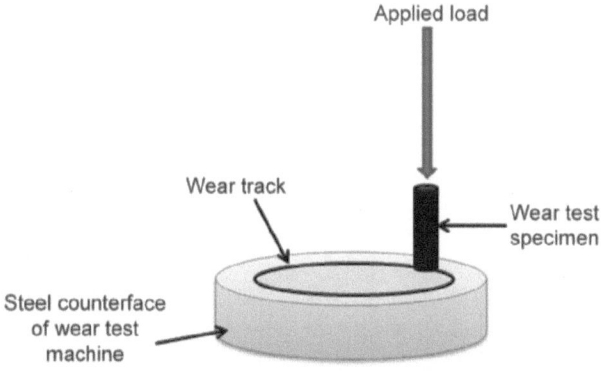

FIGURE 2.2   Schematic representation of pin-on-disc tribometer.

Sources: Chaudhary et al. (2018c), Chaudhary et al. (2020a).

polymer composites, both fiber and reinforced matrix come across the sliding action of the machine counterface. When the specimen rubs against the counterface plate, the hardened counterface wears out the specimen surface. In this action, the orientation of fibers in the polymer matrix influences the mechanism of wear of the specimen. Fibers are distributed into the polymer matrix in one of several forms: randomly particulate, single fiber, fiber mat (unidirectional mat and bi-directional fiber mat) and also twirl or twill fiber mat. All these distributions of fiber influence the wear rate of fiber-reinforced polymer composites. When the fiber phase comes in contact with the sliding surface it enhances the friction between the specimen and moving counterface surface. Wear of the fiber-reinforced composites may occur in one of the following ways.

a.   In the case of reinforcement with randomly distributed fiber particles, first the polymer matrix surface of the specimen slides against the counterface and wears out, and then randomly distributed particles take part in the sliding action and wear out or debond from the matrix phase owing to frictional force between the counter surface plate and the composite specimen.

b.   In the case of unidirectional mat reinforcement, the longitudinal direction of the fibers takes part in wear testing for anti-parallel orientation of the specimen on the counterface plate and the traverse direction of the fibers takes part in wear testing for parallel orientation of the specimen on the counterface plate.

c.   In the case of bi-directional fiber mat reinforcement, both longitudinal and traverse fibers take part in wear testing for parallel and anti-parallel orientation of the specimen on the counterface plate.

When longitudinal fibers come in contact with the counterface plate during sliding action, the longitudinal fiber generates friction and wears out and breaks owing to friction and the sliding force between surfaces. But when traverse fibers come in contact with the counterface plate during sliding action, the traverse fiber generates more friction compared to a longitudinal fiber and enhances the wear rate of the specimen.

TABLE 2.1
Recent Studies of Tribological Properties of Fiber-Reinforced Polymer Composites

Fibers	Polymer Matrix	Properties	Apparatus	Remarks	Reference
Basalt, aramid	Phenol formaldehyde	Specific wear rate and coefficient of friction	Pin-on-disc	Addition of clay and tin powder enhanced the coefficient of friction and reduced the specific wear rate.	Manoharan et al. (2019)
Eucalyptus Kraft pulp	Bio-epoxy	Specific wear rate and coefficient of friction	Pin-on-disc	Incorporation of eucalyptus Kraft pulp with bio-epoxy resulted in low wear loss as compared to neat bio-epoxy.	Barari et al. (2016)
E-glass	Epoxy	Specific wear rate and coefficient of friction	Block on roller multi-tribometer	The orientation of fiber greatly influenced the coefficient of friction and wear: parallel orientation had a higher value of wear and friction values than normal orientation.	Sen et al. (2015)
Basalt and jute fiber	Polyester/vinylester	Specific wear rate and coefficient of friction	Linear reciprocating sliding flat-on-flat tribometer	Vinylester-based composites achieved low wear loss as compared to polyester-based composites	Toth et al. (2020)
Cissus quandrangularis stem fiber	Isophthalic polyester	Specific wear rate and coefficient of friction	Pin-on-disc	Alkali- and silane-treated fiber-reinforced composites show lower wear loss compared to untreated fiber-based composites.	Naik et al. (2020)
Red mud-filled pineapple fiber	Polyester resin	Erosion rate	Pin-on-disc	Incorporation of 20% wt. of red mud as a filler with pineapple/polyester composite had a higher value of erosion wear.	Sundarakannan et al. (2020)

Material	Matrix	Test method	Measured properties	Findings	Reference
Carbon fiber grafted with TiO$_2$ nanorods	Poly(hexahydrotriazine)s (PHTs)	Rotary friction and wear testing (Huahui, MST-3000)	Wear and average friction coefficient	Incorporation of TiO$_2$ with carbon fiber as a filler enhanced the interfacial adhesion between carbon and PHT matrix, which results in low wear rate during tribological testing.	Lin et al. (2020)
Aramid, carbon, glass and basalt fibers	Polyimide and polytetra fluoroethylene	Pin on flat configuration	Wear and friction	Neat polymers showed reduced wear rate as compared to developed composites during dry and slurry erosive conditions.	R. Kumar et al. (2020)
Carbon nanorod (CNR)	Poly(vinylidenefluoride)	Ball-on-flat rigs tribometer	Specific wear rate and coefficient of friction	The CNRs were incorporated to enhance the tribological properties.	Lee et al. (2020)
Glass fiber filled with graphene	Epoxy	Pin-on-disc	Specific wear rate and coefficient of friction	Graphene-filled glass/epoxy composite showed lower value of wear loss as compared to glass/epoxy composite.	S. Kumar et al. (2020)
Particulate wood charcoal and periwinkle shell	Polyester	Pin-on-disc	Specific wear rate and coefficient of friction	Hybrid composite achieved better wear properties as compared to single-fiber-reinforced polymer composites	Edoziuno et al. (2020)
MoS$_2$, SiO$_2$, Si$_3$N$_4$, or graphite	Polymide oligomer end-capping with 4-phenylethymyl anhydride	Pin-on-disc	Specific wear rate and coefficient of friction	PI/MoS$_2$–20 wt% composite showed the lowest friction coefficient and PI/graphite-20 wt% composite showed the least wear loss.	Yang et al. (2020)
Exfoliated graphite (EG)	Silicone rubber (QM) 5060-U	Pin-on-disc	Specific wear rate and coefficient of friction	Value of coefficient of friction and specific wear rate were reduced as the exfoliated graphite content increased.	Sarath et al. (2020)
Electrolytic copper powder, graphite powder	Phenolic resin	HRS-2M high-speed reciprocating friction testing machine	Friction and wear rate	The friction coefficient and wear rate of the composites with added foam copper are smaller than those of the graphite/copper composites without foam copper.	Zhu et al. (2020)

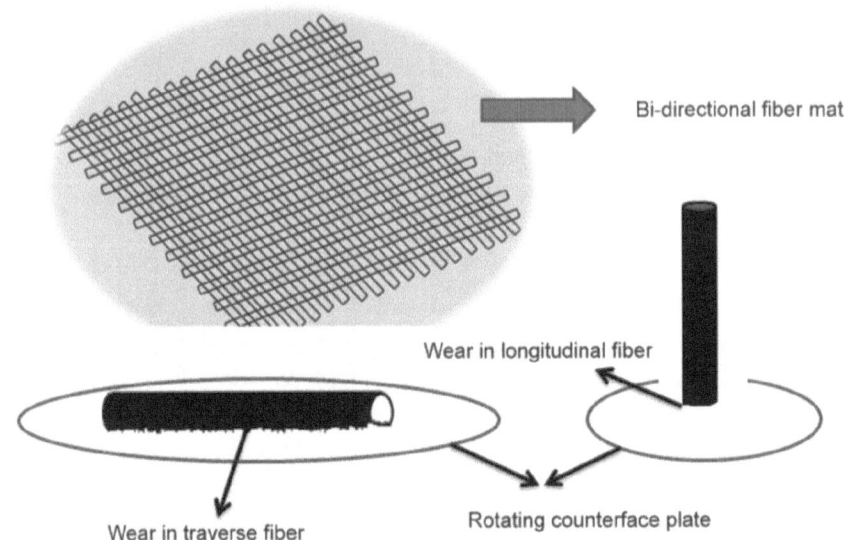

FIGURE 2.3   Wear mechanism during sliding wear analysis on pin-on-disc apparatus.

In both cases, wear takes place because of fiber breakage, fiber fracture, debonding of fiber from the matrix phase and fiber pull-out. When interfacial adhesion between fiber and matrix phase decreases, complete fiber pull-out takes place as a result of large frictional force during wear analysis. Figure 2.3 shows the arrangement of a bi-directional mat, the geometry of longitudinal fiber and traverse fiber and the mechanism of wear in fibers during sliding on pin-on-disc tribometer.

Frictional analysis and wear rate are calculated in tribological characterization of fiber-reinforced polymer composites as discussed below.

2.5   FRICTIONAL ANALYSIS AND WEAR RATE IN FIBER-REINFORCED POLYMER COMPOSITES

During tribological analysis of fiber-reinforced polymer composites, rubbing action between the specimen and the counterface of the tribo testing machine generates the friction which is recorded by using computer software. The coefficient of friction was calculated by the following equation.

$$Cofficient\ of\ friction(\mu) = \frac{Frictional\ force\left(F_f\right)}{Applied\ load\left(F_n\right)} \tag{1}$$

In the case of specific wear rate, the initial weight of the test specimen was calculated before testing and then the specimen was subjected to the wear test. The following equation was used to calculate the specific wear rate of the test sample.

$$Specific\,wear\,rate\,\left(K_S\right)$$

$$= \frac{Mass\,loss\,during\,test\,(m)}{Density\,(\rho)\times Applied\,load\,\left(F_n\right)\times Sliding\,distance\,(L)} \quad (2)$$

Various authors carried out studies on wear and friction analysis of fiber-reinforced polymer composites.

Manoharan et al. (2019) fabricated hybrid composites with recycled basalt and aramid fibers reinforcing phenol formaldehyde. The authors performed tribological testing on a pin-on-disc machine. They concluded that incorporation of basalt and aramid fiber into phenol formaldehyde polymer matrix enhanced the coefficient of friction and reduced the specific wear rate of the developed composite as compared to a neat phenol formaldehyde sample. They also concluded that 25 wt% of fiber content reinforcement in composite gives better results for tribological properties as compared to other weight percentages of fiber. Barari et al. (2016) developed eucalyptus Kraft pulp/bio-epoxy composite and performed their tribological testing on a pin-on-disc machine. The authors concluded that the weight percentage of fiber influences the tribological properties of fiber-reinforced polymer composites. They also concluded that addition of eucalyptus Kraft pulp with bio-epoxy gives low wear loss as compared to neat bio-epoxy. Toth et al. (2020) investigated the tribological properties of basalt and jute fiber–reinforced unsaturated polyester/vinylester–based thermoset composites using a sliding flat-on-flat test configuration. They blended the polytetrafluoroethylene (PTFE), polyoxymethylene (POM) and MoS_2 filler with fiber and matrix phase. The authors concluded that composites blended with PTFE showed the lowest coefficient of friction and longest service life and MoS_2-filled composites achieved the highest coefficient of friction. Naik et al. (2020) performed the tribological analysis of alkali- and silane-treated *Cissus quandrangularis* stem fiber reinforcement. The authors concluded that chemical treatment (alkali and silane) enhanced the interfacial adhesion of fiber and matrix phase, which improves the tribological performance of developed composites.

2.6 CONCLUSION

The following points can be concluded from the findings of this review:

- Fiber-reinforced polymers are widely used for bearing applications because of their good wear and frictional properties.
- Incorporation of reinforcing fiber with polymer matrix material enhances the tribological properties of polymer composites.
- Tribological properties of fiber-reinforced polymer composites are largely influenced by the orientation of fiber, volume percentage of fiber and polymer matrix, and the shape and size of reinforcing fiber.
- Strong interfacial adhesion between fiber and polymer matrix largely depends on the surface properties of the fiber, which greatly influences the tribological properties of fiber-reinforced polymer composites.

- In tribological testing, wear and tear of fiber-reinforced polymer composites for every orientation of fiber occurs as a result of fiber breakage, matrix breakage, fiber pull-out and fiber debonding.

2.7 FUTURE SCOPE

Fiber-reinforced polymer composites have a wide scope of applications in the automotive, aerospace, heavy industry, healthcare and other fields. Future research is needed to develop sustainable fiber–reinforced polymer composites that reduce environmental impact and waste compared to plastics. The future is likely to see greater adoption of sustainable fiber–reinforced polymer composites by the manufacturing as well as healthcare sectors.

REFERENCES

Agarwal, Bhagwan D., Lawrence J. Broutman, and K. Chandrashekhara. *Analysis and performance of fiber composites.* John Wiley & Sons, 2017.

Barari, Bamdad, Emad Omrani, Afsaneh Dorri Moghadam, Pradeep L. Menezes, Krishna M. Pillai, and Pradeep K. Rohatgi. "Mechanical, physical and tribological characterization of nano-cellulose fibers reinforced bio-epoxy composites: An attempt to fabricate and scale the 'Green' composite." *Carbohydrate Polymers* 147 (2016): 282–293.

Chand, Navin, and Mohammed Fahim. *Tribology of natural fiber polymer composites.* Woodhead Publishing, 2008.

Chaudhary, Vijay, Akash Kumar Rajput, and Pramendra Kumar Bajpai. "Effect of particulate filler on mechanical properties of polyester based composites." *Materials Today: Proceedings* 4, no. 9 (2017): 9893–9897.

Chaudhary, Vijay, Pramendra Kumar Bajpai, and Sachin Maheshwari. "Studies on mechanical and morphological characterization of developed jute/hemp/flax reinforced hybrid composites for structural applications." *Journal of Natural Fibers* 15, no. 1 (2018a): 80–97.

Chaudhary, Vijay, Pramendra Kumar Bajpai, and Sachin Maheshwari. "Effect of moisture absorption on the mechanical performance of natural fiber reinforced woven hybrid bio-composites." *Journal of Natural Fibers* (2018b).

Chaudhary, Vijay, Pramendra Kumar Bajpai, and Sachin Maheshwari. "An investigation on wear and dynamic mechanical behavior of jute/hemp/flax reinforced composites and its hybrids for tribological applications." *Fibers and Polymers* 19, no. 2 (2018c): 403–415.

Chaudhary, Vijay, Khushi Ram, and Furkan Ahmad. "Reprocessing and disposal mechanisms for fiber reinforced polymer composites." In *Reinforced polymer composites: Processing, characterization and post life cycle assessment*, pp. 249–261. Wiley, 2019.

Chaudhary, Vijay, and Furkan Ahmad. "Effect of moisture absorption on the wear and dynamic mechanical behavior of polymer composites." *Journal of Reinforced Plastics and Composites* 39, no. 15–16 (2020a): 572–586.

Chaudhary, Vijay, Rohit Sahu, Ankit Manral, and Furkan Ahmad. "Comparative study of mechanical properties of dry and water aged jute/flax/epoxy hybrid composite." *Materials Today: Proceedings* 25 (2020b): 857–861.

Chaudhary, Vijay, and Furkan Ahmad. "A review on plant fiber reinforced thermoset polymers for structural and frictional composites." *Polymer Testing* (2020c): 106792.

Correa, Carlos Eduardo, Santiago Betancourt, Analía Vázquez, and Piedad Gañan. "Wear resistance and friction behavior of thermoset matrix reinforced with Musaceae fiber bundles." *Tribology International* 87 (2015): 57–64.

Daniel, Isaac M., Ori Ishai, Issac M. Daniel, and Ishai Daniel. *Engineering mechanics of composite materials.* Vol. 1994. New York: Oxford University Press, 2006.

Edoziuno, Francis O., Richard O. Akaluzia, Benjamin U. Odoni, and Salifu Edibo. "Experimental study on tribological (dry sliding wear) behaviour of polyester matrix hybrid composite reinforced with particulate wood charcoal and periwinkle shell." *Journal of King Saud University-Engineering Sciences* (2020).

El-Tayeb, N. S. M., B. F. Yousif, and T. C. Yap. "An investigation on worn surfaces of chopped glass fibre reinforced polyester through SEM observations." *Tribology International* 41, no. 5 (2008): 331–340.

Koronis, Georgios, Arlindo Silva, and Mihail Fontul. "Green composites: A review of adequate materials for automotive applications." *Composites Part B: Engineering* 44, no. 1 (2013): 120–127.

Kumar, Rahul, Bastien Malaval, Maksim Antonov, and Gai Zhao. "Performance of polyimide and PTFE based composites under sliding, erosive and high stress abrasive conditions." *Tribology International* 147 (2020): 106282.

Kumar, Subhash, K. K. Singh, and Santosh Kumar. "Tribological behaviour of glass/ epoxy laminated composite reinforced with graphene and MWCNT." *Materials Today: Proceedings* 22 (2020): 2791–2797.

Lee, Jae Hun, Min Su Park, Chang Soo Lee, Tong-Seok Han, and Jong Hak Kim. "Wear-resistant carbon nanorod-embedded poly (vinylidene fluoride) composites with excellent tribological performance." *Composites Part A: Applied Science and Manufacturing* 129 (2020): 105721.

Lin, Zhe, Xiaohua Jia, Jin Yang, Yong Li, and Haojie Song. "Interfacial modification and tribological properties of carbon fiber grafted by TiO_2 nanorods reinforced novel depolymerized thermosetting composites." *Composites Part A: Applied Science and Manufacturing* 133 (2020): 105860.

Manoharan, S., R. Vijay, D. Lenin Singaravelu, S. Krishnaraj, and B. Suresha. "Tribological characterization of recycled basalt-aramid fiber reinforced hybrid friction composites using grey-based Taguchi approach." *Materials Research Express* 6, no. 6 (2019): 065301.

Mazumdar, Sanjay. *Composites manufacturing: Materials, product, and process engineering.* CRC Press, 2001.

Naik, Santosh Madeva, Vinod Moger, Veeranna Gedigeri, Ajith Kumar, and M. Sreeramulu. "Tribological properties of Cissus quandrangularis stem fiber reinforced isophthalic polyester composites." *Materials Today: Proceedings* 46, no. 16 (2020).

Nasir, R. M., M. R. A. Montaha, V. Radha, A. Y. Saad, and H. W. Gitano-Briggs. "Tribological performance of resin impregnated gunny (RIG) and resin reinforced honeycomb (RRH) material composites." *Materials & Design* 48 (2013): 34–43.

Nirmal, Umar, Jamil Hashim, and K. O. Low. "Adhesive wear and frictional performance of bamboo fibres reinforced epoxy composite." *Tribology International* 47 (2012): 122–133.

Pickering, Kim L., and Tan Minh Le. "High performance aligned short natural fibre–Epoxy composites." *Composites Part B: Engineering* 85 (2016): 123–129.

Sarath, P. S., Sohil Varghese Samson, Rakesh Reghunath, Mrituanjay Kumar Pandey, Józef T. Haponiuk, Sabu Thomas, and Soney C. George. "Fabrication of exfoliated graphite reinforced silicone rubber composites – mechanical, tribological and dielectric properties." *Polymer Testing* 89 (2020): 106601.

Sen, Mainak, Pujan Sarkar, Nipu Modak, and Prasanta Sahoo. "Woven E-glass fiber reinforced epoxy composite – preparation and tribological characterization." *International Journal of Materials Chemistry and Physics* 1, no. 2 (2015): 189–197.

Sundarakannan, R., V. Arumugaprabu, V. Manikandan, and R. Deepak Joel Johnson. "Tribo performance studies on redmud filled pineapple fiber composite." *Materials Today: Proceedings* 24 (2020): 1225–1234.

Toth, Levente Ferenc, Jacob Sukumaran, Gábor Szebényi, Ádám Kalácska, Dieter Fauconnier, Rajini Nagarajan, and Patrick De Baets. "Large-scale tribological characterisation of eco-friendly basalt and jute fibre reinforced thermoset composites." *Wear* 450 (2020): 203274.

Wei, Jianqiang, and Christian Meyer. "Utilization of rice husk ash in green natural fiber-reinforced cement composites: Mitigating degradation of sisal fiber." *Cement and Concrete Research* 81 (2016): 94–111.

Yang, Ming, Chunhong Zhang, Guangdong Su, Yanjun Dong, and Tadele Daniel Mekuria. "Preparation and wear resistance properties of thermosetting polyimide composites containing solid lubricant fillers." *Materials Chemistry and Physics* 241 (2020): 122034.

Zhu, Linying, Maozhong Yi, Liming Wang, and Shengan Chen. "Effects of foam copper on the mechanical properties and tribological properties of graphite/copper composites." *Tribology International* 148 (2020): 106164.

3 Tribological Behavior of Fiber-Reinforced Polymer Composites (FRPC)

J. Ashok Raj,[1]* L. Arulmani,[2] Thrinai Pavan Lavuluri,[3] Kadiyala B. Drupad[3]

[1]Department of Automobile Engineering, Bharath Institute of Higher Education and Research, Chennai, India

[2]Department of Mechanical Engineering, RR Institute of Technology, Bangalore, India

[3]Department of Mechanical Engineering, GITAM School of Technology, Bangalore, India

*Corresponding author: ashoktribology@gmail.com

CONTENTS

DOI: 10.1201/9781003109723-3

3.1   INTRODUCTION

Composites are materials composed of different, versatile constituents that are chemically distinct and prominently possess a distinct interface between the constituents (Rubio-López et al., 2017, Katogi et al., 2016, Mohammed et al., 2015). Two or more discontinuous phases are included in a continuous phase to assemble a hybrid composite. The discontinuous phase is known as hybrid reinforcement and is generally unbreakable and more impregnable than the continuous phase, and the continuous phase is designated as a matrix. Matrix media can be categorized as metallic, polymeric, and ceramic. Recently, the polymer matrix composites have been used for such different applications as automobile parts, interior components of airplanes, household appliances, and construction materials (Rubio-López et al., 2017, Katogi et al., 2016, Mohammed et al., 2015, Bijwe et al., 2001, Kolluri et al., 2009).

The reinforcing phase can be classified as fibrous or non-fibrous, and if the fibers are extracted from flora or some other living organism, they are termed natural fibers. Environmental problems have aroused significant interest in the development of the latest composite materials with biodegradable resources, such as natural fibers, which are low-cost and eco-friendly alternatives to synthetic fibers. These fibers have been extricated by different methods such as mechanical decorticators, water retting, and chemical retting.

The fibers that are fabricated using chemicals are called synthetic fibers and include carbon, glass, aramid, boron, and ceramic fibers. The fibers that are obtained from plants and naturally contain certain chemicals (cellulose, hemicellulose, lignin, pectin, wax, and moisture) are called natural fibers and may include jute, flax, hemp, sisal, coir, banana, agave, snake grass fiber, and others. Hybrid fibers (combinations of either natural fibers, synthetic fibers, or natural with synthetic fibers) may also be used to reinforce a polymer matrix amalgamation. The hybrid fibers in the composites can resist extreme loads compared with single-fiber reinforcements in different ways depending on the reinforcement, and the surrounding matrix sustains them in the desired position and orientation, acting as a higher load transfer medium between them (C. Singh and J. Singh, 2014, Basavarajappa and Chandramohan, 2005).

Commercially available automobile brake pads are a possible application for organic resistance materials since the matrix of these multiplex composites is made by one or more polymers (Rubio-López et al., 2017). The frictional surface materials generally contain four classes with additives: binders, reinforcements, friction modifiers, and fillers. Building a successful friction material generally requires discovering the best balance among different factors of yielding acceptable execution, capital costs, and eco-friendliness. Asbestos fiber is a superior friction fiber that enhances the polymeric composites in brake pads, brake linings, and brake couplings and so on, but it has been restricted because of dangerous environmental issues and human health complications (Mishra, 2012, Nguong et al., 2013, Shivamurthy et al., 2009). According to the regulations against hazardous ingredients in the United States and Europe, several raw materials are typically used in commercial friction materials that would have a potential negative impact on the environment. Friction compounds that are produced by chemicals such as antimony trisulfide, copper, lead, tin, potassium titanate whisker, silicon carbide whisker, and others have been widely used in the past (Katogi et al., 2016, Mohammed et al., 2015). Increased environmental awareness around the world has raised interest in natural fibers and their use in different applications.

Natural fibers are now being considered as a major replacement for synthetic fibers used in different automobile sectors. The use of natural fibers as reinforcing materials in both thermoplastics and thermoset matrix composite has a positive environmental impact with respect to ultimate degradability and premier utilization of raw materials. During the last few years, a sequence of studies has been completed on replacing the conventional synthetic fibers with natural fibers (Bijwe et al., 2001, Kolluri et al., 2009, C. Singh and J. Singh, 2014, Basavarajappa and Chandramohan, 2005, Mishra, 2012, Nguong et al., 2013, Shivamurthy et al., 2009). For example, hemp, sisal, jute, cotton, flax, and broom are the most widely used fibers to reinforce polymers such as polyolefin, polystyrene, and epoxy resins. In addition, fibers such as sisal, jute, coir, oil palm, bamboo, wheat and flax straw, waste silk, and banana have been shown to provide good and efficient reinforcement in the thermoset and thermoplastic matrices. The benefits of natural fibers over traditional reinforcing compounds such as glass fiber and carbon fiber are their specific strength properties, enhanced energy recovery, high toughness, corrosion resistance, good thermal properties, reduced tool wear, and reduced dermal and respiratory problems. Table 3.1 describes the mechanical functions of natural fibers compared to conventional reinforcing fibers. Natural fiber–reinforced composite has been reported to have similar properties to traditional synthetic fiber–reinforced composite (Rajesh et al., 2012, Basavarajappa and Chandramohan, 2005).

TABLE 3.1

Mechanical Properties of Natural Fibers Compared to Conventional Reinforcing Fibers

Fiber	Density (g/cm³)	Elongation	Tensile Strength	Young's Modulus
Cotton	1.5–1.6	7.0–8.0	287–597	5.5–12.6
Jute	1.3	1.5–1.8	393–773	26.5
Flax	1.5	2.7–3.2	345–1035	27.6
Hemp	–	1.6	690	–
Ramie	–	3.6–3.8	400–938	614–128
Sisal	1.5	–	1000	40.0
Coir	1.2	30.0	175	4.6–6.0
Viscose (cord)	–	11.4	593	11.0
Softwood Kraft	1.5	–	1000	40.0
E-glass	2.5	2.5	2000–3500	70.0
S-glass	2.5	2.8	4570	86.0
Aramid (normal)	1.4	3.3–3.7	3000–3150	63.0–67.0
Carbon (standard)	1.4	1.4–1.8	4000	230–240

Natural fibers as reinforcement have gained the attention of researchers because of their advantages over other available materials. They are eco-friendly, completely biodegradable, abundantly available, reusable, low cost, and have low density. Plant fibers are light compared to glass, carbon, and aramid fibers. The biodegradability of plant fibers is good for the environment, while their low cost and high performance fulfill the economic demands of industries. Natural fiber–reinforced plastics, by using biodegradable polymers as matrices, can be the most environmentally friendly composites. A number of automotive components previously manufactured with glass fiber composites are now being manufactured with environmentally friendly composites (Deo and Acharya, 2009).

In the United States and Europe, car manufacturers are aiming to manufacture every part of their vehicles to be either recyclable or biodegradable. Even natural fibers and their composites that are environmentally friendly and reusable have various drawbacks. They have poor wettability, incompatibility with some polymeric matrices, and high moisture absorption (Kranthi et al., 2010). Thus, composite materials that are manufactured with the use of unmodified plant fibers have exhibited unsatisfactory mechanical properties. To overcome this, in many cases, surface treatment or compatibilization agents need to be used prior to composite fabrication. The properties can be improved by both physical and chemical treatments (Gohil and Shaikh, 2010a). During the past two decades, a number of significant industries such as automotive, construction, and packaging have shown enormous interest in the development of new, natural fiber–reinforced composite materials. However, little information concerning the tribological performance of natural fiber–reinforced composite material has been available. Hence, this chapter reviews the tribological performance of natural fiber–reinforced composites (Table 3.2 and Table 3.3 later in the chapter) (Basavarajappa and Chandramohan, 2005, Deo and Acharya, 2009, Kranthi et al., 2010).

3.1.1 Natural Fiber–Reinforced Polymer Composites

The research on natural fibers has expanded greatly, motivated by environmentalism and the surging cost of synthetic materials. Natural fiber–reinforced polymer composite (NFRPC) (Figure 3.1) is an alternative to the man-made fiber–reinforced polymeric composites because natural fibers are profusely available, economical, reusable and biodegradable and, moreover, possess good mechanical strength. Thus NFRP composite research and manufacturing applications are growing rapidly. The natural fibers termed cellulosic plant fibers, like bamboo, jute, coir, banana, oil palm, and kenaf, are often used as reinforcement in a polymeric matrix (Schön, 2004, Boopathi et al., 2012, Mahapatra and Chaturvedi, 2009). Figure 3.2 describes the differentiation of polymer composites based on the type of polymer and reinforcements. During production, a strong mold or shaping tool is essential for forming the specific structure of the component. Mostly, FRP components are made with the use of a tool or a mold that can be a concave female mold, a male mold, or a mold that can entirely enclose the part with a top and bottom. Several lines of current research are the fittingness, keenness, and competence of NFRP composites (Gohil and Shaikh, 2010a, Schön, 2004, Boopathi et al., 2012, Mahapatra and Chaturvedi, 2009).

FIGURE 3.1 Life cycle of NFRP composites.

Rubio-López et al. (2017) fabricated a biodegradable composite using polylactic acid (PLA) as a biodegradable matrix and flax as a natural fiber. Flax-PLA composites have better mechanical properties than the carbon-epoxy composites. The natural fibers from different sets of plants have different chemical combinations of cellulose, hemicellulose, and lignin, resulting in different properties. Katogi et al. (2016) suggested that the bonding between fiber and matrix can be further improved by treating the fiber surfaces using various methods. The NFRPCs are generally thermally stable at about 240°C, beyond which denaturing starts. The thermal stability of natural fibers can be developed using chemical treatments to diminish the hemicelluloses, lignin, and wax contents on the surface of the fiber. Figure 3.3 shows different parameters that influence the mechanical and tribological behavior of composites (Rubio-López et al., 2017, Katogi et al., 2016, Mahapatra and Chaturvedi, 2009, Wong et al., 2010).

The staging and characteristics of NFRPCs are also controlled by the hydrophilicity of natural fiber, fiber-loading percentage, manufacturing techniques, the interaction of fiber with the matrix material, surface texture and chemical configuration of natural fiber, fiber geometry (length) and particle geometry (shape and size) (Rao et al., 2012, Majhi et al., 2012, Ishak et al., 2009). The properties of NFRP composites

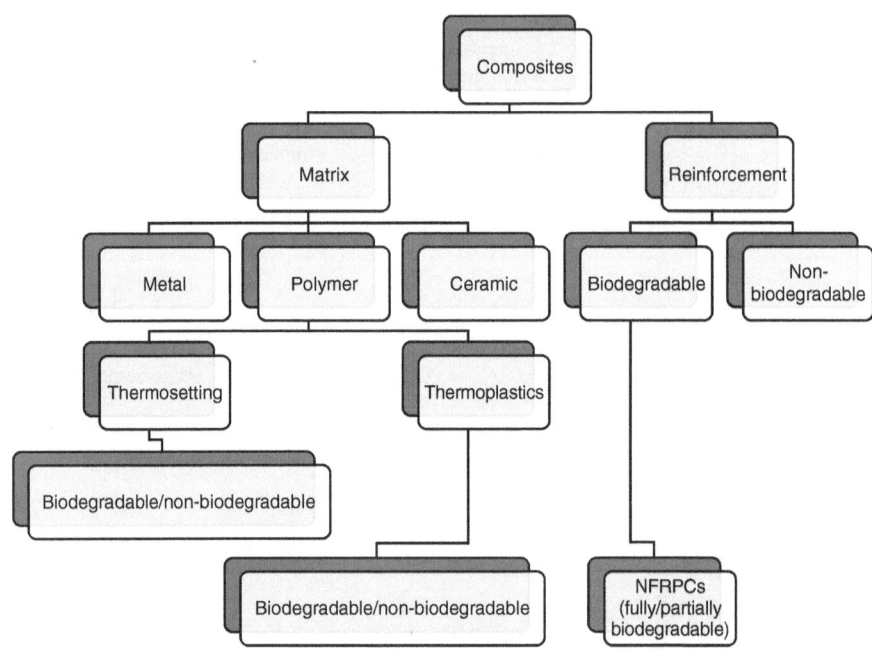

FIGURE 3.2 Classification of polymer-based composite materials.

mainly rely on the fiber volume fraction, orientation, surface quality, hydrophilicity, and manufacturing process limitations. Yousif et al. (2009) briefly appraise tensile and flexural properties of NFRPCs and their application potentials. The consequences of the above composite parameters on the tensile and flexural properties of NFRP composites are considered next.

3.1.2 Effects on Tensile and Flexural Properties

A review of the literature shows that fiber loading, surface treatments, interfacial adhesion, fiber orientation, and physical and chemical properties are the fiber properties by which the mechanical performance of NFRPCs can be projected (Rubio-López et al., 2017, Basavarajappa and Chandramohan, 2005). Process parameters such as curing temperature and pressure, methodology, and environment for the composite development also influence the mechanical properties and overall performance of NFRPCs. The mechanical characteristics of NFRPCs include tensile, compressive, impact and flexural strengths, creep resistance, and hardness. These characteristics mainly depend on the interfacial adhesion between the fiber and matrix, fiber strength, the physical properties of the fiber, fiber volume fraction, orientation and moisture uptake resistance, and properties of the constituent materials. A large interfacial adhesion smooths the stress transfer between fiber and matrix (Kolluri et al., 2009, Mishra, 2012, Nguong et al., 2013, Yousif et al., 2009).

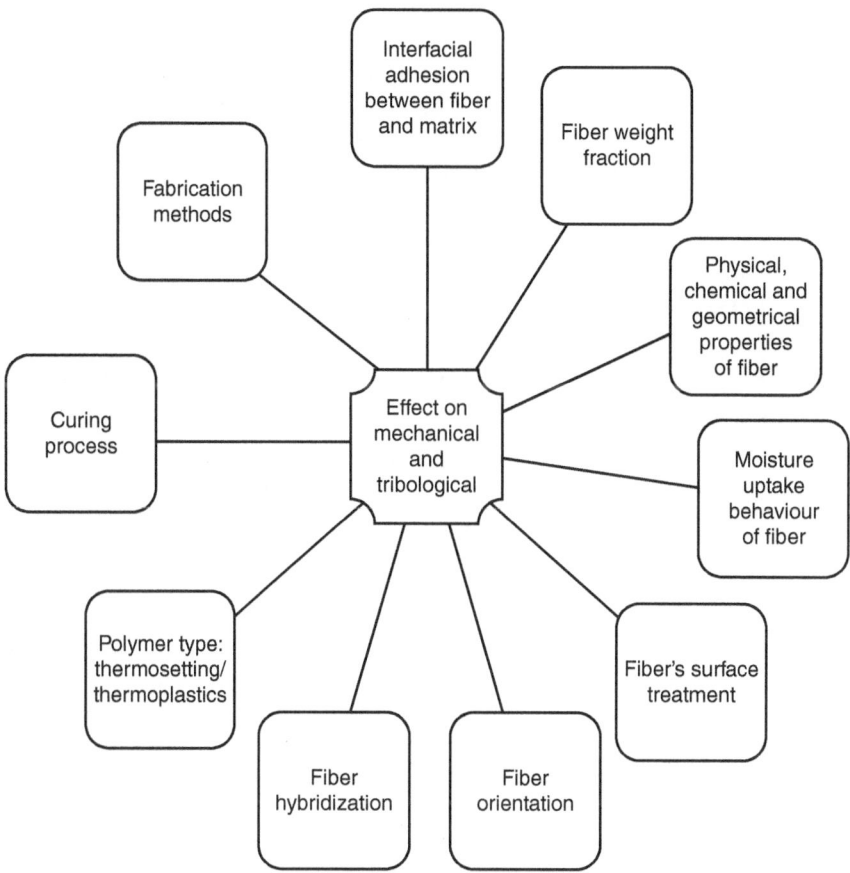

FIGURE 3.3 Factors affecting mechanical and tribological properties of natural fiber–reinforced plastic composites.

Mohammed et al. (2015) reported an increase in the stiffness, stress transfer, loss modulus, storage modulus, hardness, elongation, and fracture strength of NFRPCs with an increase in the fiber contents. In the amalgamation process, air and other volatile particles may be trapped inside the composite; these will be converted into micro-voids near the fiber-deficient region, leading to sudden failure of the composite. The existence of a void also depends on the curing and cooling temperature. Larger voids are the major cause of large differences in mechanical properties and large moisture uptake. The hydrophilic character of natural fiber diminishes the adhesion between fiber and matrix, resulting in mediocre stress transfer from the fiber to the matrix and vice versa. This drawback is reduced by different types of fiber surface treatment. Alkali treatment gives better mechanical properties than other treatments. The mechanical properties can be further improved by the surface treatment of fiber with doping compatibilizers or a coupling agent (Ishak et al., 2009, Yousif et al., 2009, Yousif et al., 2010).

3.1.3 Applications

NFRPCs have potential for increased use in agricultural equipment, automobile components, packaging products, aerospace interiors, recreation equipment, boats, safety kits, sports items, electronic goods, and other applications, thereby reducing the use of limited conventional resources. Materials consisting of NFRPCs are economical, comparably reliable, eco-friendly and reduce fuel consumption in automobiles. German auto manufacturers have broadly used NFRPCs for the improvement of different automobile components. The existing components are increasingly being replaced by NFRPCs.

3.1.4 Analyses of Tribological Properties of Synthetic Fiber–Reinforced Polymer Matrix Composites

Fiber-reinforced polymer matrix composites (FRPCs) are one of the most important classes of tribo materials. Extensive wear and friction are one of the main causes of several kinds of engineering systems failure. The friction and wear rate of the material depends on the materials chosen for reinforcement and resin, manufacturing process, operational parameters, fiber orientation, fiber volume fraction, fiber length, and surface treatments (Rubio-López et al., 2017, Katogi et al., 2016, Mohammed et al., 2015, Bijwe et al., 2001). There is no material that is ideal for all kinds of wear models. Hence it is important to study the tribo properties of composites by evaluating them in the laboratory under various operating conditions. Bijwe et al. (2001) experimented with friction and wear behavior of polyetherimide glass fiber composites under different wear modes with different percentages of fiber. From the experimentation, the authors concluded that the rate of wear resistance of composites is different for different types of wear modes and fiber percentage. Kolluri et al. (2009) investigated friction and wear of phenolic composites with varying temperature. It was found that the wear rate of the phenolic composite increased with increasing temperature irrespective of graphite particle size and loading condition. C. Singh and J. Singh (2014) have worked on Al composites prepared by powder metallurgy. Their work has proved that the wear resistance of the Al composition can be improved by adding SiC fillers. Basavarajappa and Chandramohan (2005) have examined wear for MMCs. Mishra (2012) has done experimentation on the tribological behavior of rubber dust epoxy composite by changing the rubber percentage. He found that wear resistance was maximal for specimens with 10% rubber dust (Kolluri et al. 2009, C. Singh and J. Singh, 2014, Basavarajappa and Chandramohan, 2005, Mishra, 2012).

Nguong et al. (2013) proved that the tribological properties of polymer composites are increased using nanoparticles like silica carbide and nano clay. Shivamurthy et al. (2009) worked on glass fiber epoxy composite with SiO_2 as filler; Rajesh et al. (2012) worked on Al matrix with red mud fillers; Basavarajappa and Chandramohan (2005) have worked on glass fiber epoxy with SiO_2 fillers. From all the studies it was found that fillings have a significant effect on the wear rate of composite materials and the wear rate can be controlled by varying the amount of filling. Table 3.2 shows the various research in the field of wear and friction analysis of synthetic fiber–polymer matrix composites.

TABLE 3.2
Wear Rate and Friction Analysis of Synthetic Fiber–Polymer Matrix Composites

Research	Polymer	Fiber	Fillers	Variables[1]	Manufacturing Technique	Dry Sliding Coefficient of Friction	Dry Sliding Wear Rate	Observed Wear Mechanism
Basavarajappa and Chandramohan, 2005	Epoxy	Glass	Graphite and sic	SV, L, SD	Hand lay up	—	Decrease	—
Rajesh et al. 2012	–	Al	Red mud	L, SV, FVF	Powder metallurgy	Decrease	Decrease	Adhesive
Shivamurthy et al. 2009	Epoxy	Glass	SiO$_2$	L, filler content	Cylindrical mold	Decrease	Decrease	Adhesion
Mishra, 2012	Epoxy	Rubber dust	–	Speed, L, time	Molding	–	Decrease	Abrasive
Yousif et al., 2009	Polyester	Glass	–	Speed, FO	Hand lay up	Decrease	Decrease	Back transfer polymer flim
Kolluri et al. 2009	Phenolic	Graphite	–	L, T	Compression molding	High T, high FF	High T, high wear	–
Bijwe et al. 2001	PEI	Glass	Graphite, MoS$_2$, PTFE	L	–	–	Decrease	Adhesion
Yousif and El-Tayeb, 2010	Polyester	Glass	–	SD, L	Molding	Decrease	Decrease	Plastic deformation

Note: FF, friction force; FO, foam out; FVF, fiber volume fraction; L, load; PEI, polyethylenimine; PTFE, polytetrafluoroethylene; SD, sliding distance; SiC, silicon carbide; SV, sliding velocity; T, temperature.

Extensive work has been carried out on sustainable materials from natural resources owing to their environmental appeal. Because natural fibers are biodegradable, low cost, readily available, and low weight, research on natural fibers has increased enormously in the last two decades. The major drawback of natural fibers is their biodegradability, as it is one of the main properties of celluloses. Degradation of the material takes place by oxidation as hemicelluloses absorb moisture. Oxidation of the material can be reduced by modifying fibers by chemical or physical treatments. Hence it is worth investigating tribological properties of natural fiber–reinforced composites (Basavarajappa and Chandramohan, 2005, Deo and Acharya, 2009, Kranthi et al., 2010, Gohil and Shaikh, 2010a, Schön, 2004, Boopathi et al., 2012).

3.1.5 TRIBOLOGICAL PROPERTIES OF NATURAL FIBER–REINFORCED POLYMER MATRIX COMPOSITES

Different operating parameters and material parameters affect the tribological properties of fiber-reinforced polymer matrix composites (FRPCs). The major parameters responsible for the excessive wear and friction investigated by different authors have been illustrated here. Composites are produced by combining fibers and resin materials. The orientation of the fibers has a major effect on the friction and wear behavior of the materials; in the composites, the fibers work as reinforcement and take up the maximum load. Many researchers have conducted various investigations of fiber orientation like normal, parallel, anti-parallel, and random and have listed the results. Tribological characteristics have been studied by various investigators on composite materials from combining different natural fibers with polymer resins, such as jute and linen with unsaturated polyester resin, bamboo fibers with epoxy, kenaf fiber with epoxy resin, bagasse fiber with epoxy resin, and powdered bamboo fiber with polyester reinforcement. Investigations of these composites using different models have given the same conclusion that normal orientation gives a minimal wear rate while the random orientation gives a maximal wear rate. Investigators revealed wear rate is significantly affected by fiber orientation. These wear rates are based on different operating parameters and could not be directly used to draw conclusions about the fiber (Basavarajappa and Chandramohan, 2005, Deo and Acharya, 2009, Kranthi et al., 2010, Gohil and Shaikh, 2010a, Schön, 2004).

3.1.6 FIBER VOLUME FRACTION

Fiber volume fraction is related to load-carrying capacity; hence it contributes to taking up a maximum load in the composite material. Investigators found that a higher fiber volume fraction increases the load-carrying capacity up to a limit; after that, an increase in fiber is responsible for debonding between fiber and resin. Several studies have been conducted on natural fiber volume fractions: Deo and Acharya (2009) worked on *Lantana camara* fiber with polymer matrix; Kranthi et al. (2010) have investigated different fiber weight fraction of composite material formed from pine wood dust with epoxy resin. Gohil and Shaikh (2010a) have experimented on banana fiber–reinforced epoxy composite and studied how mechanical characteristics of composition are affected by fiber volume fraction. Schön (2004) worked on carbon fiber with fiber volume fractions of 3–9% with epoxy resin. All these studies concluded that a high fiber volume fraction leads to poor bonding between the fiber and the resin, decreasing the wear and friction properties.

3.1.7 FIBER LENGTH

Short fiber length results in reduced load-carrying capacity leading to a high wear rate. Fiber length contributes to creating interface bonding between fiber and matrix. Up to a limit, long fibers increase interface bonding, thereby increasing load-carrying capacity and reducing wear rate. Fiber length plays a vital role in creating interfacial bonding between fiber and matrix (Kolluri et al., 2009, C. Singh and J. Singh, 2014, Basavarajappa and Chandramohan, 2005). Beyond a certain fiber length, wear rate increases because excessive length results in easy pull-out of the fibers. Several experimental works have been carried out by different researchers: Boopathi et al. (2012) analyzed the wear rate for *Borassus* fruit fiber with various fiber lengths with epoxy resin. Mahapatra and Vedansh (2009) have experimented on sugarcane fiber with different fiber lengths. From all the investigations it was evident that fibers of optimum length provided the best wear and frictional properties over too-short and too-long fibers.

3.1.8 SURFACE TREATMENT

Cellulose is responsible for the degradation of material and cellulose is the major content of natural fibers. Combining cellulosic fibers and matrix forms a very weak interface bond as the fibers are hydrophilic whereas the matrix is hydrophobic. A weak bond results in a higher wear rate. To overcome this problem many investigations were done. Yousif and El-Tayeb (2010) experimented on the effect of NaOH treatment on oil palm polyester composites. Wong et al. (2010) analyzed the effect of alkaline treatment on bamboo fibers. Rao et al. (2012) and Majhi et al. (2012) investigated benzoline treatment of coir and rice husk fibers respectively; the fibers were then used with epoxy resin. Ishak et al. (2009) experimented on the effect of seawater treatment on sugarcane fiber. Yousif et al. (2009) registered a patent for the effect of surface treatment on coir fiber–reinforced polymer matrix composites. All these investigations showed that surface treatment improves adhesion between the fiber and the matrix.

3.1.9 OPERATING PARAMETERS

Yousif et al. (2010) conducted wear analysis on betelnut, and Navin and Fahim (2009) performed wear analysis of sisal fiber polyester composites under different loading conditions. N. Singh et al. (2011) investigated kenaf fiber under different load and sliding distance conditions. Gohil and Shaikh (2010a) have identified the effect of various operating parameters on glass fiber epoxy composites. From all these studies it can be concluded that the wear performance of the composites depends on the different operating parameters. In this chapter, several works are reviewed and listed in Table 3.3 for the wear rate and friction coefficient of different types of natural fiber–polymer matrix composites.

TABLE 3.3
Wear and Friction Analysis of Natural Fiber–Polymer Composites

Research	Polymer	Fiber	Treatment/ Fillers	Variables	Manufacturing Technology	Dry Sliding		Observed Wear Mechanism
						Coefficient of Friction	Wear Rate	
Nguong et al. 2013	Epoxy	Recycled cellulose fiber	Nano SiC, nano clay	–	–	–	Decrease	–
N. Singh et al. 2011	Polyurethane (PE)	Kenaf	Alkaline	FO, L, SD	Hand lay-up	Decrease APO	Decrease for NO	Plastic deformation
Rao et al. 2012	Epoxy	Coir	Benzolylation	FVF, L, speed	Cylindrical mold	–	Decrease	Abrasive
Majhi et al. 2012	Epoxy	Rice husk	Benzolylation	L, FVF	Cylindrical mold	–	Decrease	Abrasive
Yousif and El-Tayeb, 2010	Polyester	Untreated and traced oil palm	Alkaline	SD, L	Cylindrical mold	Decrease	Decrease	Debonding
Boopathi et al. 2012	Epoxy	Borassus fruit fiber	Alkaline	L, FV, FL	Cylindrical mold	Decrease	Decrease	Micro cracks
Yousif et al. 2009	Polyester	Coir	Alkaline	SD, L	Hand lay-up	Increase	Decrease	Adhesion
Yousif et al. 2009	Polyester	Betelnut fruit fiber	–	SD, L	Hand lay-up	Increase	Decrease	Macro and micro cracks, debonding and fiber pullout
Kranthi et al. 2010	Epoxy	Pine wood dust	–	FVF, SD, SV, L	–	–	Decrease	–
Deo and Acharaya, 2009	Epoxy	*Lantana camara*	–	FVF, L, SD	Molding	Decrease	Decrease	Abrasion

Note: FO, foam out; FVF, fiber volume fraction; L, load; SD, sliding distance; SiC, silicon carbide; SV, sliding velocity.

3.2 NATURAL-FIBER SELECTION AND PREPARATION

Natural fibers can be grouped based on their bast (strong woody) qualities, like those of jute, hemp, kenaf, and flax. These fibers are produced from fibrous sheaves of dicotyl plants or vessel sheaves of monocotyl plants, fibers derived from the seed (cotton), hard fibers from the leaf (sisal, pineapple), and several others. All these fibers have distinct mechanical and physical properties (Tables 3.1 and 3.4). So far, wood from trees is considered to be the most plentiful fiber in the world with an annual world production of 1.75×10^9 tonnes per year from well over 10,000 species. Cotton production, by comparison, is 18.5×10^6 tonnes per year, while kenaf, flax, and hemp are 9.7×10^5, 8.3×10^5, and 2.1×10^5 tonnes per year, respectively. Even though there are a large variety of plant fibers in nature available in great quantity, only a few are suited for automotive applications. Cellulose is the main component of natural fibers. Different fibers have different amounts of pure cellulose, hemicellulose, pectin, lignin, and other extractives. For structural composites produced from natural fiber for value-added applications, the most prominent fibers in use are flax, kenaf, and hemp, primarily owing to their fiber strength (Gohil and Shaikh, 2010a, Schön, 2004, Boopathi et al., 2012, Mahapatra and Vedansh, 2009, Yousif et al., 2009).

Table 3.4 shows the tensile strength and elastic modulus of the major natural and man-made fibers. The table shows calculations of specific strength (tensile strength)

TABLE 3.4
Properties of Selected Natural and Synthetic Fibers

Fiber	Density (g/cm³)	Elongation (%)	Tensile Strength (MPa)	Young's Modulus (GPa)	Reference
Cotton	1.5–1.6	7.0–8.0	287–597	5.5–12.6	C. Singh and J. Singh (2014), Basavarajappa and Chandramohan (2005)
Jute	1.3	1.5–1.8	393–773	26.5	C. Singh and J. Singh (2014)
Flax	1.5	2.7–3.2	345–1035	27.6	Bijwe et al. (2001)
Hemp	1.45	2–4	690	70	Bijwe et al. (2001)
Ramie	–	3.6–3.8	400–938	614–128	Bijwe et al. (2001)
Sisal	1.5	–	1000	40.0	Mishra (2012)
Coir	1.2	30.0	175	4.0–6.0	Mishra (2012)
Viscose (cord)	–	11.4	593	11.0	Nguong et al., (2013)
Softwood Kraft pulp	1.5	–	1000	40.0	Nguong et al., (2013)
E-glass	2.5	2.5	2000–3500	70.0	Nguong et al., (2013)
S-glass	2.5	2.8	4570	86.0	Nguong et al., (2013)
Aramid (normal)	1.4	3.3–3.7	3000–3150	63.0–67.0	Nguong et al., (2013)
Carbon (standard)	1.4	1.4–1.8	4000	230–240	Nguong et al., (2013)

and modulus (modulus of elasticity/density) of different natural fibers. Comparing all the fibers to E-glass, with respect to specific strength it is evident that E-glass is superior to all fibers, although flax fiber is very competitive (3,500 vs. 1,035, respectively). However, the specific modulus value of hemp is comparatively favorable to that of E-glass. From the data in the table, it can be noted that bast fiber characteristics are comparable or in some cases exceed those of glass fibers. Natural fibers like kenaf, jute, and hemp can be planted and harvested two to three times annually, and almost all can be planted and harvested annually. Kenaf grows to a height of 4 meters in four to five months and can yield two or three harvests a year in tropical climates. Kenaf is native to Africa and widely cultivated in the United States. Jute is grown in India, China, Bangladesh, and other countries and can be grown in four to six months. The cultivation of jute has declined in recent years owing to the development of synthetic fibers. Agronomically, jute and kenaf have advantages in their resistance to climatic extremes, pests, and diseases. Hemp has a history of more than 10,000 years as a source of rope, cloth, and other textiles, and it grows as a yearly crop in most climates (Mishra, 2012, Nguong et al., 2013, Shivamurthy et al., 2009, Rajesh et al., 2012, Basavarajappa and Chandramohan, 2005).

All of these plants clean the air by consuming large quantities of carbon dioxide (CO_2) as they have a high rate of assimilation of CO_2, which is the main cause of the greenhouse effect. Sisal can produce fibers for up to 20 years after planting, until the plant blooms and dies. All the plant bast fibers after harvesting undergo retting for separation of useful fibers from the rest of the plant. Fibers are separated by softening and partial rotting, which is done by moistening or soaking. There are several methods to accomplish this, such as creating a moist environment, using microorganisms, or chemical breakdown. These methods help to unbind the fiber and non-fiber portions, during which time the retting process removes the hemicellulose and lignin components. The following are the most prominently used retting processes these days, and each has certain advantages (Nguong et al., 2013, Majhi et al., 2012, Gohil and Shaikh, 2010b).

In the dew retting method for fiber separation, the stalks are left in the field and rain, dew, or irrigation is used to keep the stems moist. Light-brown, coarse fibers are produced by this process, which may take up to five weeks. In water retting, the stems are submerged in water for about 10 days, promoting the growth of bacteria that break down the pectin. Water retting results in good-quality fibers. In warm-water retting, bundles are soaked for 24 hours, after which the water is replaced. For the next two or three days, heat is applied to warm the batch, which results in a uniform, clean fiber. In green retting, used when the fiber is needed for textiles, paper, or fiberboard products, the components are separated using a mechanical process. In chemical retting, pectin is dissolved using chemicals, which allows the components to be separated. This process takes around 48 hours and produces a high-quality product (Basavarajappa and Chandramohan, 2005, Rao et al., 2012, N. Singh et al., 2011).

Even though the chemical retting process separates the components in a short time, the major disadvantage of this process is it affects several properties, including a loss in tenacity, color, and luster. Natural retting processes take longer but the resulting fibers have many desirable characteristics. Bacterially retted fibers have better properties than those retted chemically (Kolluri et al., 2009, C. Singh and J. Singh,

2014, Basavarajappa and Chandramohan, 2005, Mishra, 2012, N. Singh et al., 2011. The properties of natural reinforcing fibers can be altered using a wide range of physical and chemical methods such as techniques to correct for fiber deficiencies and to improve bonding and adhesion of the fibers, dimensional stability, and thermoplasticity. Calendaring, stretching, thermo-treatment, and weaving or integration are the physical methods to produce yarns from natural fibers. Physical methods do not change the chemical composition of the fiber but change the structural and surface properties of the fiber and thereby influence its mechanical bonding to polymers. Surface modification of natural fibers can be used to optimize the properties of the fiber-matrix interface (Rubio-López et al., 2017, Katogi et al., 2016, Mohammed et al., 2015, Bijwe et al., 2001).

3.3   AUTOMOTIVE APPLICATIONS

Interest in bio-based materials, and particularly, natural fiber–reinforced composites, coincides not with laws or regulations in large markets such as the European Union but with the interest of numerous major automakers in worldwide sustainability. Sustainability here relates to corporate responsibility, an automaker's responsibility to its workers and customers and beyond. For instance, DaimlerChrysler's sustainability endeavors attempted phenomenal innovation advancement and innovation exchange activities including the use of bio-based materials within the Philippines, South America, and South Africa. DaimlerChrysler went one step further, distinguishing bio-based materials as one of the two key parts of its efforts to form a worldwide sustainability network. Another key portion was the use of renewable energies as the alternative and replacement for conventional fuels, in an effort to achieve a bio-based car supply chain involving suppliers from the farmer to the car wholesaler (Katogi et al., 2016, Mishra, 2012, Basavarajappa and Chandramohan, 2005).

Global automotive suppliers such as Honda embarked on using natural-fiber materials, such as wood-fiber parts in the floor area of the Pilot sport utility vehicle (SUV), a decision that was driven by engineering considerations as well as corporate philosophy. Overall, the variety of bio-based automotive parts currently in production is astonishing; Daimler is the biggest proponent with up to 50 components in its European vehicles being produced from bio-based materials (Figure 3.2). Manufacture of a number of automotive parts with natural-fiber reinforcement is feasible . Flax, sisal, and hemp are processed into door cladding, seatback linings, and floor panels. Seat bottoms, back cushions, and head restraints are made using coconut fiber, while cotton helps to provide soundproofing, and wood fiber is used in seatback cushions. Under-floor body panels are made from abaca (Kolluri et al., 2009, C. Singh and J. Singh, 2014, Basavarajappa and Chandramohan, 2005, Mishra, 2012, Nguong et al., 2013).

The wide scope of natural fibers has also driven manufacturers to incorporate natural materials into their cars (Table 3.5). For example, the BMW Group in 2004 incorporated 10,000 tonnes of natural fibers, including a considerable amount of renewable raw materials, into its vehicles. At General Motors, a kenaf and flax mixture has been used for the door panel and package tray inserts for Saturn L300s and European-market Opel Vectras, while wood fiber has been used for seatbacks in

TABLE 3.5
Example of Interior and Exterior Automotive Parts
Produced from Natural Materials

Vehicle Part	Materials Used
Interior glove box	Wood/cotton fibers molded, flax/sisal
Door panels	Flax/sisal with thermoset resin
Seat coverings	Leather/wool backing
Seat surface/backrest	Coconut fiber/natural rubber
Trunk panel	Cotton fiber
Trunk floor	Cotton with polypropylene/poly(ethylene terephthalate) (PP/PET) fibers
Insulation	Cotton fiber
Exterior floor panels	Flax mat with polypropylene

the Cadillac DeVille and in the cargo area floor of the GMC Envoy and Chevrolet Trailblazer. In Europe, Ford mounts Goodyear tires that are made with corn on its fuel-sipping Fiestas. Goodyear has found that in comparison to traditional tires, corn-infused tires have lower rolling resistance, so they provide better fuel economy. Wood fiber was used in sliding door inserts for the Ford Freestar. Toyota has an interest in using kenaf to make Lexus package shelves, and the body structure of Toyota's i-unit (single-seat) and i-foot (single-seat, with two legs for stair climbing) vehicles are made using kenaf (Mishra, 2012, Deo and Acharya, 2009, Rao et al., 2012).

Currently, there is a good deal of global research into the applications for natural-fiber composites, and automakers are generating prototypes that offer a look into the destiny of manufacturing. For example, a wide variety of agricultural products have been developed by the U.S. Agricultural Research Service. Groups such as the United Soybean Board and the National Corn Growers Association that concentrate on researching and advancing new markets for members' crops and waste items are assisting research into new applications. In addition, tier 1 suppliers, those who manufacture final products for a brand, are actively engaged in developing proto-type parts: Visteon has developed a system for manufacturing flax-based instrument panels; a process has been developed by Composite to produce door panels from flax; Findlay Industries, one of the suppliers of cargo area floors for the GM and Honda SUVs and the package shelves for Saturn and Opel, also manufactures and supplies headliners for Mack Trucks which are produced from hemp, flax, kenaf, and sisal mixture; and soy-resin body panels have been developed that are currently used on John Deere tractors (Katogi et al., 2016, Mishra, 2012, Basavarajappa and Chandramohan, 2005, Kolluri et al., 2009.

As mentioned earlier, at present several experimental parts of complicated geometries are in either the prototype or production stages. Figure 3.4 and Figure 3.5 illustrate the demanding applications that can be met with natural-fiber composites. Figure 3.6 depicts the front-end grill reinforcement for a Ford Montageträger produced

FIGURE 3.4 Flax, hemp, sisal, wool, and other natural fibers are used to make 50 Mercedes Benz E-Class components.

FIGURE 3.5 (a) Harvester Works combines, in which John Deere has replaced steel gull-wing doors with (b) soy-resin body panels.

Source: Photos courtesy of Richard Wool, University of Delaware, with permission.

from a hemp-polypropylene composite, and Figure 3.5 depicts the underbody panel compression molded from flax-polypropylene for an A-Class DaimlerChrysler automobile (Basavarajappa and Chandramohan, 2005, Mishra, 2012, Nguong et al., 2013, Shivamurthy et al., 2009.

FIGURE 3.6 A front-end grill opening reinforcement for the Ford Montageträger.

REFERENCES

Basavarajappa, S., and G. Chandramohan. "Wear studies on metal matrix composites: A Taguchi approach." *Journal of Materials Science and Technology* 21, no. 6 (2005): 845–850.

Bijwe, Jayashree, J. Indumathi, J. John Rajesh, and M. Fahim. "Friction and wear behavior of polyetherimide composites in various wear modes." *Wear* 249, no. 8 (2001): 715–726.

Boopathi, L., P. S. Sampath, and K. Mylsamy. "Influence of fiber length in the wear behaviour of borassus fruit fiber reinforced epoxy composites." *International Journal of Engineering Science and Technology* 4, no. 09 (2012): 4119–4129.

Deo, Chittaranjan, and S. K. Acharya. "Solid particle erosion of lantana camara fiber-reinforced polymer matrix composite." *Polymer-Plastics Technology and Engineering* 48, no. 10 (2009): 1084–1087.

Gohil, Piyush, and A. A. Shaikh. "Unidirectional banana-epoxy-reinforced composite: Experimentation and theoretical estimation." *Composites: Mechanics, Computations, Applications: An International Journal* 1, no. 3 (2010a).

Gohil, Piyush P., and A. A. Shaikh. "Experimental evaluation for mechanical property of unidirectional banana reinforced polyester composites." In *Advanced Materials Research*, vol. 123, pp. 1147–1150. Trans Tech Publications Ltd, (2010b).

Ishak, M. R., Z. Leman, S. M. Sapuan, M. Y. Salleh, and S. Misri. "The effect of sea water treatment on the impact and flexural strength of sugar palm fibre reinforced epoxy composites." *International Journal of Mechanical and Materials Engineering* 4, no. 3 (2009): 316–320.

Katogi, H., K. Takemura, and R. Sebori. "Fatigue property of natural fiber after alkali treatment." In *Proceedings of the 2nd International Conference on High Performance and Optimum Design of Structures and Materials (EHPSM 2016)*, pp. 343–350. (2016).

Kolluri, Dilip, Anup K. Ghosh, and Jayashree Bijwe. "Analysis of load-speed sensitivity of friction composites based on various synthetic graphites." *Wear* 266, no. 1–2 (2009): 266–274.

Kranthi, Ganguluri, Rajlakshmi Nayak, Sandhyarani Biswas, and Alok Satapathy. "Wear performance evaluation of pine wood dust filled epoxy composites." Paper presented at the International Conference on Advancements in Polymeric Materials APM 2010, CIPET, Bhubaneswar, India, February 20–22, 2010 (2010).

Mahapatra, S. S., and Vedansh Chaturvedi. "Modelling and analysis of abrasive wear perform-ance of composites using Taguchi approach." *International Journal of Engineering, Science and Technology* 1, no. 1 (2009): 123–135.

Majhi, Sudhakar, S. P. Samantarai, and S. K. Acharya. "Tribological behavior of modified rice husk filled epoxy composite." *International Journal of Scientific & Engineering Research* 3, no. 6 (2012): 180–184.

Mishra, Antaryami. "Dry sliding wear behavior of epoxy-rubber dust composites." *International Journal of Mechanical and Mechatronics Engineering* 6, no. 7 (2012): 1218–1223.

Mohammed, Layth, Mohamed NM Ansari, Grace Pua, Mohammad Jawaid, and M. Saiful Islam. "A review on natural fiber reinforced polymer composite and its applications." *International Journal of Polymer Science* 2015 (2015).

Navin, C. and Mohammed Fahim. *Tribology of Natural Fiber Polymer Composites*. Woodhead Publishing, 2009.

Nguong, C. W., S. N. B. Lee, and D. Sujan. "A review on natural fibre reinforced polymer composites." *International Journal of Materials and Metallurgical Engineering* 7, no. 1 (2013): 52–59.

Rajesh, S., S. Rajakarunakaran, and R. Sudhakara Pandian. "Modeling and optimization of sliding specific wear and coefficient of friction of aluminum based red mud metal matrix composite using Taguchi method and response surface methodology." *Materials Physics and Mechanics* 15, no. 2 (2012): 150–166.

Rao, CH Chandra, S. Madhusudan, G. Raghavendra, and E. Venkateswara Rao. "Investigation in to wear behavior of coir fiber reinforced epoxy composites with the Taguchi method." *International Journal of Engineering Research and Industrial Applications* 2 (2012): 2248–9622.

Rubio-López, A., J. Artero-Guerrero, Jesús Pernas-Sánchez, and C. Santiuste. "Compression after impact of flax/PLA biodegradable composites." *Polymer Testing* 59 (2017): 127–135.

Schön, Joakim. "Coefficient of friction and wear of a carbon fiber epoxy matrix composite." *Wear* 257, no. 3–4 (2004): 395–407.

Shivamurthy, B., Siddaramaiah, and M. S. Prabhuswamyc. "Influence of SiO_2 fillers on sliding wear resistance and mechanical properties of compression moulded glass epoxy com-posites." *Journal of Minerals and Materials Characterization and Engineering* 8, no. 07 (2009): 513.

Singh, Charanjit, and Jagteshwar Singh. "Synthesis of Al-SiC composite prepared by mech-anical alloying." *Journal of Mechanical and Civil Engineering* 11, no. 3 (2014): 12–17.

Singh, Narish, B. F. Yousif, and Dirk Rilling. "Tribological characteristics of sustainable fiber-reinforced thermoplastic composites under wet adhesive wear." *Tribology Transactions* 54, no. 5 (2011): 736–748.

Yousif, Belal, Ong B. Leong, Low K. Ong, and Wong K. Jye. "The effect of treatment on tribo-performance of CFRP composites." *Recent Patents on Materials Science* 2, no. 1 (2009): 67–74.

Yousif, B. F., and N. S. M. El-Tayeb. "Wet adhesive wear characteristics of untreated oil palm fibre-reinforced polyester and treated oil palm fibre-reinforced polyester com-posites using the pin-on-disc and block-on-ring techniques." *Proceedings of the Institution of Mechanical Engineers, Part J: Journal of Engineering Tribology* 224, no. 2 (2010): 123–131.

Yousif, B. F., Saijod TW Lau, and S. McWilliam. "Polyester composite based on betelnut fibre for tribological applications." *Tribology International* 43, no. 1–2 (2010): 503–511.

Wong, K. J., B. F. Yousif, and K. O. Low. "The effects of alkali treatment on the interfacial adhesion of bamboo fibres." *Proceedings of the Institution of Mechanical Engineers, Part L: Journal of Materials: Design and Applications* 224, no. 3 (2010): 139–148.

4 Effect of Reinforcements on the Tribological Properties of Polymer Composites

N. Thangapandian,[1]* M. Satthiyaraju[2a]

[1]Department of Mechanical Engineering, St. Joseph's Institute of Technology, Chennai – 600 119, India.

[2]Department of Mechanical Engineering, National Institute of Technology, Tiruchirappalli – 620015, India.

*Corresponding author: erpandian@gmail.com

Other email ID: [a]nittsathya@gmail.com

CONTENTS

4.1 INTRODUCTION

Tribology is the study of surfaces in contact, which becomes an inevitable area of study among materials researchers owing to the heavy loss caused by tribological contacts, including around 20% of the world's total energy use (Friedrich, 2018). The direct cost of wear failures includes replacement of damaged components, reduced operating efficiency and loss of productivity, and wear also has adverse effects on the environment and human health (Patnaik et al., 2010). Therefore, numerous researchers have tried to reduce wear and tear by different techniques including coating, lubrication and incorporating different reinforcements. In all these techniques, the basic principles

DOI: 10.1201/9781003109723-4

are the reduction of adhesion between the mating surfaces and enhancing a material's tensile strength, hardness, stiffness and compression strength (Friedrich et al., 2002, and Sahin and Mirzayev, 2015). With a wide variety of reinforcements, polymer and its composites replace conventional heavy metals in the industries. Polymer composites are preferred in industries owing to their high strength-to-weight ratio, higher shock-absorbing capacity, self-lubrication and low cost. Because most high performance polymer composites are used in extreme applications such as aircraft wings and turbine blades, critical issues like friction, wear and heat retardancy must be taken care of while working with polymer composites. In such applications, the material undergoes a higher wear rate owing to the impact of foreign particles. The major failures caused by wear are (1) matrix wear, (2) fiber sliding wear, (3) fiber fracture and (4) fiber-matrix interfacial debonding (Cirino et al., 1988). These failures depend on the properties of matrix and reinforcing materials. Novel nanoparticles also significantly improves the tribological properties of polymer composites. Conventional coarse-grained reinforcing materials tend to pull out or abrade their counterparts. Also, the failure to form a protective layer can be ignored while using the nano-reinforcing materials over the coarse-grained reinforcements. These nanocomposites exhibit synergetic wear resistance greater than that of the individual components. Thus, understanding the tribological properties of the individual constituent material is essential to better use polymer composites in heavy load conditions.

Different types of wear (Figure 4.1) originate in an insignificant manner and lead to major energy loss. Adhesive wear occurs owing to the adhesive interaction between the molecules of two different surfaces while one rubs over another (Belyi et al., 1977). The material undergoes failure whenever the adhesive force while rubbing exceeds the

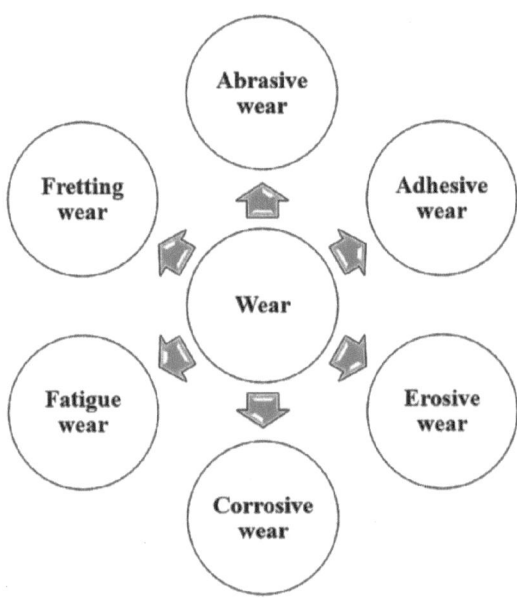

FIGURE 4.1 Major forms of wear.

internal molecular interactions within the composites. Adhesive wear is predominant in polymer composites and can be controlled by the proper selection of reinforcement.

Abrasive wear accounts for 50% of total wear loss (Bijwe et al., 2007), and it occurs when the harder asperity hits the relatively softer counterpart, digging into the mating surface of the material (Mohan et al., 2011). It can occur as two-body abrasion, three-body abrasion or a combination of both. Improving the hardness and strength of the material significantly lower this loss, since more energy is required to abrade the particle from the matrix.

As the name implies, fatigue wear is due to the cyclic loading in polymer matrix composites (PMCs). The increase in temperature during repeated loading causes the matrix to soften. A small crack initiation followed by fatigue failure is dependent on parameters like number of cycles, load, coefficient of friction and elasticity of the mating material. Fretting wear in the composite material results from the oscillation or reciprocating movement between two surfaces in contact. A corrosive or oxidative environment also causes wear when two surfaces are in contact; this is known as corrosive wear. The impingement of foreign particles with significant velocity causes the loss of wear from the solid surface, known as erosive wear.

The conventional techniques such as coating, lubrication and strong reinforcing materials are still practiced for reducing the wear and tear in polymer composites. In all these techniques, the adhesion between the counterparts is reduced and improved material properties (such as hardness, stiffness and compression strength) for reducing the abrasive wear are achieved. The most common and best-known technique to improve tribological properties is coating. In addition to improving scratch and abrasion resistance, coatings are also custom-made to improve the chemical, mechanical and electrical properties.

Lubrication can be applied internally or externally in the polymer composites to reduce the surface interaction and thus the wear. External lubrication can be classified into boundary lubrication and hydrodynamic lubrication. These lubrication types result in reduced adhesion of polymer composites to the mating surface.

This chapter discusses these three techniques – coating, reinforcement and lubrication – for improving tribological properties of polymer composites. In addition, the effect of various reinforcement parameters such as the orientation, volume fraction and the size and shape of reinforcements are discussed.

4.2 COATING TECHNIQUES FOR POLYMER COMPOSITE MATERIALS

Poor abrasion resistance restricts the use of polymer composites for different applications, especially high temperature applications. Coatings are considered as an engineering solution to improve thermal conductivity, electrical conductivity, protection from wear and resistance to thermal softening (Zhou et al., 2011). Good adhesion, low porosity and greater wear resistance characterize effective coating. Different processes have been employed (Table 4.1) for the surface coating over the polymer matrix composites.

Among all the listed coating and metallization processes, thermal spray coating was cheap and suitable for producing thick coating on the polymer-based composite

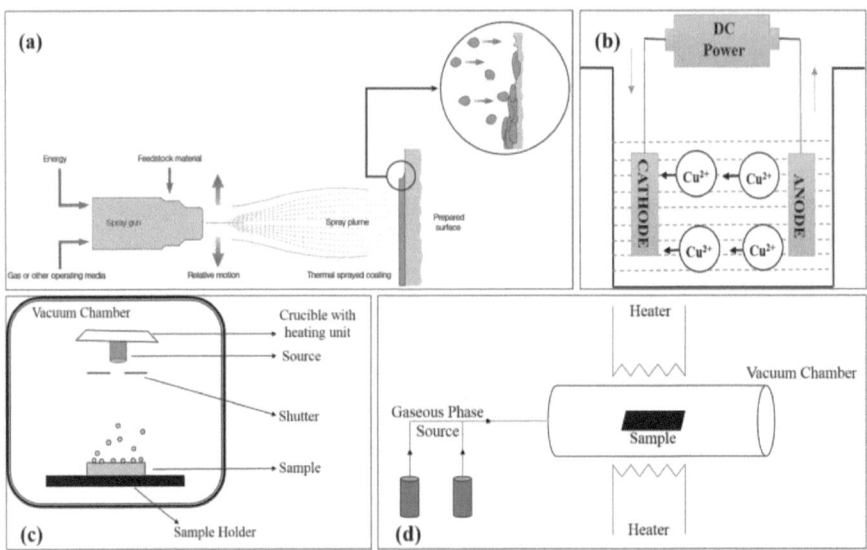

FIGURE 4.2 Different coating techniques for polymer composites: (a) thermal spray coating, (b) electroplating, (c) physical vapour deposition, (d) chemical vapour deposition.

Source: (a) www.griekspoorthermalcoatings.com/techniques.

substrate (Therrien, 2013). Thermal degradation and char formation of matrix may occur during the thermal spray coating owing to the temperature of the metal particles used and the high temperature carrier gases. This coating process uses thermal energy to melt the coating material into powder that is spurted over the substrate using a carrier gas or plasma (Figure 4.2).

The coating adhesion can be improved by different methods such as sand blasting, chemical treatment, usage of fillers and preheating of substrate. Even after the surface modification, adhesion strength was not up to the mark to meet industrial needs. Also, lubrication of PMCs with molybdenum disulfide, silicon carbide (SiC) and graphite were not successful in reducing the wear in polymer composites. However, researchers are doing many experiments with reinforcements and their parameters for the successful fabrication of composites to meet the needs of high wear applications.

4.3   EFFECT OF REINFORCEMENT ON THE WEAR PROPERTIES OF POLYMER COMPOSITE MATERIALS

The properties of the polymer composite materials are strongly reliant on the reinforcing material characteristics such as volume, interfacial bonding, size and shape. Recently, nanosized reinforcing materials have also gained researchers' attention because of their exceptional properties.

The diameter of reinforcement material may range from a few nanometers to microns. The composites are classified into three types as shown in Figure 4.3: particle

TABLE 4.1

The Properties of PMCs after Different Coating Processes

Process	Base Material	Coating Material	Properties	References
Physical vapour deposition, chemical vapour deposition	PMMA/silicate glass	Ti, TiN, TiO$_2$	Adhesion was good immediately after coating	Straumal et al., 2001
	Carbon fiber–based material	Carbon nanotube, (CNT)	Less porous	Delfini et al., 2017
	GFRP	Alumina, NiCrAlY and silicon nitride	Alumina offered good resistance to wear; NiCrAlY adhesion was good but delaminated	Harding et al., 1994
Electroplating	Carbon fiber–reinforced epoxy composites	Cu-SiO$_2$	Required layer formed, increase in hardness and decrease in electrical conductivity	Li et al., 2009
	Melamine and phenolic-based conducting PMC	Silver-coated CNT	Homogeneously dispersed CNT layer with improved interfacial bond	Oh et al., 2008
Cold spray technique	Carbon fiber–reinforced PMC	Al, Al-Cu	Less porous, excellent bonding	Zhou et al., 2011
	PC/ABS, polyamide-6, polypropylene, polystyrene and glass fiber composites	Cu, Al, Sn powder	Good electrical conductivity, erosion due to carrier gas velocity	Lupoi & O'Neill, 2010
Thermal spray techniques	Carbon epoxy PMC	Zn powders	Good adherence	Robitaille et al., 2009
	GFRP epoxy composites	Al-12Si coating	Improved adherence after gridding	Therrien, 2013

Note: Acrylonitrile butadiene styrene (ABS); glass fiber–reinforced polymer (GFRP); polycarbonates (PC).

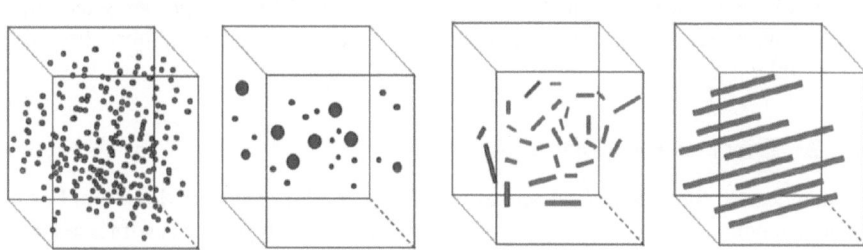

FIGURE 4.3 Types of reinforcements in PMCs.

reinforced (nano- and micron-sized), whiskers reinforced and continuous fiber composites.

Though dry sliding wear characteristics of particle-reinforced composites depend on the reinforcement's properties and operating characteristics, this chapter deals with the reinforcements' properties alone. This section on the effect of reinforcement is divided and presented as follows:

1. Effect of reinforcement volume percent on wear properties.
2. Effect of orientation of fiber.
3. Effect of reinforcement size and shape on wear properties.

4.3.1 EFFECT OF REINFORCEMENT VOLUME

The effects of volume percentage on the wear characteristics have been studied extensively for glass fiber (Zhang et al., 2007), carbon fiber (Bijwe et al., 2004), graphite and MoS$_2$ (Bijwe et al., 1990), aramide (Bijwe et al., 2004), carbon nanotubes (Sakka et al., 2017, and Zoo et al., 2004), nano-Al$_2$O$_3$ (Meena et al., 2019) and natural fiber (Pujari and Srikiran, 2019).

In general, the addition of nanoparticle reinforcements like carbon nanotube (CNT) and graphite greatly improved the tribological properties when compared to their unfilled composites (Zoo et al., 2004). The problem when using a nanosized reinforcement is greater surface area, which restricts the use of greater amounts of reinforcement. During wear, nanosized reinforcements usually pulled out of the matrix and formed an interfacial layer, which greatly reduced wear. Lack of wear grooves in the scanning electron micrographs (Figure 4.4) endorses the effectiveness of CNT reinforcement in improving the tribological properties. Also, the excessive fibrillation of fibers in PMC is helpful in absorbing the maximum amount of shear stress to the counterpart during the abrasive action (Bijwe et al., 2004).

Ultra-high molecular weight polyethylene (UHMWPE) composites with different weight percentages of CNT reinforcement were tested for their tribological properties using a ball-on-disc wear tester. Figure 4.5 shows the reduction in wear loss and the increase in coefficient of friction with the increase in weight percentage of CNTs (Zoo et al., 2004) due to improvement in shear stress offered by CNTs.

In general, an increase in weight percentage or volume percentage of the reinforcement will improve tribological properties up to a certain limit. Beyond that limit the interfacial bonding between the matrix and reinforcement will decrease significantly, which worsens the tribological and other properties. Thus, the proper selection of reinforcing material and volume percentage is tedious and a multi-criteria optimization problem.

4.3.2 EFFECT OF FIBER DIRECTION

In many cases, the tribological properties also depend on the fiber direction. For the same percentage of fiber volume, a significant change in properties is witnessed when the fiber direction changes. Initially most of the polymer composites were fabricated as unidirectional composites, later as bidirectional, and finally randomly oriented polymer composites were developed to overcome the anisotropy.

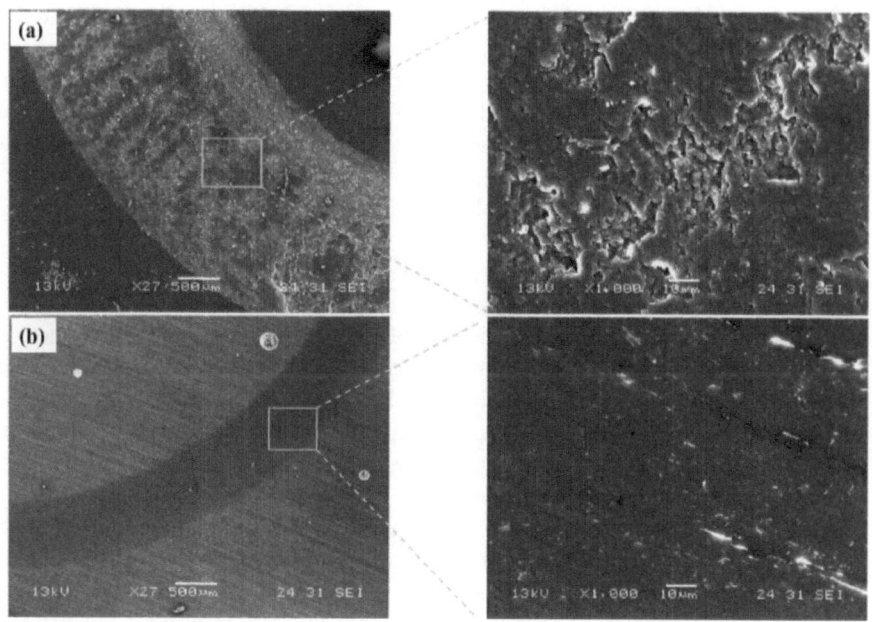

FIGURE 4.4 SEM images of worn surfaces for (a) neat epoxy and (b) epoxy-treated CNT composite.

Source: Sakka et al., 2017.

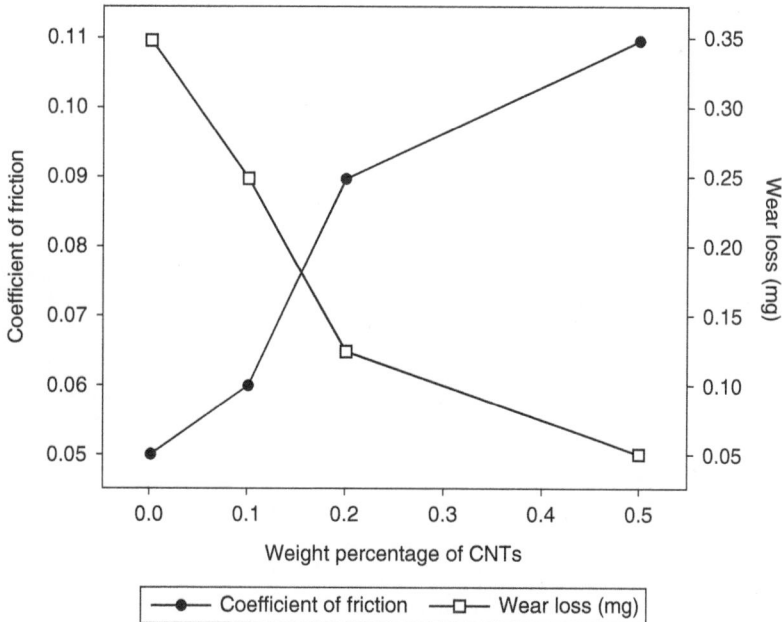

FIGURE 4.5 Effect of weight percentage in tribological properties of UHMWPE-CNT composites.

The effects of fiber direction on wear properties of PMCs were studied by a few researchers, with little difference in the results. Composites with three different orientations (P, N, AP – parallel, normal and antiparallel) were tested for their wear properties (Lhymn, 1987). The coefficient of friction is highest in P orientation, followed by AP and N (perpendicularly) oriented composites. The specific wear rate on the natural fiber composites with these three orientations was measured using a pin-on-disc friction sliding wear machine with the following formula (Dwivedi and Chand, 2009):

$$\text{Specific wear rate } k_o = w/\rho DL$$

where w = weight loss (grams), ρ = density of sample, D = sliding distance (meters) and L –= applied load (newtons). Similar experiments were carried out on the three differently oriented composites. As shown in Figure 4.6 the results confirm the wear rate increased in the order of normal, antiparallel and parallel direction. Also, the researchers concluded that the increased interfacial bonding, flexibility and shock-absorbing capacity of the sisal fiber increased the resistance to wear (Panda, 2016).

Much of the work reported used the same three orientations for different fibers such as glass, carbon and natural fibers. Comparing those results with randomly oriented fiber composites is difficult, since most of the randomly oriented composites are particulate composites. The particulate composites were reinforced for different

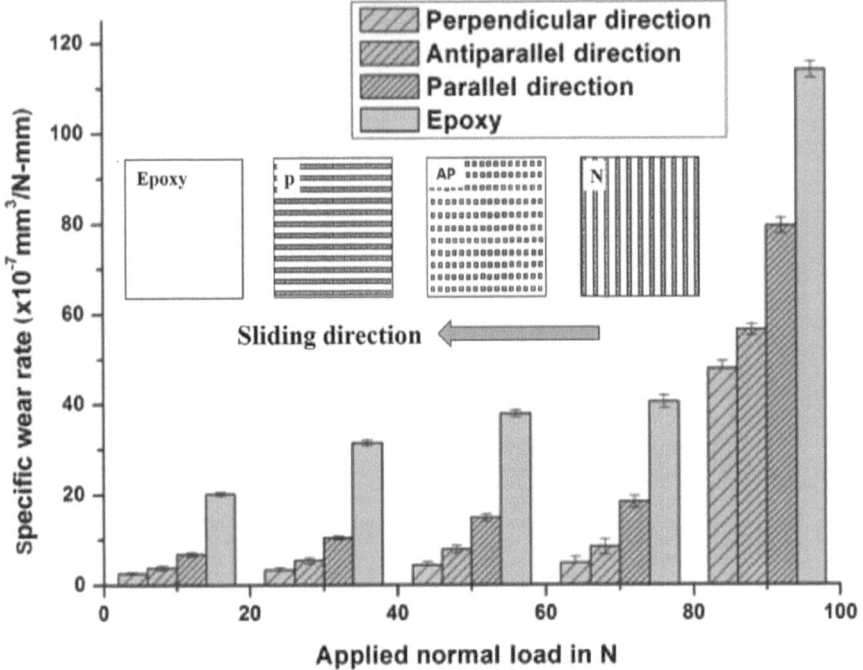

FIGURE 4.6   Three different fiber orientations and their specific wear rates.

Source: Panda, 2016.

shapes and directions than fiber-reinforced composites, whose properties depend on the size and shape of the reinforcement.

4.3.3 Effect of Reinforcement Size and Shape on Wear Properties

4.3.3.1 Nano- and Micron-Sized Particle Reinforcement

The reinforcements for composite materials have evolved many dimensional changes due to technology development. Continuous fibers, discontinuous fibers, micron-sized particles (spherical and irregular) and nano particles are some of the major PMC reinforcements. Different metal and ceramic powders such as Cu, SS, a mixture of Cu-SS, Si, TiO_2, SiC, SiO_2, WC and tantalum niobium carbide have been used for tribological applications. Researchers also used carbon nanotubes (CNTs) for wear applications after their invention by Iijima in 1992. These fillers provide excellent wear properties when even a small amount is added to the matrix. Moreover, dissimilar properties can be achieved with respect to the locations, for example, excellent wear properties on the outer shell and good toughness and damage tolerance on the inner side (Rezzoug et al., 2019). Appropriate procedure is mandatory to avoid an agglomeration of nanoparticles.

The scratchiness of harder nanoparticles decreases wear significantly owing to the reduction in angularity, compared to micron particle–reinforced PMC. The zero-dimensional nanofillers serve as nanolevel polishing agents that move freely between the two contact surfaces. They also reduce the coefficient of friction and temperature on mating regions of polymer nanocomposites. Typically, the nanolevel particles promote better wear properties in comparison with bigger particles. This was confirmed with the wear test results for PEEK reinforced with nano-SiC particle, micron SiC particle and whiskers SiC by Xue and Wang (1997).

The effect of nanoparticles on the frictional coefficient and contact temperature of the polyamide 6 (PA6) composites reinforced with graphite nanoparticles is shown in Figure 4.7. During the initial stage, both composites exhibit similar performance

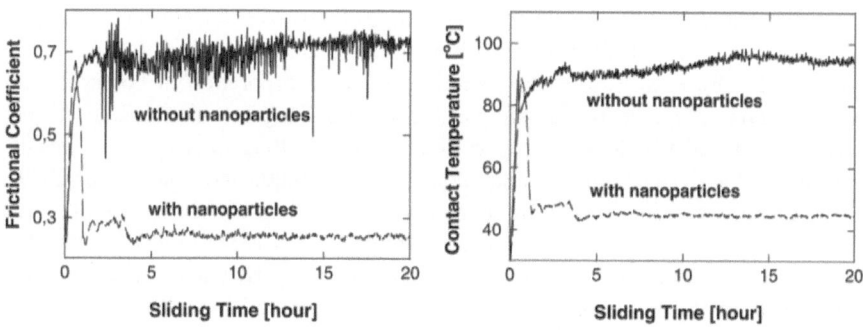

FIGURE 4.7 Wear behavior of PA6 polymer composites reinforced with graphite nanoparticles or short carbon fibers (parameters: time = 20 h, sliding velocity = 1 m/s, applied pressure = 4 MPa).

Source: Friedrich et al., 2005.

TABLE 4.2
**Least Wear Rates of Different Polymer/
Nanoparticle Composites**

Polymer/Nanoparticle Composites	Lowest Wear Rate ($\times 10^{-6}$ mm^3/Nm)
PEEK/Si$_3$N$_4$	1.3
PEEK/SiO$_2$	1.4
PEEK/SiC	3.4
PEEK/ZrO$_2$	3.9
PPS/Al$_2$O$_3$	10.4

against wear, but later the nanocomposite shows exceptional reduction in contact temperature and frictional coefficient due to the action of nanoparticles (Friedrich et al., 2005).

Researchers have been working with different nanoparticles such as SiC, SiO$_2$, Si$_3$N$_4$, CuO, CaCO$_3$, ZnO, TiO$_2$ and Al$_2$O$_3$ in polytetrafluoroethylene (PTFE), poly ether etherketone (PEEK), epoxy, polymethyl methacrylate (PMMA) and polyphenylene sulfide (PPS) matrices in efforts to enhance wear properties. The least wear rates of various nanoparticle reinforcements of polymeric materials (as shown in Table 4.2) have been discussed elsewhere in detail (Sawyer et al., 2003). Also, they suggested PTFE for the highest loading of nanofillers.

4.3.3.2 Short and Long Fiber Reinforcement

The fiber-reinforced polymer composites can be incorporated with short fiber or continuous fiber in random or aligned orientation of distribution. The short fibers, or chopped fibers, are used in most mechanical components such as bearing elements, gears and cams. These short fibers carry nearly all the forces. This enhances the wear resistance of the composite above that of polymer composites incorporated with nanofillers. The short fiber–based polymer composites also enhance creep resistance, impact resistance and stiffness even in high temperature applications (Cirino et al., 1988). Sudhakara et al. (2013) did experiments on incorporation of borassus fruit fiber into epoxy polymer to examine wear properties. They revealed that short fiber provides enhanced tribological properties compared to long fibers. Similar results were obtained with sugarcane fiber composites by Mahapatra and Chaturvedi (2009).

Additionally, hybrid fiber reinforcement with nanofillers could improve the wear behavior of polymeric composites. Recently, Singh et al. (2018) examined hybrid fiber reinforcements with epoxy resin polymer to enhance the performance in tribological applications. In this study, the wear results were compared for epoxy composites filled with glass fiber, carbon fiber and multi-walled carbon nanotubes (MWCNTs) in a dry and inert environment.

Mechanical behavior and resistance to wear are enhanced with the addition of continuous fibers in PMC. However, processing of such continuous fiber is difficult and cannot be accomplished by traditional methods such as extrusion and injection molding. Continuous fibers contribute wear resistance in both parallel and normal orientation to the direction of sliding. This is the main reason that polymeric materials reinforced with continuous fiber are used in biomedical and mechanical applications.

Many reports are available on continuous fiber composites with specific orientations which provide better improvements to wear resistance, compared to others. Different orientations are highly favorable to obtain appropriate wear rates of the polymer composites. Moreover, the appropriate orientation and kind of fiber could have a synergetic effect leading to additional improvements. Also, the natural fibers such as coir, jute, kenaf, sisal and rice husk are incorporated into the polymeric composites to improve wear performance. The results reveal better tribological performance compared to the neat polymer.

In summary, the nanofillers, short fibers and continuous fibers are attractive to reinforce the polymer composites to replace conventional materials with high wear performance. Fiber pull-out, debonding and delamination by adhesive wear are the main reasons for the failure of polymer matrices reinforced with continuous fiber, whereas debonding and matrix cracking are the failure mechanisms in polymeric matrices reinforced with particulates (such as nano and micron-size metal powders, oxides and ceramics). Additionally, better wear performance can be achieved by incorporating both short and long fibers. Because producing continuous fibers is difficult, most researchers have experimented with the short fibers such as glass fiber, carbon fiber and aramid fiber. The volume content, temperature, applied loads, fiber orientation and chemical modification of fibers can influence tribological performance.

The polymer composites have been widely used in automotive applications including panels, dashboards, ball joints, and others. A recent study reveals that the use of carbon nanotubes (CNTs) in the automotive body significantly reduced vibration (Joy et al., 2020). Different sporting goods have also been made with polymer-based composites. Improving the tribological properties will increase the life of the composites, and thus critical environmental problems of disposal of worn-out materials and resources to produce replacements are greatly reduced.

4.4 CONCLUSION

The effects of different types of reinforcement parameters such as orientation of fiber, fiber size, and fiber volume on the wear mechanism have been discussed in detail. Of the different reinforcement directions, PMCs with P, or parallel, orientation exhibited a higher coefficient of friction. The composites with short fibers and particulates demonstrate superior mechanical and tribological properties, and such reinforcements can be handled with ease compared to long fibers. The effects of different metal and ceramic particles have been discussed, along with the need to develop newer reinforcements for better tribological properties. The presented results pave the way for better understanding and provide new insight into improving the tribological properties of existing polymeric materials. These reinforced polymeric composites can be used in many mechanical parts that experience high wear and tear.

REFERENCES

Belyi, V. A., A. I. Sviridyonok, V. A. Smurugov, and V. V. Nevzorov. "Adhesive Wear of Polymers." *Journal of Tribology* 99, no. 4 (1977): 396–400.

Bijwe, J., C. M. Logani, and U. S. Tewari. "Influence of Fillers and Fibre Reinforcement Abrasive Wear Resistance of Some Polymeric Composites." *Wear* 138 (1990): 77–92.

Bijwe, J., S. Awtade, B. K. Satapathy, and A. Ghosh. "Influence of Concentration of Aramid Fabric on Abrasive Wear Performance of Polyethersulfone Composites." *Tribology Letters* 17, no. 2 (2004): 187–194.

Bijwe, J. Ã., Rekha Rattan, and M. Fahim. "Abrasive Wear Performance of Carbon Fabric Reinforced Polyetherimide Composites: Influence of Content and Orientation of Fabric." *Tribology International* 40 (2007): 844–854.

Crino, M., K. Friedrich, and R. B. Pipes. "Evaluation of Polymer Composites for Sliding and Abrasive Wear Applications." *Composites* 19, no. 5 (1988): 383–392.

Delfini, A., A. Vricella, R. Bueno Morles, R. Pastore, D. Micheli, F. Gugliermetti, and M. Marchetti. "CVD Nano-Coating of Carbon Composites for Space Materials Atomic Oxygen Shielding." *Procedia Structural Integrity* 3 (2017): 208–216.

Dwivedi, U. K., and Navin Chand. "Influence of Fibre Orientation on Friction and Sliding Wear Behaviour of Jute Fibre Reinforced Polyester Composite." *Applied Composite Materials* 16, no. 2 (2009): 93–100.

Friedrich, K. "Polymer Composites for Tribological Applications." *Advanced Industrial and Engineering Polymer Research* 1, no. 1 (2018): 3–39.

Friedrich, K., R. Reinicke, and Z. Zhang. "Wear of Polymer Composites." *Proceedings of the Institution of Mechanical Engineers, Part J: Journal of Engineering Tribology Wear of Polymer Composites* 216 (2002): 415–426.

Friedrich, K., Zhong Zhang, and Alois K. Schlarb. "Effects of Various Fillers on the Sliding Wear of Polymer Composites." *Composites Science and Technology* 65 (2005): 2329–2343.

Harding, David R., James K. Sutter, Marla A. Schuerman, and Elizabeth A. Crane. "Oxidation Protective Barrier Coatings for High-Temperature Polymer Matrix Composites." *Journal of Materials Research* 9, no. 6 (1994): 1583–1595.

Joy, A., Susy Varughese, Anand K. Kanjarla, S. Sankaran, and Prathap Haridoss. "Effect of the Structure and Morphology of Carbon Nanotubes on the Vibration Damping Characteristics of Polymer-Based Composites." *Nanoscale Advances* 2, no. 3 (2020): 1228–1235.

Lhymn, C. "Tribological Properties of Unidirectional Polyphenylene Sulfide-Carbon Fiber Laminate Composites." *Wear* 117, no. 2 (1987): 147–159.

Li, Hao, Yizao Wan, Hui Liang, Xiaolei Li, Yuan Huang, and Fang He. "Composite Electroplating of Cu-SiO$_2$ Nano Particles on Carbon Fiber Reinforced Epoxy Composites." *Applied Surface Science* 256, no. 5 (2009): 1614–1616, www.griekspoorthermalcoatings.com/techniques.

Lupoi, R., and W. O'Neill. "Deposition of Metallic Coatings on Polymer Surfaces Using Cold Spray." *Surface and Coatings Technology* 205, no. 7 (2010): 2167–2173.

Mahapatra, S. S., and Vedansh Chaturvedi. "Modelling and Analysis of Abrasive Wear Performance of Composites Using Taguchi Approach." *International Journal of Engineering, Science and Technology* 1, no. 1 (2009): 123–135.

Meena, Anoj, Harlal Singh Mali, Amar Patnaik, and Shiv Ranjan Kumar. "Investigation of Wear Behavior of Nanoalumina and Marble Dust-Reinforced Dental Composites." *Science and Engineering of Composite Materials* 1 (2019): 84–96.

Mohan, N., S. Natarajan, and S. P. Kumaresh Babu. "Abrasive Wear Behaviour of Hard Powders Filled Glass Fabric-Epoxy Hybrid Composites." *Materials and Design* 32 (2011): 1704–1709.

Oh, Youngseok, Daewoo Suh, Youngjin Kim, Eungsuek Lee, Jee Soo Mok, Jaeboong Choi, and Seunghyun Baik. "Silver-Plated Carbon Nanotubes for Silver/Conducting Polymer Composites." *Nanotechnology* 19, no. 49 (2008): 4539–4545.

Panda, Rajesh. "Investigation of Thermal, Mechanical, and Tribological Behaviour of Biobased-Resin–Crosslinked, Sisal-Fibre–Reinforced Epoxy Composites." Ph.D Dissertation, University of Toronto, 2016.

Patnaik, Amar, Alok Satapathy, Navin Chand, N. M. Barkoula, and Sandhyarani Biswas. "Solid Particle Erosion Wear Characteristics of Fiber and Particulate Filled Polymer Composites: A Review." *Wear* 268 (2010): 249–263.

Pujari, Satish, and S. Srikiran. "Experimental Investigations on Wear Properties of Palm Kernel Reinforced Composites for Brake Pad Applications." *Defence Technology* 15, no. 3 (2019): 295–299.

Rezzoug, Amine, Said Abdi, Samir Mouffok, Fares Djematene, and Boubekeur Djerdjare. "Tribological Investigation of Carbon Fiber-Epoxy Composite Reinforced by Metallic Filler Layer." *Indian Journal of Engineering & Materials Sciences* 26 (2019): 334–341.

Robitaille, F., M. Yandouzi, S. Hind, and B. Jodoin. "Metallic Coating of Aerospace Carbon/Epoxy Composites by the Pulsed Gas Dynamic Spraying Process." *Surface and Coatings Technology* 203, no. 19 (2009): 2954–2960.

Sahin, Yusuf, and Husseyn Mirzayev. "Wear Characterization of Polymer Based Composites." *Mechanics of Composite Materials* 51, no. 5 (2015): 543–554.

Sakka, M. M., Z. Antar, K. Elleuch, and J. F. Feller. "Tribological Response of an Epoxy Matrix Filled with Graphite and/or Carbon Nanotubes." *Friction* 5, no. 2 (2017): 171–182.

Sawyer, W. Gregory, Kevin D. Freudenberg, Praveen Bhimaraj, and Linda S. Schadler. "A Study on the Friction and Wear Behavior of PTFE Filled with Alumina Nanoparticles." *Wear* 254 (2003): 573–580.

Singh, K. K., Jiban Jyoti Kalita, and Nisha Sharma. "Investigation of Friction and Wear Behavior of MWCNTs Filled HFRP Composites under Different Sliding Parameters and Different Environment." *Materials Today: Proceedings* 5, no. 14 (2018): 28347–28353.

Straumal, B. B., N. F. Vershinin, A. Cantarero-Saez, M. Friesel, P. Zieba, and W. Gust. "Vacuum Arc Deposition of Protective Layers on Glass and Polymer Substrates." *Thin Solid Films* 383, no. 1–2 (2001): 224–226.

Sudhakara, P., Dani. Jagadeesh, YiQi Wang, C. Venkata Prasad, A. P. Kamala Devi, G. Balakrishnan, B. S. Kime, and J. I. Song. "Fabrication of *Borassus* Fruit Lignocellulose Fiber/PP Composites and Comparison with Jute, Sisal and Coir Fibers." *Carbohydrate Polymers* 98, no. 1 (2013): 1002–1010.

Therrien, David Stuart. "Heat Transfer Analysis of Flame-Sprayed Metal-Polymer Composite Structures." Master's thesis, University of Alberta, 2013.

Xue, Qun Ji, and Qi-Hua Wang. "Wear Mechanisms of olyetheretherketone Composites Filled with Various Kinds of SiC." *Wear* 213 (1997): 54–58.

Zhang, Huibo, W. Li, Xujie Yang, Yongchun Zhang, and Yadong Chen. "Microstructural Characterizations and Mechanical Behavior of Polyurethane Elastomers Strengthened with Milled Fiberglass." *Journal of Materials Processing Technology* 190, no. 1–3 (2007): 96–101.

Zhou, X. L., A. F. Chen, J. C. Liu, X. K. Wu, and J. S. Zhang. "Preparation of Metallic Coatings on Polymer Matrix Composites by Cold Spray." *Surface and Coatings Technology* 206, no. 1 (2011): 132–136.

Zoo, Yeong Seok, Jeong Wook An, Dong Phil Lim, and Dae Soon Lim. "Effect of Carbon Nanotube Addition on Tribological Behavior of UHMWPE." *Tribology Letters* 16, no. 4 (2004): 305–309.

5 Mechanical and Tribological Behaviour of Particulate–Reinforced Metal Matrix Composite

A. S. Vivekananda,[1] J. Jenix Rino[2a]

[1]Department of Mechanical Engineering, College of Engineering, Guindy, Chennai-25, Anna University, India

[2]Associate Professor and Head, Department of Mechanical Engineering, Stella Mary's College of Engineering, Kanyakumari District, Tamil Nadu, India.

*Corresponding author: lance.vivek@gmail.com

[a]Other email ID: jenixproject@gmail.com

CONTENTS

5.1   INTRODUCTION

Metal matrix composites (MMCs) are replacing the metals and alloys in structural and non-structural applications, especially in the aircraft, automobile and electronics industries (Kaczmar and Pietrzak, 2000). The reasons are the high strength-to-weight ratio, high thermal and chemical stability, and higher resistance to wear and corrosion of MMCs compared with metals and alloys. Figure 5.1 shows the advantages of composite over metals and alloys. Earlier, continuous reinforced metal matrix composites had been used in aircraft industries, as they have higher strength than particulate composite in a given direction. However, the fabrication of continuous reinforced metal

DOI: 10.1201/9781003109723-5

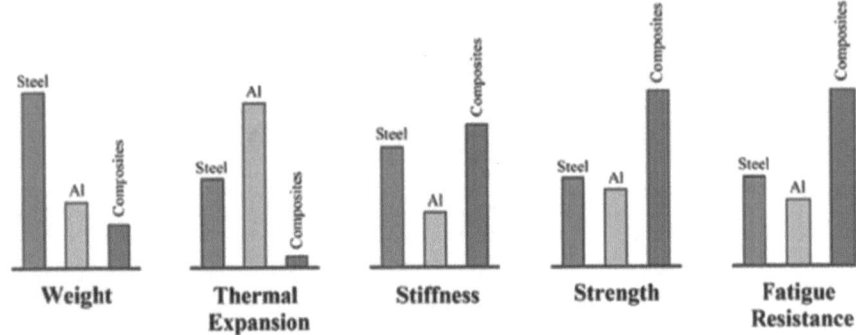

FIGURE 5.1   Comparison of properties between metals and composites.

Source: Deutsch (1978).

matrix composites is complex. Interfacial reaction between matrix and reinforcement fiber arises anisotropically and affects the properties of composites, and it is the major reason for introducing new processing techniques and developing newer composites. Usage of MMCs in various areas has expanded after the development of particulate reinforcement of metal matrix composites (PRMMCs). In the last two decades, the applications of PRMMCs have increased due to the ease of the fabrication process, which lowers cost, and the isotropic nature of these composites. Cost and applications are major deciding factors in the choice of the processing method with ceramic reinforcement. Generally, light metals like aluminium, magnesium and titanium alloys are used as matrix material for making the metal matrix composite. Therefore, the discussion in this chapter is restricted to only the mechanical behaviour of PRMMCs.

Higher strength with lower weight is the major reason for using MMCs in structural applications. Normally, direct and indirect methods are used for strengthening the composite. An effective load transfer from matrix to reinforcement is a direct strengthening method. The presence of hard reinforcement particles in matrix material improves the strength of composite, and reinforcement also restricts the dislocation of matrix material under loading conditions. But a lower amount of reinforcement cannot contribute significantly to strengthening the particle-reinforced composite. The coefficient of thermal expansion mismatch between matrix and reinforcement is very high while fabricating the composite, which initiates the dislocations around the particles through inducement of residual plastic stresses. This thermal mismatch plays an inevitable role in strengthening of composite indirectly through an increase of dislocation density. Decrease of particle size is also an indirect strengthening mechanism of composite and is called Orowan strengthening. The mobility of dislocations will be restricted by hard nano particles through this Orowan pinning mechanism. Similarly, grain refinement strengthening is also an indirect strengthening method. It may be achieved through addition of an alloying element, faster cooling rate, or other means to activate more nucleate sites. In recent years, secondary processing methods such as extrusion, equal channel angular pressing and accumulative roll bonding have been used for refining the grain size significantly.

Normally, mechanical properties of MMCs can be altered by the processing route, size of particle reinforcement, type of reinforcement, nature of the interface, wettability between matrix and reinforcement, and weight or volume fraction of reinforcement. For example, in a system of Al/SiC$_p$ (where p indicates particles), the formation of Al$_4$C$_3$ affects the wettability between the matrix and reinforcement, followed by degrading the strength of the composite (Ren et al., 2007). Hence, the aforementioned parameters have to be considered while making the MMCs. We have restricted ourselves here to discussing an overview of the variation in mechanical properties (tensile strength, creep and fatigue) of metal matrix composite with respect to different parameters.

5.2 TENSILE BEHAVIOUR

Tensile strength and Young's modulus of MMCs are higher than in unreinforced material. This is because the presence of hard ceramic particles arrests the dislocations and coalescence of voids on the matrix material during loading conditions. In addition, the reinforcement induces stress on the matrix material which increases the work-hardening rate. An increase in the work-hardening rate eventually increases the strength of the matrix material. The strength of a few particulate-reinforced metal matrix composites are given in Table 5.1.

TABLE 5.1
Strength of Various Al PRMMCs

Composite System (Matrix/ Reinforcement)	Reinforcement (wt.% or vol.%)	Manufacturing rocess	Tensile Strength	References
Al/AlN	50 vol.%	Squeeze casting	154 MPa	Zhang et al., 2003
2020Al/Al$_2$O$_3$	30 wt.%	Stir casting	100–120 MPa	Kok, 2005
A356/Al$_2$O$_3$	1.5 vol.%	Stir casting	265 MPa	Karbalaei Akbari et al., 2013
AA7075/TiB$_2$	0, 3, 6 and 9 wt.%	Stir casting	275–300 MPa	Michael Rajan et al., 2013
Al6061/WC	3 wt.%	Powder metallurgy	200–250 MPa	Kumar & Pramod, 2015
Al/SiC	10 wt.%	Stir casting	163 MPa	Jen et al., 2015
Al7075/SiC	6 wt %	Stir casting	290 MPa	Balaji et al., 2015
Al7075/SiC	9.5 wt.%	Stir casting	230 MPa	Ram Prabhu, 2017
Al/SiC	15 wt.%	Centrifugal casting	145 MPa	El-Galy et al., 2017
LM13 Al/ZrB$_2$-TiC particles	5 wt.% of TiC and 0–10 wt.% of ZrB$_2$	Stir casting	238 MPa	Arunbharathi et al., 2019

However, size and volume fraction of particles, processing methods and micro-structural changes of matrix material must be considered in fabricating MMCs because these factors affect the tensile strength of the composite. For example, stress is concentrated on the particle-agglomerated regions when the particles are not dis-tributed uniformly. The increased brittleness of those agglomerated regions could easily initiate cracks, which could then propagate drastically in particle-absent re-gions on matrix material.

Several studies reported that the size and volume fraction of particles affect the strength of composite significantly (Kok, 2005, Michael Rajan et al., 2013, Zhong et al., 2007, Hassan and Gupta, 2006). Reinforcement with finer particles enhances strength more effectively than coarse particles (Zhong et al., 2007, Sree Manu et al., 2017, Hassan and Gupta, 2006) even at lower volume fractions. Therefore, the strength of composite is decreased on increasing the particle size.

The effect of volume fraction and size of particles in Al-Cu-Mg (2080)/SiC$_p$-T8 composite have been explained by Chawla and Shen (2001). Their results revealed that at constant volume fraction, a decrease in particle size increased the strength of composite. But handling the finer particles is challenging. They have poor wettability into matrix material and also favour particle agglomeration. Kok (2005) observed the same results in an Al/Al$_2$O$_3$ system, where the strength of composite increased on decreasing the particle size and increasing the weight percentage of particles.

Sree Manu et al. (2017) observed the reduced strength of composite above a cer-tain limit of volume fraction. Reinforcement with Al$_2$O$_3$ nano particles has increased the strength of Al matrix material through trapping the dislocation between par-ticles. Tensile strength is significantly improved at lower volume fraction (Al/0.25 wt.% Al$_2$O$_3$). But the strength of the composite is slightly reduced after 0.5 wt.% reinforcement. This is because a higher surface area and surface energy of nano particles allows them to easily cluster in matrix material. Similarly, an increase of volume fraction increases the strength of composite only up to a certain limit. The number of particles increase along with their the increased volume fraction in the composite. Therefore the particles have occupied a larger area, which is important in restricting dislocation when applying load. Limitation of increased strength as a result of increased volume fraction is due to increases in porosity and in stresses on the composite material system which affect the mechanical properties of the composite.

Ductility of composite is reduced on increasing the volume fraction of reinforce-ment. This is also a major reason for limiting the volume fraction of reinforcement. At higher volume fraction, too much brittleness and stress concentration will be induced on the matrix material, which favours the crack initiation process when the load is ap-plied. But in particulate nano composite, the localised ductility is effectively retained for the matrix material (Sree Manu et al., 2017, Hassan and Gupta, 2006, Mazahery and Shabani, 2012) through trapping the dislocation between particles, also known as Orowan strengthening. At the same time, distributions of particles have to be con-sidered while fabricating PRMMCs because non-uniform distribution of particles creates a deleterious effect on the tensile strength, flow stress and fatigue strength (Kennedy and Wyatt, 2000, Hong et al., 2003). Agglomeration in the composites pro-duces brittleness and also promotes void nucleations.

In recent years, in-situ reactions have been used to incorporate fine particles into matrix material to avoid the drawbacks of incorporating finer reinforcement particle in ex-situ methods. These composites have better bonding strength and more uniform distribution of particles than ex-situ composites. In-situ methods are now commonly used for Al- and Mg-based composites (Kuruvilla et al., 1990).

5.3   CREEP BEHAVIOUR

Creep property is evaluated for metals and alloys being considered for various applications, especially at elevated temperature conditions. Normally, creep rate is increased on increasing dislocation and grain boundary sliding in all types of materials systems. The three stages of creep are considered while investigating the creep rate of materials. In the primary stage, strain rate is decreases with increased time. In the secondary stage, strain rate is constant, also known as steady state creep rate. Here the strain varied linearly with respect to time. But in the tertiary stage, strain rate is drastically increased with time. In this final stage, the material is going to be fractured through void nucleation and cavitation.

Generally, the steady state creep rate is expressed by the following power law equation:

$$Creep\,rate\,(\dot{\varepsilon}_s) = A\sigma^n \exp(-Q\,/\,RT) \tag{1}$$

where A is a constant, σ is the applied stress, T is the temperature in kelvins, Q is the activation energy, R is the gas constant and n is the stress exponent. From this equation it is easy to identify the stress exponent and activation energy during creep. The values are compared with the monolithic matrix material to help identify the improvement in the creep behaviour of the composite materials system. Apart from this equation, various approaches are using to analyse the creep behaviour of a material system based on interface characteristics in composite, dispersion strengthening of alloys with dislocation interactions, load transferring efficiency, and so on (Park and Mohamed, 1995). The discontinuous composite system acts like a dispersion-strengthened alloy. A few investigations have incorporated the threshold stress into equation 1 (Nardone and Strife, 1987) and observed that the creep deformation is driven by an effective stress instead of the applied stress.

In MMCs, an incorporation of hard reinforcements has restricted the dislocation and grain boundary sliding. Therefore MMCs have superior creep resistance compared with metals and alloys (Park and Mohamed, 1995). However, the creep characteristic of MMCs is an important property because of the usage of MMCs in structural and non-structural applications subjected to constant loading for a longer time, especially at elevated temperature conditions. Creep behaviour of MMCs depends on occurrence of steady state condition, microstructural changes and deformation mechanisms, which are highly responsible for deciding stress exponent and activation energy. Many studies have investigated the creep behaviour of MMCs in detail (Nieh, 1984, Tjong and Ma, 1999). Nieh (1984) compared the creep resistance of Al/SiC$_p$ and Al/SiC$_w$ (where w indicates whiskers as opposed to p, particles) with unreinforced Al alloy. Al composite has higher creep resistance than Al alloy. More deformation is induced on Al alloy than composite during creep testing. The stress exponent is very high and apparent activation energy is three times higher than self-diffusion of Al alloy. Comparatively, Al/SiC$_w$ composite has better creep resistance than particulate composite. This is because of the shape, effective load transfer ability and higher strength of single crystalline whiskers compared with polycrystalline particles. Al/SiC composite has a higher stress exponent than Al alloy, which means the composite has good load-bearing capacity.

Load transfer mechanism is one of the important parameters to enhance the creep resistance of composite. It differs with respect to the type of reinforcement, processing route, microstructural changes, size of particles and so on. As mentioned, creep resistance is lower in particulate composite than in whisker composite owing to lower load transfer ability and lower strength. Li and Langdon (1998) have investigated the effect of processing route, such as powder metallurgy (PM) and ingot metallurgy (IM), on the creep behaviour of composite. Rate of controlling the creep in PM and IM by dislocation climb and viscous glide is studied respectively. Powder metallurgy composite has better creep resistance than ingot composite, and PM composite also has a higher threshold stress level, which means that below that stress level, creep cannot be possible.

Size of particles and volume fraction also significantly affect the creep rate of composite. In an Al/SiC$_p$ composite system, smaller particles (3.5 µm) have better creep resistance than unreinforced Al alloy, and the creep resistance of Al alloy reinforced with coarser particles (10 and 20 µm) is almost the same as that of unreinforced Al alloy (Tjong and Ma, 1999). Creep resistance increases on increasing the reinforcement volume fraction of an Al/SiC$_p$ composite system (Krajewski et al. 1997).

5.4 FATIGUE BEHAVIOUR

Metals and alloys are subjected to cyclic load when used in various applications. As a result of their lower fatigue resistance against the applied alternating stresses, metals and alloys fail easily, which is a driving force for the development of metal matrix composites. An addition of hard ceramic particles by tailoring into a soft monolithic matrix material increases the fatigue resistance substantially over unreinforced alloy within an acceptable cost range. In recent years, the use of MMCs in structural applications, especially in the aircraft and automobile industries, is very high, so the enhancement of fatigue resistance has become a significant mechanical property.

Normally, fatigue occurs as a result of cyclic mechanical or thermal loads or a combination of the two. Basically mechanical fatigue is classified as high cycle fatigue (HCF) and low cycle fatigue (LCF) based on the region of the stress strain curve where the repetitive application of load is taking place. Usually HCF is a stress-controlled fatigue and this applied stress lies in the elastic region. So the test sample takes more cycles to fail, hence the name high cycle fatigue. LCF is a strain-controlled fatigue, and while the test sample is cyclically strained the deformations lie mostly in the plastic region. HCF induces a work-hardening effect in the test sample owing to its high number of cycles, which leads to a very short duration of crack propagation even though the time for crack initiation is greater. But in LCF, the duration of crack propagation is comparatively greater as there is no straining due to low cycles.

The volume fraction of reinforcement, particle size, distribution of reinforcement, grain size, microstructure, presence of inclusions or defects that arise from processing and testing environment are the major deciding factors of the fatigue resistance of MMCs (Chawla et al., 1998, Chawla et al., 2000). The increase in fatigue resistance of a metallic material with the addition of reinforcement is in accordance with the load transfer from the soft matrix to particles with a high shear modulus. However, the bonding strength at the interface of matrix and reinforcement is very important. If the interface is weak and readily delaminates, the weak interfaces will act as the sites of fatigue crack initiation. Added reinforcement also acts as nucleation sites on solidification, which controls the grain growth and thus increases the strength of the material. Addition of reinforcement also reduces the stress concentration around the inclusions and precipitates, which minimises the sites for fatigue crack initiation and increases fatigue life. The bigger particle size as reinforcement may cause particle cracking during secondary processing before the material is used in an application. The particle cracking will act as a new energy site for fatigue crack initiation. Hence the MMCs of smaller particle size possess more fatigue life. Figure 5.2a and b are S-N (stress versus number of cycles) diagrams

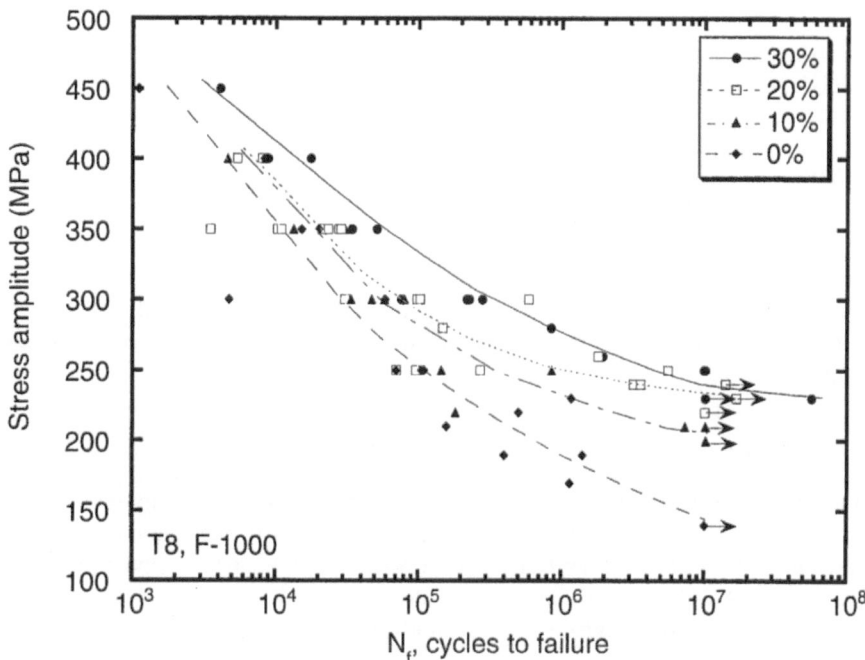

FIGURE 5.2   (a) Effect of SiC reinforcement volume fraction on fatigue life.

Source: Chawla et al. (1998).

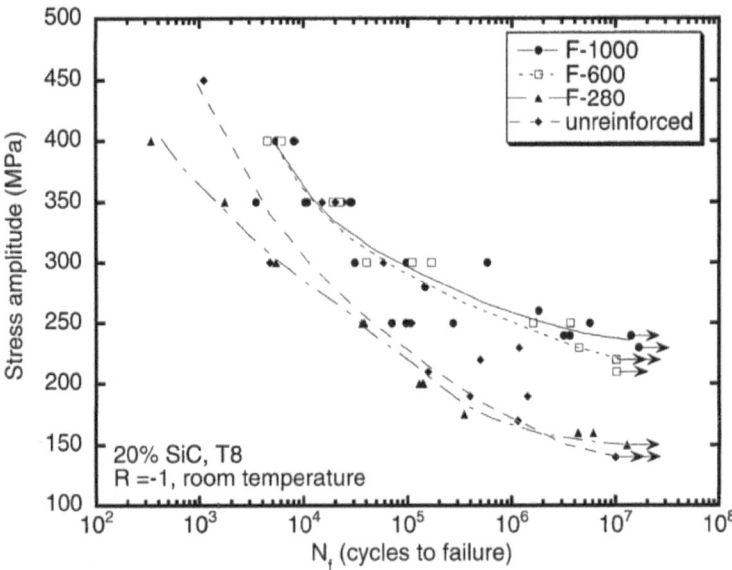

FIGURE 5.2   (b) Effect of reinforcement particle size on fatigue life.

Source: Chawla et al. (1998).

showing the effect of SiC reinforcement volume fraction on fatigue life and the effect of reinforcement particle size on fatigue life reported by Chawla et al. (1998).

The intermetallic inclusion in the matrix and the clusters, or agglomeration, of particles also affect fatigue resistance as shown in Figure 5.3a and 5.3b, respectively.

MMCs are usually made by combining hard ceramic particles with the metals by a liquid metallurgical or a powder metallurgical route. The resulting MMCs have an anisotropic property at macroscopic scale. The addition of particles substantially increases the stiffness and yield strength of matrix material compared to the unreinforced metal or alloys, and also the presence of hard particles reduces the ductility of matrix material. Hence, the volume fraction of the reinforcement is generally limited to 30%.

5.5   WEAR

Higher wear resistance is a driving force for the use of MMCs in automobile industries. Al-based MMCs are mainly used in piston areas in the automobile sector. The auto industry has successfully applied Al-based particulate composites, mostly SiC_p/Al and Al/Al$_2$O$_3$, in pistons, engine blocks, disc rotor brakes, drums, calipers, connecting rods, drive shafts, snow tire studs and other components (Lindroos and Talvitie, 1995). Progressive loss of material, also known as wear, occurred during either sliding or relative motion between contact surfaces. An incorporation of hard reinforcements into matrix material has improved the wear resistance and thermal stability of MMCs. A number of studies have reported the tribological behaviour of

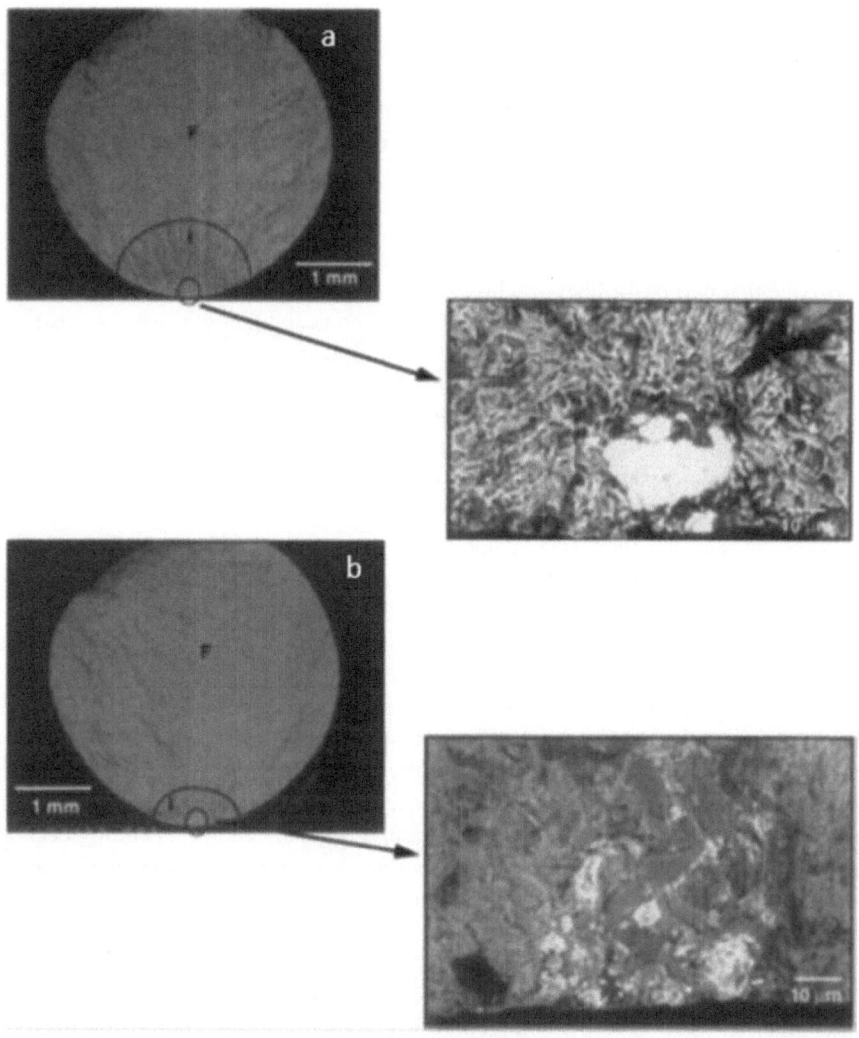

FIGURE 5.3 SEM micrograph of (a) intermetallic inclusion and (b) particle cluster that served as a fatigue failure initiation site. Fatigue initiation and fast fracture regions are marked I and F, respectively.

Source: Chawla et al. (1998).

MMCs and their successful implementation for many applications (Bai et al., 1992, Kukutschová et al., 2009, Bauri and Surappa, 2008). For example, the presence of hard graphite particles in Al matrix material improved the wear resistance of the composite because the graphite acts as a solid lubricant during sliding (Rohatgi et al., 1992). Normally, abrasion, adhesion, fatigue and oxidative wear play major roles in the failure of MMCs. Different factors can affect the wear resistance of MMCs. These factors determine the transition from mild to severe wear during sliding. Parameters

such as internal factors (size of reinforcement, volume fraction, type of reinforcement, addition of an alloying element and distribution of particles) and external factors (sliding distance, sliding speed, applied load, etc.) could control the tribological
properties of MMCs. The following sections discuss variation in the wear resistance
and coefficient of friction of MMCs with respect to a few of these parameters.

5.5.1 INTERNAL FACTORS

For the fabrication of MMCs, various parameters like type of reinforcement in the
form of either particle or fiber or whisker, size of reinforcement, volume fraction, and
processing techniques are considered. These factors indirectly affect the wear rate
of the composite by way of wear parameters. For example, non-uniform distribution
of particles concentrates stress on particular regions with particles, which does not
enhance the wear resistance of composite well. In addition, particles easily plough
out from matrix material during the sliding if an interface bonding between matrix
and reinforcement is weak, and the rolling particles add wear, a situation called three-
body wear.

Type of reinforcement is a major factor in the wear resistance of composite.
Particle-reinforced composites have better wear resistance than whisker- and fiber-reinforced composites (Miyajima and Iwai, 2003). Because the diameter of a particle is
greater than those of whisker and fiber, a particle has superior load-bearing capacity
during sliding.

Transition from mild wear to severe wear is not only dependent on reinforcement
load-bearing capacity (Alpas and Zhang, 1994). In the Al/SiC$_p$ composite system, the
transition of wear regime is also affected by using different counterface materials.
Mullite counterface material against composite has induced higher wear of the composite than SAE 52100 steel. Large wear debris is generated during sliding between
Al/SiC$_p$ and mullite. This is because of the absence of a transfer layer on the sample
surface while sliding against mullite, which proves that the composite with steel
counterface has higher wear resistance.

Apart from type of reinforcement and type of counterface material, size and volume
fraction of reinforcement have played major roles in the variation of wear resistance
of composite. Many investigations report that an increase in particle size leads to increased wear resistance of composite (Alpas and Zhang, 1994, Mandal et al., 2004).
Finer particles are easily pulled out from matrix material during sliding. This occurs
because fine particles are not deeply embedded in matrix material, and after pulling
out they generate three-body wear on the material system. Fine particles have poor
wettability with matrix material and also easily get clustered because of their higher
surface area and surface energy (Mahdavi and Akhlaghi, 2011). The effect of particle
size on wear rate of composite is shown in Figure 5.4. Some investigations reported a
decrease in wear rate due to reduction of particle size. These results revealed that the
variation of wear occurs with respect to size of particles depending on the different
testing methods (Liang et al., 1995).

Similarly, an increase in volume fraction reduced the wear rate of composite
(Madhan Kumar et al., 2016). The increased number of particles prevents dislocation in larger areas of matrix material. Finally, in most of the investigations, a higher

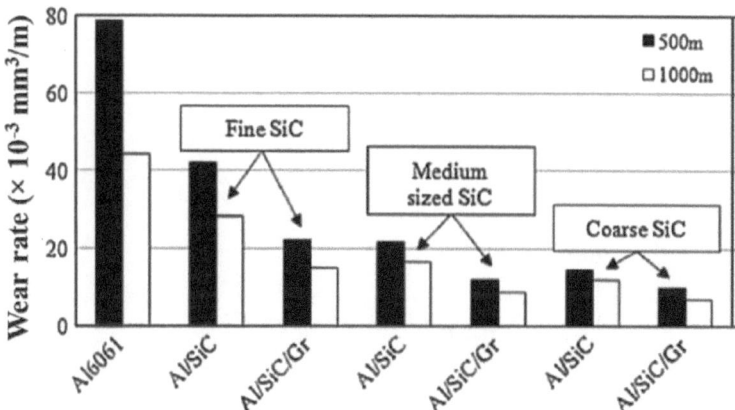

FIGURE 5.4 Variation of the wear rates of Al/SiC and Al/SiC/Gr composites containing different sized SiC particles at the sliding distances of 500 and 1000 m.

Source: Mahdavi and Akhlaghi (2011).

volume fraction with uniform distribution of particles improved the wear resistance of composite (Gopalakrishnan and Murugan, 2012). The coefficient of friction of composite is always lower than that of matrix material because ceramic particles separate the contact asperity between pin and disc. The coefficient of friction slightly increased above a certain volume fraction but was still lower than for unreinforced alloy.

5.5.2 EXTERNAL FACTORS

The normal load, sliding distance, sliding speed, temperature, lubrication effect and so on externally affect the wear resistance of composite. These factors are dominant in determining the transition from mild to severe wear. A number of studies reported the effect of wear process parameters on wear rate of composite (Gopalakrishnan and Murugan, 2012, Tjong and Lau, 2000, Watson et al., 2005, Yigezu et al., 2013, Baradeswaran and Perumal, 2013). Generally, an increase of normal load decreases the wear resistance of composite as shown in Figure 5.5. This is due to an increase of stresses in contact regions, followed by an increase in the interface temperature (Tjong and Lau, 2000). An Al/TiB$_2$ composite system has induced a similar kind of load effect (Mandal et al., 2007). The occurrence of wear is smaller at lower load conditions (19.6 N and 39.2 N in the study). This is due to the presence of TiB$_2$ particles and the formation of a mechanical mixed layer (MML) separating the pin and the disc. Additionally, the presence of TiB$_2$ particles significantly improves the hardness of composite. Only shallow grooves appeared on the worn surface, which indicates the inducement of small plastic deformation of Al/TiB$_2$ composite. Increasing the load increases the stresses between contact asperities of pin on disc followed by increased interface temperature, which increased wear drastically. The formation of a transfer layer is destroyed completely by increasing the load; therefore matrix material came into earlier contact with counterface material. Delaminations appeared on the worn

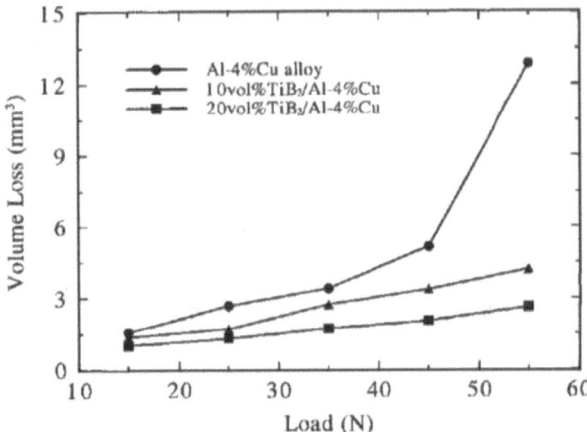

FIGURE 5.5   Effect of load on wear rate of Al-Cu/TiB$_2$ composite variation of volume loss with applied normal loads for Al–4Cu alloy and MMCs tested for sliding distance of 2500 m and sliding velocity of 1 ms^{-1}.

Source: Tjong and Lau (2000).

surface of Al/TiB$_2$ composite that revealed severe wear through inducement of sub-surface plastic deformation. Natarajan et al. (2006) compared the effect of load on the tribological behaviour of cast iron and A356/SiC$_p$ against friction lining material. Transfer film was fully destroyed at a faster rate on increasing the load, which reduced the wear resistance of cast iron and Al/SiC$_p$ composite. Comparatively, A356/ SiC$_p$ composite had higher wear resistance than cast iron while sliding against a friction lining pin.

However, applied load is not the sole deciding parameter for wear behaviour of composite. Sliding distance is another dominant factor for deciding the wear resistance of composite. Surface roughness, surface chemistry and surface-related properties play a major role in increasing the sliding distance. In an Al/SiC$_p$ composite system, wear rate increased on increasing the sliding distance (5000 m) (Rao and Das, 2011a). Seizure of the specimen was noticed on Al alloy when the sliding distance varied from 4000 m to 5000 m. But this cannot occur in composite on account of the presence of SiC$_p$ particles. In an Al/TiC system, sliding distance was severely affected when the load reached peak condition (Yigezu et al., 2013). Figure 5.6 shows that the weight loss and coefficient of friction increased almost linearly while increasing the sliding distance irrespective of weight fraction. Micro softening occurred while increasing the sliding distance, which decreased bonding strength between matrix and reinforcement. With weakening of bonds, the particles are ploughed out easily from the matrix material, which induces three-body wear, increasing the wear rate of the composite.

On the other hand, wear resistance of Al MMCs increased on increasing the sliding speed during the sliding wear test (Ravikiran and Surappa, 1997, Shorowordi et al., 2004). Shorowordi et al. (2004) compared the effect of sliding speed on wear behaviour

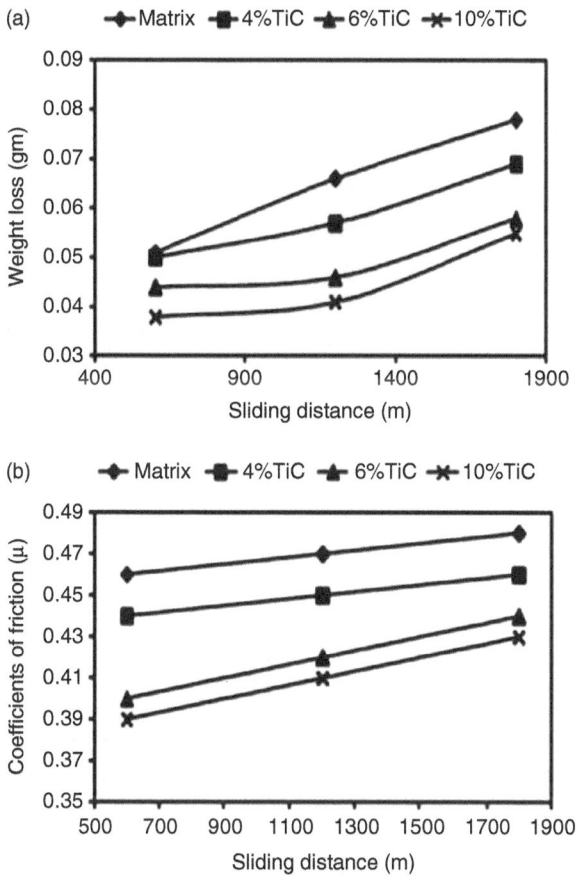

FIGURE 5.6   Effect of sliding distance on the weight loss and coefficients of friction of Al alloy and Al/TiC composite.

Source: Yigezu et al. (2013).

of Al-B$_4$C, Al/SiC$_p$ and unreinforced Al alloy. They observed that the composite has higher wear resistance than Al alloy and also that the wear resistance of composite decreased on increasing the sliding speed as shown in Figure 5.7. Delaminations appeared on the worn surface of Al alloy while increasing the sliding speed, indicating severe wear. But in MMCs, only very fine marks and small grooves were observed indicating mild wear. Other studies also support this result for increasing the sliding speed (Ravikiran and Surappa, 1997, Tjong and Lau, 2000).

A few studies have reported that the wear rate of MMCs increased on increasing the sliding speed (Kwok and Lim, 1999, Ranganath et al., 2001, Qin et al., 2008, Rao and Das, 2011b, Rao et al., 2013), but as usual the wear rate was lower than Al alloy. Kwok and Lim (1999) observed in Al/SiC composite that an increase of interface temperature softened the matrix material during the sliding at higher speed. At lower speed (less than 3 m/s), occurrence of wear is much smaller and in addition no

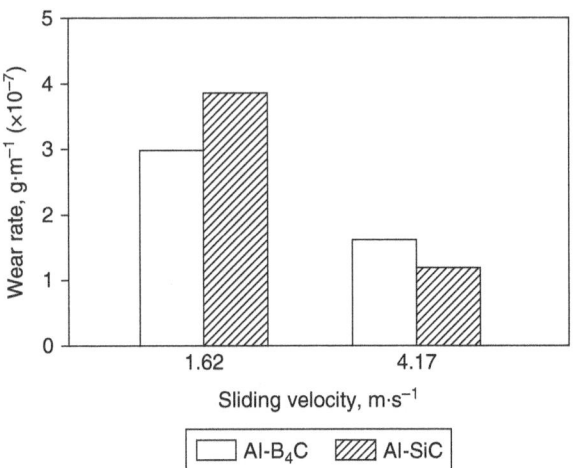

FIGURE 5.7   Wear rate of Al-B₄C and Al-SiC_p composite at different velocities.

Source: Shorowordi et al. (2004).

failure was observed. At intermediate speed (3 to 8 m/s), wear is increased sharply, followed by failure on the composite surface. But sudden failure occurred when the sliding speed exceeded 8 m/s, indicating the occurrence of severe wear. In an Al/Mg$_2$Si system, the same sliding speed effect was reported by Qin et al. (2008). Composite with modified Mg$_2$Si (modified by phosphorus) had superior wear resistance to others.

This observation of contrary results revealed that the load and sliding distance have played roles with an increase of sliding speed. Catastrophic failure has occurred through an increase of interface temperature when the speed is higher with critical load. Therefore, other factors have to be considered while increasing the sliding speed. Figure 5.8 explains the variation of wear mechanisms with respect to load and sliding speed.

5.6   SUMMARY

This chapter clearly demonstrates that the particle size, volume fraction of reinforcement, nature of bonding and processing routes affect the tensile strength, creep fatigue and wear behaviour of composite. The addition of hard ceramic particles as reinforcement in the soft matrix material improves the strength and performance of matrix material by increasing the ability to withstand higher stresses than unreinforced matrix material. These particles could improve such key properties of a metallic material as stiffness, hardness, tensile strength and thermal stability of the matrix material. However, the particle reinforcement size, volume fraction, distribution of particles and wear parameters are directly connected to the frictional components which predominantly influence the strength and wear resistance of composites. At the same time, toughness of composite is reduced at a higher volume fraction of

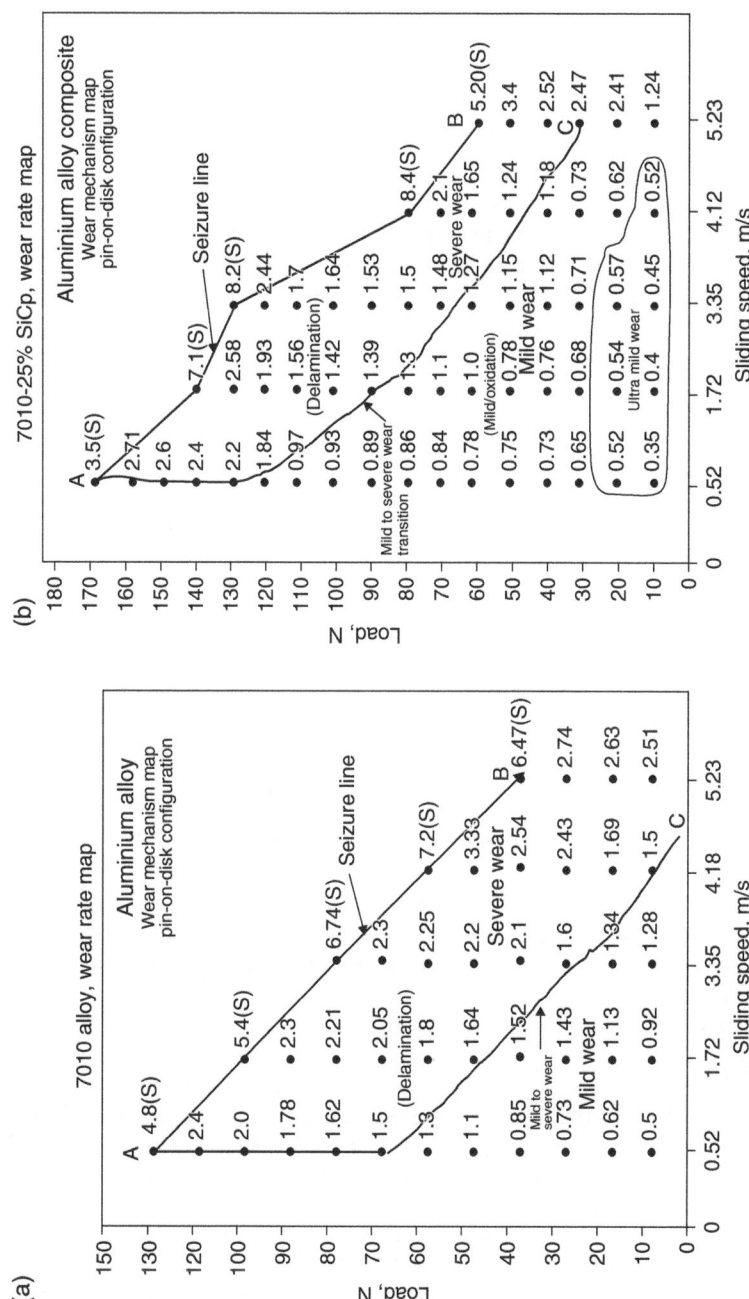

FIGURE 5.8 Effect of sliding speed versus load on wear transition for (a) Al 7075 alloy, (b) Al 7010/25%SiC$_p$.

Source: Rao and Das (2011b).

reinforcement, which also affects the wear resistance at a higher loading condition. Normally, reinforcement with finer particles increases the strength and also makes the material more difficult to deform or fragment than material with coarser particles, and inducement of cracks is easy with coarse particle reinforcement. Fine reinforcement particles improve the strength of composite significantly through their ability to minimize micro voids by pinning the triple junctions in grain boundaries, being harder to fragment by stress concentrations and restricting crack propagation, and inducing an improved grain refinement as the reinforcements act as active sites for nucleation during solidification. However, finer particles could easily plough out of matrix material due to their low penetration and less mechanical interlocking at the interface of composite material. Thus, the importance of interfacial bonding is also a key factor to be considered. Regarding the fatigue behaviour of PRMMCs, the fatigue crack is initiated from voids and inclusions such as weak interphases. Hence, the PRMMC processing route should be chosen with care to suit the desired properties of the end material and to minimise voids, pores and weak interphases.

5.7   SCOPE OF PRMMCS

PRMMCs provide metallic material with novel properties to suit the new technologies, advancing engineering applications and future industrial developments. The possibilities to manipulate the properties of metals with the different composition of matrix and reinforcement phases, different processing routes and processing parameters, and post processing combinations host opportunities for materials and design engineers. The existing barriers are being addressed through various research efforts. The literature indicates the future of PRMMCs in industrial and commercial applications is bright. There is a huge interest in research to overcome the lack of design data and property modelling, as well as the high cost for processing and post treatments. However, the advancements in nano particle reinforcement, in-situ reinforced composite processing, novel rheocasting and functionally graded PRMMCs have been technological breakthroughs with promising signs. These technologies drastically improved the properties of PRMMCs for high temperature applications with enhanced wear corrosion and fatigue resistance. PRMMCs potentially offer new roles and research opportunities in automotive applications.

REFERENCES

Alpas, A.T., and J. Zhang. "Effect of microstructure (particulate size and volume fraction) and counterface material on the sliding wear resistance of particulate-reinforced aluminum matrix composites." *Metallurgical and Materials Transactions A* 25 (1994): 969–983, https://doi.org/10.1007/BF02652272.
Arunbharathi Ramaswamy, Ashoka Varthanan Perumal, and Samson Jerold Samuel Chelladurai. "Investigation on mechanical properties and dry sliding wear characterization of stir cast LM13 aluminium alloy-ZrB$_2$-TiC particulate hybrid composites." *Materials Research Express* 6, no. 6 (2019): 066578, https://doi.org/10.1088/2053-1591/ab0ef8.
Bai, B.N.P., B.S. Ramasesh, and M.K. Surappa. "Dry sliding wear of A356-Al-Sic composites." *Wear* 57, no. 2 (1992): 295–304, https://doi.org/10.1016/0043-1648(92)90068-J.

Balaji, V., N. Sateesh, and M. Mansoor Hussain. "Manufacture of aluminium metal matrix composite (AL 7075-SiC) by stir casting technique." *Materials Today: Proceedings* 2, no. 4–5 (2015): 3403–3408, https://doi.org/10.1016/j.matpr.2015.07.315.

Baradeswaran, A., and A.E. Perumal. "Influence of B₄C on the tribological and mechanical properties of Al 7075 – B₄C composites." *Composites Part B*, 54 (2013): 146–152, https://doi.org/10.1016/j.compositesb.2013.05.012.

Bauri, R., and M.K. Surappa. "Sliding wear behavior of Al – Li – SiC$_p$ composites." *Wear* 265, no. 11–12 (2008): 1756–1766. https://doi.org/10.1016/j.wear.2008.04.022.

Chawla, N., and Y.L. Shen, "Mechanical behaviour of particle reinforced metal matrix composites." *Advanced Engineering Materials* 3, no. 6 (2001): 357–370, https://doi.org/10.1002/1527-2648(200106)3:6<357::AID-ADEM357>3.0.CO;2-I.

Chawla, N., C. Andres, J.W. Jones, and J.E. Allison. "Effect of SiC volume fraction and particle size on the fatigue resistance of a 2080 Al/SiC$_p$ composite." *Metallurgical and Materials Transactions A: Physical Metallurgy and Materials Science* 29, no. 11 (1998): 2843–2854, https://doi.org/10.1007/s11661-998-0325-5.

Chawla, N.N., U. Habel, Y.L. Shen, C. Andres, J.W. Jones, and J.E. Allison. "The effect of matrix microstructure on the tensile and fatigue behaviour of SiC particle-reinforced 2080 Al matrix composites." *Metallurgical and Materials Transactions A: Physical Metallurgy and Materials Science* 31 (2000): 531–540, https://doi.org/10.1007/s11 661-000-0288-7.

Deutsch, S., "Automotive Applications for Advanced Composite Materials." In *23rd National SAMPE Symposium*, pp. 34–60, (May 1978).

El-Galy, I.M., M.H. Ahmed, and B.I. Bassiouny. "Characterization of functionally graded Al-SiCp metal matrix composites manufactured by centrifugal casting." *Alexandria Engineering Journal* 56, no. 4 (2017): 371–381, https://doi.org/10.1016/j.aej.2017.03.009.

Gopalakrishnan, S., and N. Murugan. "Production and wear characterisation of AA 6061 matrix titanium carbide particulate reinforced composite by enhanced stir casting method." *Composites Part B* 43, no. 2 (2012): 302–308, https://doi.org/10.1016/j.comp ositesb.2011.08.049.

Hassan, S.F., and M. Gupta. "Effect of particulate size of Al₂O₃ reinforcement on microstructure and mechanical behaviour of solidification processed elemental Mg." *Journal of Alloys and Compounds* 419, no. 1–2 (2006): 84–90, https://doi.org/10.1016/j.jall com.2005.10.005.

Hong, S.J., H.M. Kim, D. Huh, C. Suryanarayana, and B.S. Chun. "Effect of clustering on the mechanical properties of SiC particulate-reinforced aluminum alloy 2024 metal matrix composites." *Materials Science and Engineering: A* 347, no. 1–2 (2003): 198–204, https://doi.org/10.1016/S0921-5093(02)00593-2.

Jen, P.C., D.R. Parhi, G. Pohit, et al. Crack assessment by FEM of AMMC beam produced by modified stir casting method. *Materials Today: Proceedings* 2 (2015): 2267–2276.

Kaczmar, J.W., and K. Pietrzak. "The production and application of metal matrix composite materials." *Journal of Materials Processing Technology* 106, no. 1–3 (2000): 58–67. https://doi.org/10.1016/S0924-0136(00)00639-7.

Karbalaei Akbari, M., O. Mirzaee, and H.R. Baharvandi. "Fabrication and study on mechanical properties and fracture behavior of nanometric Al₂O₃ particle reinforced A356 composites focusing on the parameters of vortex method." *Materials and Design* 46 (2013): 199–205, https://doi.org/10.1016/j.matdes.2012.10.008.

Kennedy, A.R., and S.M. Wyatt. "The effect of processing on the mechanical properties and interfacial strength of aluminium/TiC MMCs." *Composites Science and Technology* 60, no. 2 (2000): 307–314, https://doi.org/10.1016/S0266-3538(99)00125-6.

Kok, M. "Production and mechanical properties of Al_2O_3 particle-reinforced 2024 aluminium alloy composites." *Journal of Materials Processing Technology* 161, no. 3 (2005): 381–387, https://doi.org/10.1016/j.jmatprotec.2004.07.068.

Krajewski, P.E., J.E. Allison, and J.W. Jones. "Effect of SiC particle reinforcement on the creep behavior of 2080 aluminum." *Metallurgical and Materials Transactions A: Physical Metallurgy and Materials Science* 28 (1997): 611–620, https://doi.org/10.1007/s11661-997-0046-1.

Kukutschová, J., V. Roubíček, K. Malachová, Z. Pavlíčková, R. Holuša, J. Kubačková, V. Mička, D. MacCrimmon, and P. Filip. "Wear mechanism in automotive brake materials, wear debris and its potential environmental impact." *Wear* 267, no. 5–8 (2009): 807–817, https://doi.org/10.1016/j.wear.2009.01.034.

Kumar, G.B., and R. Pramod. "Influence of WC particulate reinforcement on the mechanical properties and sliding wear of Al6061 alloys." *Applied Mechanics and Materials* 813–814 (2015): 67–73.

Kuruvilla, A.K., K.S. Prasad, V.V. Bhanuprasad, and Y.R Mahajan. " Microstructure-property correlation in $Al/TiB_2(Xd*)$ composites." *Scripta Metallurgica et Materiala* 24, no. 5 (1990): 873–878, https://doi.org/10.1016/0956-716X(90)90128-4.

Kwok, J.K.M., and S.C. Lim. "High-speed tribological properties of some Al / SiC_p composites: I . Frictional and wear-rate characteristics." *Composites Science and Technology* 59, no. 1–2 (1999): 55–63. https://doi.org/10.1016/S0266-3538(98)00055-4.

Li, Yong, and Terence G. Langdon. "An examination of the effect of processing procedure on the creep of metal matrix composites." *Materials Science and Engineering: A* 245, no. 1 (1998): 1–9, https://doi.org/10.1016/S0921-5093(97)00712-0.

Liang, Y.N., Z.Y. Ma, S.Z. Li, S. Li, and J. Bi. "Effect of particle size on wear behaviour of SiC particulate-reinforced aluminum alloy composites." *Journal of Materials Science Letters* 14 (1995): 114–116. https://doi.org/10.1007/BF00456563.

Lindroos, V.K., and M.J. Talvitie. "Recent advances in metal matrix composites." *Journal of Materials Processing and Technology* 53, no. 1–2 (1995): 273–284, https://doi.org/10.1016/0924-0136(95)01985-N.

Madhan Kumar, N., S.S. Kumaran, and L.A. Kumaraswamidhas. "Aerospace application on Al 2618 with reinforced – Si_3N_4, AlN and ZrB_2 in-situ composites." *Journal of Alloys and Compounds* 672 (2016) 238–250, https://doi.org/10.1016/j.jallcom.2016.02.155.

Mahdavi, S., and F. Akhlaghi. "Effect of the SiC particle size on the dry sliding wear behavior of SiC and SiC–Gr-reinforced Al6061 composites." *Journal of Materials Science* 46 (2011): 7883–7894, https://doi.org/10.1007/s10853-011-5776-1.

Mandal, A., R. Maiti, M. Chakraborty, and B.S. Murty. "Effect of TiB_2 particles on aging response of Al-4Cu alloy." *Materials Science and Engineering: A.* 386, no. 1–2 (2004): 296–300, https://doi.org/10.1016/j.msea.2004.07.026.

Mandal, A., M. Chakraborty, and B. S. Murty. "Effect of TiB_2 particles on sliding wear behaviour of Al–4Cu alloy." *Wear* 262, no. 1–2 (2007): 160–166. https://doi.org/10.1016/j.wear.2006.04.003.

Mazahery, A., and M.O. Shabani. "Characterization of cast A356 alloy reinforced with nano SiC composites." *Transactions of Nonferrous Metals Society of China* 22, no. 2 (2012): 275–280, https://doi.org/10.1016/S1003-6326(11)61171-0.

Michael Rajan, H.B., S. Ramabalan, I. Dinaharan, and S.J. Vijay. "Synthesis and characterization of in situ formed titanium diboride particulate reinforced AA7075 aluminum alloy cast composites." *Materials and Design* 44 (2013): 438–445, https://doi.org/10.1016/j.matdes.2012.08.008.

Miyajima, T., and Y. Iwai. "Effects of reinforcements on sliding wear behaviour of aluminum matrix composites." *Wear* 255, no. 1–6 (2003): 606–616, https://doi.org/10.1016/S0043-1648(03)00066-8.

Nardone, V.C., and J.R. Strife. "Analysis of the creep behaviour of silicon carbide whisker reinforced 2124 Al(T4)." *Metallurgical Transactions A* 18 (1987): 109–114, https://doi.org/10.1007/BF02646227.

Natarajan, N., S. Vijayarangan, and I. Rajendran. "Wear behaviour of A356 / 25SiC$_p$ aluminium matrix composites sliding against automobile friction material." *Wear* 261, no. 7–8 (2006): 812–822, https://doi.org/10.1016/j.wear.2006.01.011.

Nieh, T.G. "Creep Rupture of a Silicon Carbide Reinforced Aluminum Composite." *Metallurgical and Materials Transactions A: Physical Metallurgy and Materials Science* 15 (1984): 139–146, https://doi.org/10.1007/BF02644396.

Park, K.T., and F.A. Mohamed. "Creep strengthening in a discontinuous SiC-Al composite." *Metallurgical and Materials Transactions A* 26 (1995): 3119–3129. https://doi.org/10.1007/BF02669441.

Qin, Q.D., Y.G. Zhao, and W. Zhou. "Dry sliding wear behavior of Mg$_2$ Si / Al composites against automobile friction material." *Wear* 264, no. 7–8 (2008): 654–661, https://doi.org/10.1016/j.wear.2007.05.008.

Ram Prabhu, T. "Processing and properties evaluation of functionally continuous graded 7075 Al alloy/SiC composites." *Archives of Civil and Mechanical Engineering* 17 (2017): 20–31, https://doi.org/10.1016/j.acme.2016.08.004.

Ranganath, G., S.C. Sharma, and M. Krishna. "Dry sliding wear of garnet reinforced zinc / aluminium metal matrix composites." *Wear* 251, no. 1–12 (2001): 1408–1413, https://doi.org/10.1016/S0043-1648(01)00781-5.

Rao, R.N., and S. Das. "Effect of SiC content and sliding speed on the wear behaviour of aluminium matrix composites." *Materials and Design* 32, no. 2 (2011a): 1066–1071, https://doi.org/10.1016/j.matdes.2010.06.047.

Rao, R.N., and S. Das. "Effect of sliding distance on the wear and friction behaviour of as cast and heat-treated Al – SiC$_p$ composites." *Materials and Design* 32, no. 5 (2011b): 3051–3058, https://doi.org/10.1016/j.matdes.2011.01.033.

Rao, R.N., S. Das, D.P. Mondal, G. Dixit, and S.L.Tulasi Devi." Dry sliding wear maps for AA7010 (Al–Zn–Mg–Cu) aluminium matrix composite." *Tribology International* 60 (2013): 77–82, https://doi.org/10.1016/j.triboint.2012.10.007.

Ravikiran, A., and M.K. Surappa. "Effect of sliding speed on wear behaviour of A356 Al-30 wt.% SiCp MMC." *Wear*, 206, no. 1–2 (1997): 33–38, https://doi.org/10.1016/S0043-1648(96)07341-3.

Ren, S., X. He, X. Qu, I.S. Humail, and Y. Li. "Effect of Si addition to Al – 8Mg alloy on the microstructure and thermo-physical properties of SiC$_p$ / Al composites prepared by pressureless infiltration." *Materials Science and Engineering: B* 138, no. 3 (2007): 263–270, https://doi.org/10.1016/j.mseb.2007.01.023.

Rohatgi, P.K., S. Ray, and Y. Liu. "Tribological properties of metal matrix-graphite particle composites." *International Materials Reviews* 37, no. 1 (1992): 129–149, https://doi.org/10.1179/imr.1992.37.1.129.

Shorowordi, K.M., A.S.M.A. Haseeb, and J.P. Celis. "Velocity effects on the wear, friction and tribochemistry of aluminum MMC sliding against phenolic brake pad." *Wear* 256, no. 11–12 (2004): 1176–1181, https://doi.org/10.1016/j.wear.2003.08.002.

Sree Manu, K.M., S. Arun Kumar, T.P.D. Rajan, M. Riyas Mohammed, and B.C. Pai. "Effect of alumina nanoparticle on strengthening of Al-Si alloy through dendrite refinement, interfacial bonding and dislocation bowing." *Journal of Alloys and Compounds* 712 (2017): 394–405, https://doi.org/10.1016/j.jallcom.2017.04.104.

Tjong, S.C., and K.C. Lau. "Dry sliding wear of TiB$_2$ particle reinforced aluminium alloy composites." *Materials Science and Technology* 16 (2000): 99–102, https://doi.org/10.1179/026708300773002717.

Tjong, S.C., and Z.Y. Ma. "High-temperature creep behaviour of powder-metallurgy alu-minium composites reinforced with SiC particles of various sizes." *Composites Science and Technology* 59, no. 7 (1999): 1117–1125, https://doi.org/10.1016/S0266-3538(98)00151-1.

Watson, I.G., M.F. Forster, P.D. Lee, R.J. Dashwood, R.W. Hamilton, and A. Chirazi. "Investigation of the clustering behaviour of titanium diboride particles in aluminium." *Composites Part A: Applied Science and Manufacturing* 36, no. 6 (2005): 1177–1187, https://doi.org/10.1016/j.compositesa.2005.02.003.

Yigezu, B.S., P.K. Jha, and M.M. Mahapatra. "Effect of sliding distance, applied load, and weight percentage of reinforcement on the abrasive wear properties of in situ synthesized Al – 12 % Si / TiC composites." *Tribology Transactions* 56, no. 4 (2013): 546–554, https://doi.org/10.1080/10402004.2013.767401.

Zhang, Q., G. Chen, G. Wu, Z. Xiu, and B. Luan. "Property characteristics of a AlN$_p$/Al composite fabricated by squeeze casting technology." *Materials Letters* 57, no. 8 (2003): 1453–1458, https://doi.org/10.1016/S0167-577X(02)01006-6.

Zhong, X.L., W.L.E. Wong, and M. Gupta. "Enhancing strength and ductility of magnesium by integrating it with aluminum nanoparticles." *Acta Materiala* 55, no. 8 (2007): 6338–6344, https://doi.org/10.1016/j.actamat.2007.07.039.

6 Tribological Properties of Metal Matrix Composites

S. Darius Gnanaraj,[1] T. Ram Prabhu[2]
[1]Vellore Institute of Technology, India.
Email ID: dariusgnanaraj.s@vit.ac.in
[2]Defence R&D Organization, India.
Email ID: ramprabhu.t@gmail.com

CONTENTS

DOI: 10.1201/9781003109723-6

6.1   INTRODUCTION

Composite is a material that consists of two or more distinct phases (matrix and reinforcement phases) which are mixed to have a new material that is significantly different in properties compared to those of its constituents. The composite material is different from pure metals or conventional materials in terms of the properties. Conventional materials have a few limitations like poor stiffness, strength, and density and are affected by the nature of loading and working environment. Composites meet the requirements in terms of the loading and working environment and are capable of replacing conventional materials. Metal matrix composites (MMCs) are widely used among the different types of composites due to their significant properties, namely, high specific strength, high specific modulus, damping capacity, and wear resistance. Demand for MMC has been increasing owing to its low density and the low cost of reinforcements.

MMCs show exceptional chemical inertness and thermomechanical properties compared with conventional materials, according to Hu and Cong (2018), Li and Wang (2010), Y. Zhang et al. (2019), and Niu et al. (2015). MMCs are also suitable for placing under high load, high temperature, and operating conditions leading to high wear, according to Koczak and Premkumar (1993) and Emamian et al. (2014). To understand the wear mechanism, scanning electron microscope (SEM) characterization techniques are used while conducting tribological research on MMCs. Figure 6.1 shows the braking energy density and operating speeds of modern vehicles used in transportation. Over the last two or three decades, the requirements have increased manyfold, and they are expected to increase in the future. Newer composites with enhanced braking capacity at high temperatures must be found to meet future requirements. The following sections present the current state of research in this field in an effort to provide readers with information that supports creating novel composites with excellent tribological properties.

6.2   CLASSIFICATIONS OF MMC AND APPLICATIONS

Composites are classified as follows:

- Based on the matrix material
  - Metal matrix composites
  - Ceramic matrix composites
  - Polymer matrix composites
- Based on the filler material
  - Particulate composite
  - Fibrous composite
  - Laminate composite

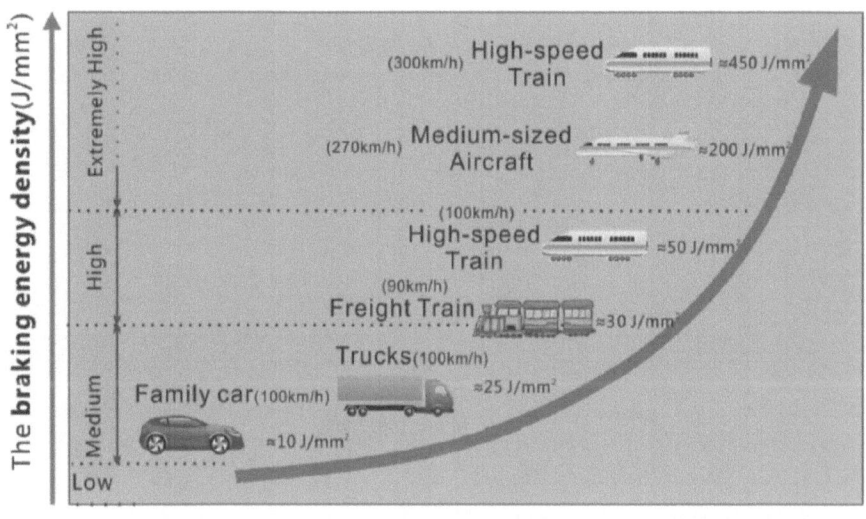

FIGURE 6.1 Braking energy density required for different loads and operating speeds.

Source: Zhou et al. (2018).

6.2.1 METAL MATRIX COMPOSITES (MMCS)

MMC consists of two or more distinct phases distributed uniformly to provide physical and chemical properties that are not possible to obtain from any of its phases. The following are commonly used categories of MMCs:

- Aluminium-based composites
- Magnesium-based composites
- Titanium-based composites
- Copper-based composites

6.2.1.1 Aluminum-Based Composites

Figure 6.2 presents the proportion of MMCs using various matrix materials; aluminium is the most widely used matrix material in MMCs. The potential of magnesium, cobalt, and other matrix materials are to be exploited to bring out the excellent properties possessed by such MMCs.

Aluminium is widely used in MMCs as the matrix material. Aluminium matrix composites (AMCs) have been extensively studied since 1920 and are widely used in electronics packing, sporting goods, aerospace, and automotive applications owing to their physical properties, high strength due to age hardening, resistance to corrosion, and thermal and electrical conductivity. Reinforcement can be easily added to aluminium. AMCs offer a large variety of mechanical properties. They are generally reinforced with materials like Al_2O_3, SiC, and carbon. Normally, AMCs are fabricated by squeeze casting, powder metallurgy, and stir casting methods AMC has numerous

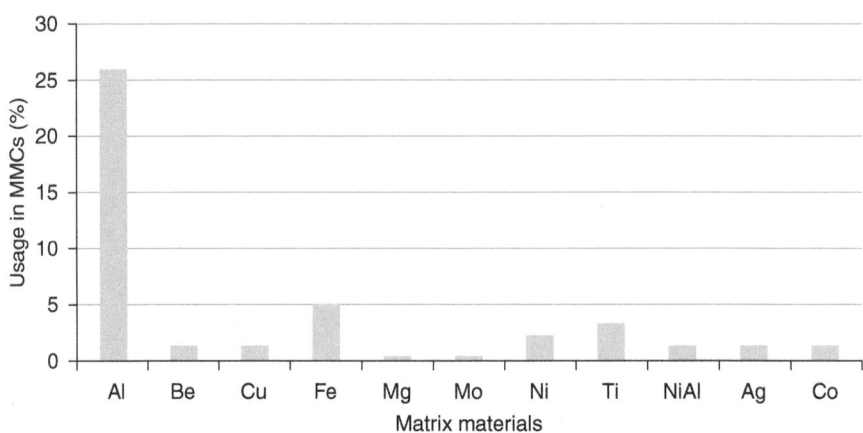

FIGURE 6.2 Matrix materials usage in MMCs.

Source: Adapted from Adebisi et al. (2011).

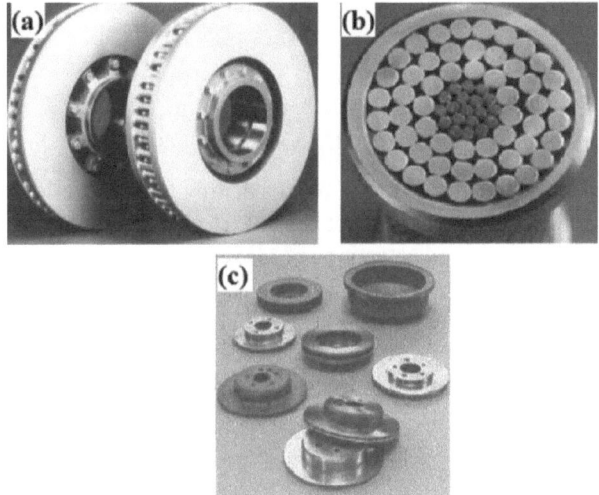

FIGURE 6.3 Industrial applications of AMCs: (a) brake rotors for a high-speed train, (b) cores for high-voltage electrical wires, and (c) automotive braking systems.

industrial applications, a few of which are shown in Figure 6.3. Figure 6.3a shows the brake rotors of a high-speed train which is made with aluminium alloy reinforced with particulates. This rotor could save approximately 120 kg/piece compared with conventional parts generally made with cast iron. AMC is also suitable for fabricating wires for the core of electrical conductors (Figure 6.3b). It gives significant performance when compared with steel-reinforced conductors. Braking systems including discs, back-plate, and drums were made with aluminium alloy reinforced with particulates for the Lupo from Volkswagen (Figure 6.3c).

6.2.1.2 Magnesium-Based Composites

Magnesium-matrix composites are seldom preferred because they are difficult to fabricate and have low thermal conductivity. Magnesium alloys have low density, and magnesium-matrix composites are used in the space industry.

6.2.1.3 Titanium-Based Composites

Titanium alloys have high corrosion resistance due to titanium's stable passive oxide film formation. They are stable at high temperatures and hence are preferred as matrix materials for MMC. Titanium is preferred over aluminium for its ability to retain strength at high temperatures, and so titanium MMCs are preferred for making aircraft parts as well as missiles that operate at very high speeds. But titanium poses problems by reacting with other reinforcement materials present in the MMC. Aircraft structures are made using fiber-reinforced titanium composites.

6.2.1.4 Copper-Based Composites

Compared to other MMCs, copper-matrix composites have high thermal conductivity and they retain strength at high temperatures. Gas turbine blades operating at high speeds and temperatures are manufactured using copper-based superalloys.

6.2.2 Filler Materials

The shape and size of the filler material is the basis for the classification of composites. Particulate composites consist of small particles distributed uniformly in the matrix. These particulates could be either in flakes or in powder form. The fiber-reinforced composites consist of fibers surrounded by the matrix. These fibers could be short or long. In laminated composites, layers of laminates are stacked one by one with different fiber orientations.

6.2.2.1 Powder or Particulate Fillers

Patil et al. (2020) reported that fly ash is a particulate form known for its density and for its economic viability as a filler material. MMCs having fly ash are considered to be an important class of materials. A content of 10 wt% fly ash increases the hardness of aluminium by 75%, and 15% fly ash increases the tensile strength of aluminium by 58%. Mishra et al. (2014) conducted tribological investigations on ZA-27 metal matrix composite (MMC) reinforced with silicon carbide (SiC) particle having different weight percentages ranging from 0 to 9 wt% in steps of 3%. They found that the wear rate decreased when they increased the filler content in the MMC.

6.2.2.2 Short and Long Fiber Fillers

Ashwanth Kumar et al. (2018) reported that hardness and tensile properties of Al-Si-fly-ash composites decreased when the percentage of fly ash goes beyond 10%. The reinforcement of basalt short fiber improved mechanical, damping, tribological and wear properties as a result of grain refinement. Isotropic properties are exhibited by MMCs having short fibers. MMCs having nanofibers have good mechanical and tribological properties compared to MMCs having microfibers. Kumar et al.

(2020) studied the dry sliding wear behavior of aluminium matrix reinforced with nickel-coated short carbon fibers (NCSCFs). The results indicated that Al/NCSCF composite had lower wear rates than that of Al alloy. The wear rates of the composites decreased with the increase of NCSCFs content. However, wear rates increased with the increase in normal load and rotating speed in both composites and Al alloy.

6.2.2.3 Laminated Composites

Qi et al. (2012) reported that high-performance self-lubricated alumina composites are regarded as potential candidates for space applications. Al_2O_3/Mo composites with a laminated structure were prepared, and their mechanical and tribological properties were studied. The results show that Al_2O_3/Mo laminated composites have excellent self-lubricating and mechanical properties. Also, the friction coefficient of the material can be reduced to 0.34 at 800°C, approximately 62.6% less than the monolithic Al_2O_3 ceramics. D.-Y. Zhang et al. (2013) studied the self-lubricating laminated composites (SLCs). The matrix zones were silicon steel sheets and the filled zones were polymer matrix filled with MoS_2 and graphite, respectively. Compared to the control specimen, the friction coefficient and wear rate of SLC was reduced by 57%. Scanning electron microscopy (SEM) images show that the lubricating mechanism of SLC was that solid lubricants embedded in filled zones expanded and smeared a layer of transfer film on the sliding path to lubricate the surface.

6.3 WEAR AND FRICTION BEHAVIOR OF MMC

MMCs are continuously replacing conventional materials in aerospace, automobile, sports, and marine applications since they have better mechanical properties and high strength-to-weight ratios. Desired tribological and mechanical properties can be incorporated in MMCs by adding suitable reinforcing materials. In this section, the effect of shape, size, type, and volume fraction of different reinforcements in various MMCs are discussed. Also, the effects of load, speed, temperature, and environmental conditions on the tribological properties of MMCs are considered. With an increase in the quantity of reinforcements, the tribological properties of MMCs generally increase. Wear rates of MMCs usually increase with the increase in load and speed, and the friction values decrease. Up to a certain temperature, friction and wear of MMCs decrease and beyond that limit they increase. The following sections present the friction and wear behavior of selected MMCs.

6.3.1 FRICTION AND WEAR BEHAVIOR OF AL MATRIX COMPOSITES

Al-Cu, Al-Si, Al-Zn, and Al-Mg-Si are used as matrix and Al_2O_3 and SiC are used as reinforcements. The influence of graphite, pressure, and temperature at different sliding conditions (dry and lubricated conditions) were analyzed. The matrix and reinforcements are used to improve the friction and wear performance. The coefficient of friction can be increased by the addition of either metal, non-metal oxides or carbides such as SiO_2, Al_2O_3, SiC, and B_4C. MoS_2, a natural solid lubricant phase, may also act as a better lubricating material by improving the composite's anti-rubbing property compared to graphite or hexagonal boron nitride (h-BN) as lubricants. With an increase in the reinforcements, the ratio of composite wear rate to matrix wear rate (normalized wear rate) decreases as seen in Figure 6.4. Aluminium matrix composite reinforced with Mg/ZrO_2 showed better wear resistance compared to other reinforcements. With an increase in reinforcements, the normalized coefficient of friction decreases as shown in Figure 6.5.

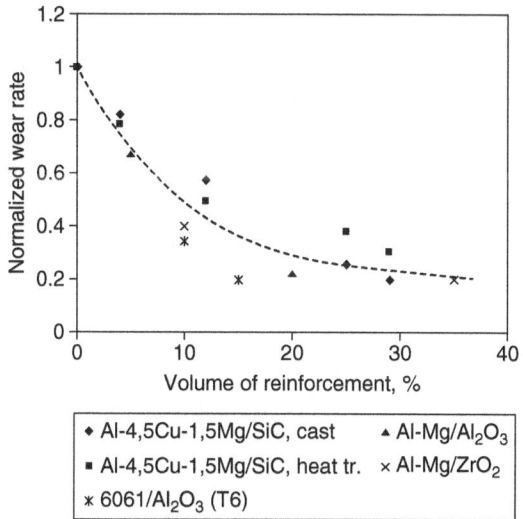

FIGURE 6.4 Wear rates of Al-based composite with reinforcements

Source: Vencl et al. (2004).

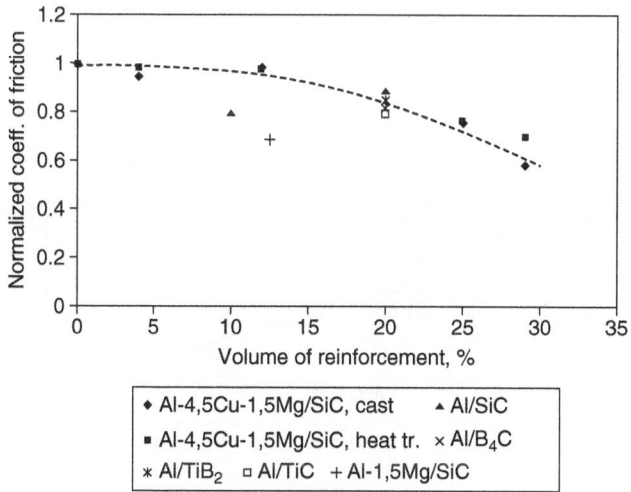

FIGURE 6.5 Friction coefficient of Al-based composites with reinforcements.

Source: Vencl et al. (2004).

The specific wear rate of composites and friction coefficient were evaluated by using the test rig. The wear resistance was higher compared to the matrix and increased with increasing reinforcements in composites. The heat energy converted from kinetic energy results in the rise of temperature, which in turn affects the wear and friction coefficient. In such a case, the surface temperature greatly increased. The secondary hard particles embedded in a soft matrix phase protect the dual phase material from sliding abrasive damages as shown in Figure 6.6. The maximum protection is attained in the dual phase material by adding the right proportion of secondary hard particles having the optimum size and shape.

Since different authors work with different conditions, comparing their results to arrive at a conclusion is difficult. Wear maps are useful to consolidate the results obtained by many researchers. The wear map in Figure 6.7 consolidates the results of operating in different zones. With the increase in load and speed, seizure takes place

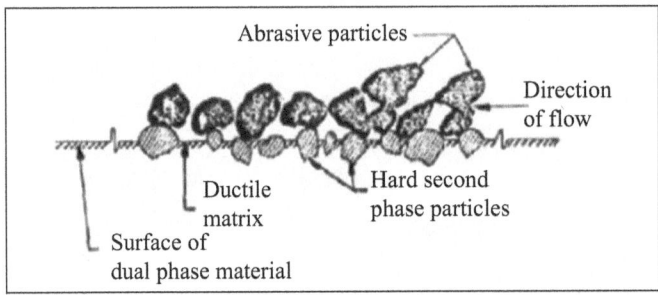

FIGURE 6.6   Protection offered by hard particles to ductile matrix against abrasion.

Source: Rohatgi et al. (2013).

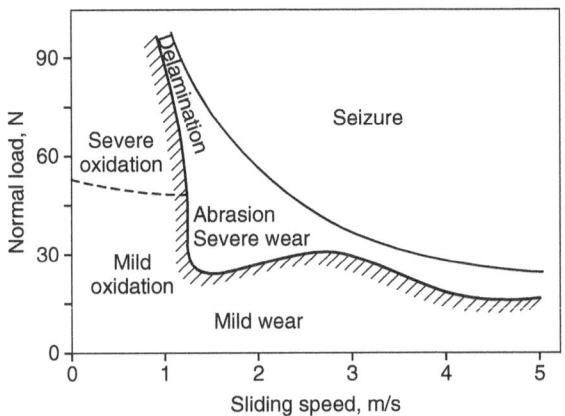

FIGURE 6.7   Wear map of Al alloy with silicon carbide.

Source: Vencl et al. (2004).

and it is not advisable to operate in those conditions. If the load is less than 15 N and speed is high, only mild wear takes place. Under low speeds and up to a load of 50 N, mild oxidation takes place. Severe oxidation takes place if the load exceeds 50 N at speeds less than 1 m/s. If the speed is higher than 1 m/s, delamination takes place. If the operating conditions fall in the region bounded by the two curves, then abrasion and severe wear take place.

6.3.2 Friction and Wear Behavior of Mg Composites

Magnesium matrix composites have better physical and mechanical properties and a low density that make them a good fit for aerospace and defense applications. The density of Mg is approximately two-thirds that of Al, which makes it a more suitable material for energy conservation by reducing fuel consumption. Magnesium matrix composites may also be used in the near future as an alternative for aluminium matrix composites because of their high wear resistance, strength, and low density. Mg alloys have a low resistance to high temperature, creep, and wear. When compared with Al matrix composites, modulus and ductility are low. Suitable reinforcements must be added to improve these properties. The reinforcements added to the magnesium matrix improve the fatigue, creep, wear, stiffness, and damping properties of the resulting MMC. This section presents the details of reinforcements used in Mg matrix composites and their relative advantages and disadvantages. Commonly used reinforcements are titanium carbide, silicon carbide, aluminium oxide, carbon nanotube, boron carbide, and fibers. Mg with Si particles in structural components showed better properties including low density, high hardness, strength, and elastic modulus. Guo et al. (2016) mentioned that Mg matrix composites showed positive results to a certain extent. When environmental temperature reaches above 150°C, the ultimate tensile strength and wear resistance decrease.

The addition of sub-micron silicon carbide to Mg matrix improves micro-hardness, elastic modulus, yield strength, and thermal stability. SiC bonds well with Mg without interfacial activity, leading to improved yield strength, tensile strength, and enhancement of elongation before fracture. Lim et al. (2003) investigated the wear properties of Mg/SiC composite under dry sliding conditions. The composite offers better wear resistance under a loading condition of 30 N at a sliding speed less than 5 m/s. Jiang et al. (2003) investigated Mg composite reinforced with titanium carbide particles and found that the tensile strength, wear resistance, and hardness increased compared to the unreinforced Mg alloy. Similarly, addition of aluminium oxide shows significant improvement in compressive strength and creep resistance of the composite. Addition of carbon nanotube (CNT) in Mg composites improves bonding and tensile strength. Lu et al. (2013) investigated the influence of adding nano Al_2O_3 and CNT to Mg matrix in different proportions and found the tribological behavior of the composites. They found that compared to AZ31 alloy all other composites have more wear resistance. The wear resistance of Mg composite reinforced with CNT alone is higher than that of nano Al_2O_3 reinforcement. The hardest element, boron carbide, increases the hardness and wear resistance by improving the flexural strength and interfacial bonding strength of Mg composites. In AZ91 alloy matrix composite having MgO, magnesium nitrate, and Al_2O_3, the ultimate strength, hardness, and strain hardening were increased by hard oxides (Bhingole et al. 2014). Application of ultrasonic vibrations to molten Mg MMC increases the uniformity in the distribution of particulates, decreases porosity, and avoids agglomeration. Among all the Mg MMCs, AZ91-6.5-UST has the highest resistance to wear as a result of the stronger interface and well-dispersed hard oxides. Mindivan et al. (2014) studied the addition of Al and CNT in magnesium composites and found that wear rate, hardness, and corrosion resistance increased but the compression strength decreased. Jiang et al. (2005) investigated the influence of boron carbide in magnesium matrix and found that MgO and MgB_2 are formed in the composites, which increased hardness and wear resistance when B_4C particulates are increased from 10% to 20%. Fiber addition influences the distribution of alloying elements in Mg matrix, increasing the tensile strength and decreasing the ductility.

6.3.3 FRICTION AND WEAR BEHAVIOR OF TITANIUM MATRIX COMPOSITES

Hayat et al. (2019) reported that titanium matrix composites are widely used materials in applications like aircraft engines and airframe manufacturing. In the past, titanium and its alloys were not generally considered the best materials for wear-resistant applications because of its low hardness. The hardness of titanium matrix can be enhanced by incorporating hard ceramic reinforcement in particle or fiber form. Prakash et al. (2018) reported that titanium alloys find their applications in aerospace and automotive areas owing to good machinability, high strength, low thermal expansion, and stiffness. Boron carbide ceramic particles (B_4C) are used as reinforcements in Ti-6Al-4V alloy, and the effects of B_4C on corrosion, wear, and mechanical properties were investigated. Addition of B_4C increased wear resistance, corrosion resistance, and hardness and decreased the density. The friction coefficient increased with applied load and reinforcement content and decreased with sliding velocity. Liu et al. (2007) investigated the tribological properties of Ti matrix alloy and TiC/Ti composite in dry sliding conditions. It was found that the wear properties of TiC/Ti composite are better than the Ti matrix alloy. Table 6.1 compares the wear properties of Ti matrix alloy and TiC/Ti composite.

Thotsaphon et al. (2008) studied the friction and wear behavior of titanium composite reinforced with multi-walled carbon nanotube (MWCNT). The composite was manufactured using spark plasma sintering and hot compression. Increasing the MWCNT content and applied load decreased the coefficient of friction of the composite. Figure 6.8 depicts the influence of applied load on the friction coefficient of titanium matrix composites. The homogeneous distribution of TiC compounds and

TABLE 6.1
Comparison of Wear of Ti Matrix Alloy and TiC/Ti Composite

Materials	Rigidity (HRC)	Average Friction Coefficient	Wear Rate (g/h)
Ti Matrix alloy	37.8	0.2878	0.97
TiC/Ti composite	42	0.2851	0.18

Source: Liu et al. (2007)

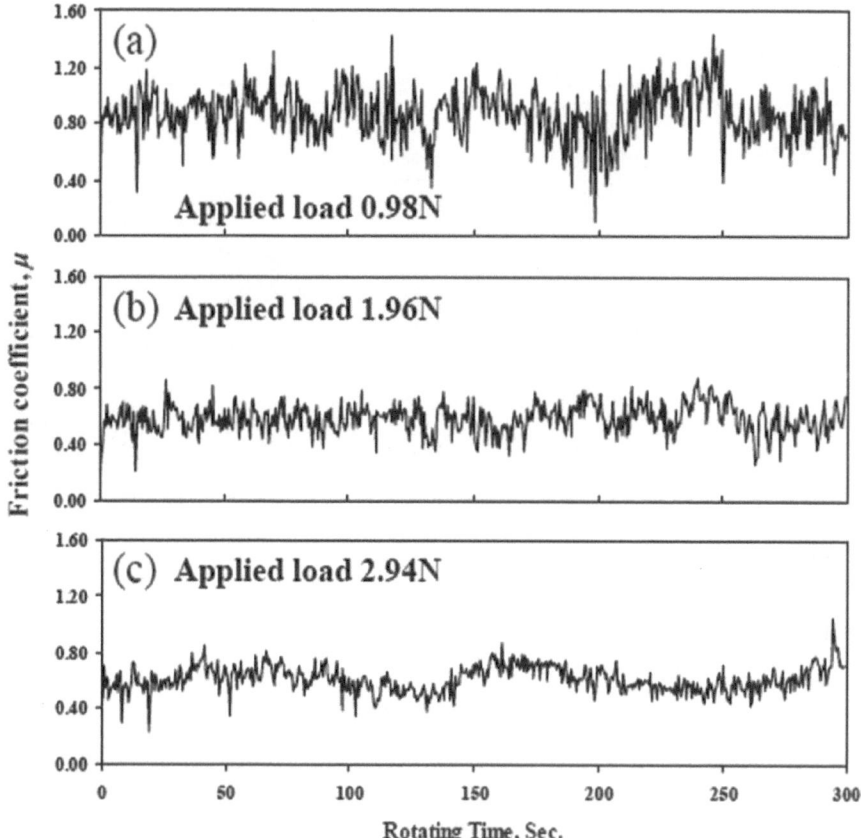

FIGURE 6.8   Friction coefficient profiles of Ti composites under different loads.

Source: Thotsaphon et al. (2008).

MWCNT enhanced the hardness to a remarkable level. Self-lubricating properties of MWCNT reduced friction. Adhesive wear took place at lower loads and abrasive wear at higher loads.

6.3.4   FRICTION AND WEAR BEHAVIOR OF METAL MATRIX COMPOSITES

The tribological behavior of composites depends on surface conditions and wear mechanisms. Worn surface, wear debris, and cross section are analyzed to understand the wear mechanism. There are five kinds of mechanisms of wear: adhesive wear, abrasive wear, severe plastic deformation, oxidation-delamination, and delamination.

6.3.4.1   Adhesive Wear and Abrasive Wear

Adhesive wear is due to the strong adhesive force involved at the sliding interface, and it leads to localized transfer of material between the sliding surfaces. Subsurface cracks may not be seen in the analysis of a cross section. Surface plastic deformation

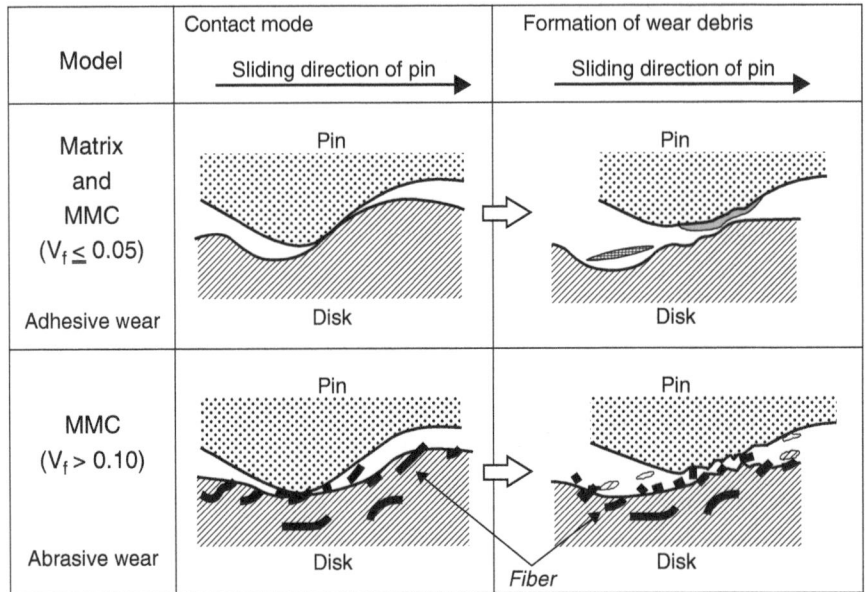

FIGURE 6.9   Schematic of wear mechanism models.

Source: Iwai et al. (2000).

and low hardness promote adhesive wear. Materials that resist thick oxide layer formation at the sliding surface will have poor adhesive wear resistance. In a sliding
system, the detached fragments (asperities) are involved in shearing action with the
relative surfaces subjected to abrasion. During abrasion, the asperities from the surface may form loose wear debris. These loose wear particles get trapped between
the sliding surfaces and induce third-body abrasion. The hard particle debris stuck
between rubbing surfaces acts as a third body and promotes abrasive wear.

Iwai et al. (2000) suggested a schematic of wear mechanism models as shown
in Figure 6.9. Model 1 shows the adhesive wear mechanism, which is predominant
in MMC having a volume fraction of reinforcements less than 0.05% ($V_f < 0.05$).
Model 2 shows the abrasive wear mechanism, which is predominant in MMC having
a volume fraction of reinforcements more than 0.1% ($V_f > 0.1$). During abrasive wear,
hard fibers are exposed at the worn surface and loose fragments are present in the
interfaces.

6.3.4.2   Plastic Deformation

During the initial sliding, the asperities at the surface interface get plastically deformed. The plastic deformation is based on the load, tangential force, yield strength,
and temperature. The plastic deformation of the asperities increases the contact area
at the junction which accommodates large tangential stresses under sliding condition.
The accumulation of tangential stress increases the friction coefficient. When the tangential stress exceeds the yield limit of the material, a sudden fracture of asperities

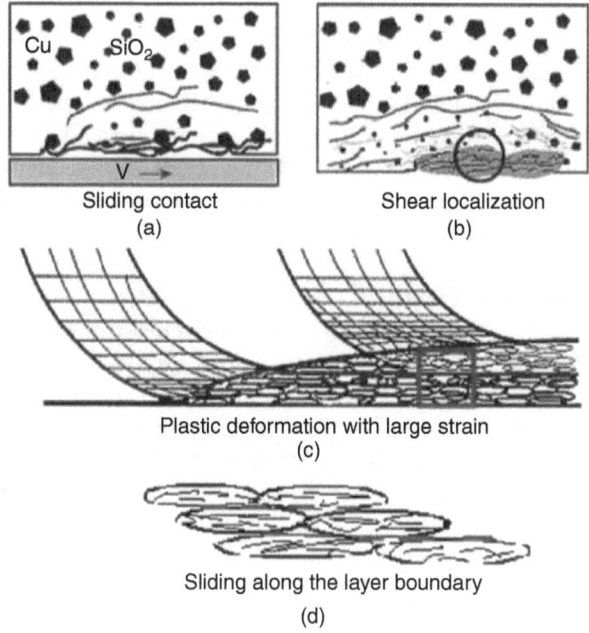

FIGURE 6.10 Plastic deformation with large strain: (a) initial sliding contact, (b) formation of shear localization, (c) distortion and elongation of grains with large plastic strain in marked area in (b), and (d) sliding along the grain boundary in the marked area in (c).

Source: Shang et al. (2012).

take place. This type of deformation takes place when a material of low hardness is tested under heavy loads. The worn surface will look wavy if severe plastic deformation takes place. The matrix deforms along the direction of sliding. A severe plastic deformation layer (PDL) similar to the worn surface will be found in the subsurface also. The wear debris due to this type of wear has a thin flake shape. When PDL hardens, it will become brittle and generate cracking leading to a large amount of wear. Shang et al. (2012) showed with the help of a schematic diagram the formation of laminar structure under the contact surface of Cu–SiO$_2$ sliding against 1045 steel as shown in Figure 6.10.

6.3.4.3 Oxidation-Delamination

Oxidation wear is the predominant wear mechanism involved in high temperature dry sliding conditions. Oxide film plays a critical role in oxidation wear. The oxide film generally reduces metal-metal contact in a sliding system. The friction coefficient and wear rate are reduced by oxide film formation. However, the oxide film gets ruptured or delaminates from the surface after attainment of critical thickness. The rate of attaining critical thickness depends on the oxidation kinetics of the material. The process of oxidation and delamination is repeated at regular intervals and results in wear loss. The amount of wear increases as a result of more frictional heat produced by hard braking. Wear debris are ground to smaller particles by repeated shear stresses combined with high pressure. When the smaller particles join together through sintering, they form a mechanical mixture layer (MML) on the friction surface which prevents further wear. Oxidation of MML takes place owing to high temperature and delamination of MML takes place owing to high shear stress. After one layer peels off, smaller particles will gather inside the pit and form a fresh MML. Oxidation-delamination wear is shown in Figure 6.11.

6.3.4.4 Delamination Wear

Delamination wear is due to subsurface crack nucleation and propagation parallel to the sliding surface. As a result of continuous sliding, there will be an accumulation of dislocations at the subsurface. Over a period of time, the accumulation of dislocations leads to the formation of voids. The voids primarily form around the hard-secondary phases in the matrix. The subsurface voids merge together to form a crack which is parallel to the sliding direction. When the subsurface crack attains sufficient length, the portion surrounded by the cracks delaminates as a sheet or flake. When cracks propagate and extend parallel to the sliding direction and when spalling takes place, delamination of a layer takes place. In the particle-matrix interface, micro-cracks are available if the bonding is not strong. In a brake pad, the action of repeated compressive stress and shearing stress causes the cracks to enlarge and join together until they reach the surface of the brake pad, resulting in delamination. A large area of pits appears on the pad leading to the formation of flake-like debris as depicted in Figure 6.12.

Figure 6.13 shows the wear map of copper-iron MMC. This map relates different iron content and braking energy density (BED) to the resulting wear mechanisms. Low BED with low content of Fe results in adhesive wear. Abrasive wear (ploughing) occurs with moderate BED and Fe content ranging from 0 to 20%. Severe plastic deformation occurs when BED is high and Fe content is low. Oxidation-delamination is observed at around 10% Fe and high BED. Delamination results when BED is high and also Fe content is about 20%.

Figure 6.14 shows the 3-D maps of the wear rate and coefficient of friction (CoF) of Cu-Fe MMC. The content of Fe and BED form the horizontal axes in both maps, and wear rate is represented by the vertical axis in part (a). The wear resistance of the MMC is high when Fe content is high and BED is low. Part (b) shows that CoF is low when Fe content is low as well as high and BED is high. CoF is high when Fe content is around 16% and BED is low.

6.4 RELATIONSHIP BETWEEN THE SURFACE HARDNESS OF THE MMC LAYER AND SPECIFIC WEAR

Tomida et al. (2001) studied the formation of the metal matrix composite layer on aluminium alloy with TiC-Cu powder by the laser surface alloying process. Figure 6.15 shows the relationship between the surface hardness of the MMC layer and specific wear.

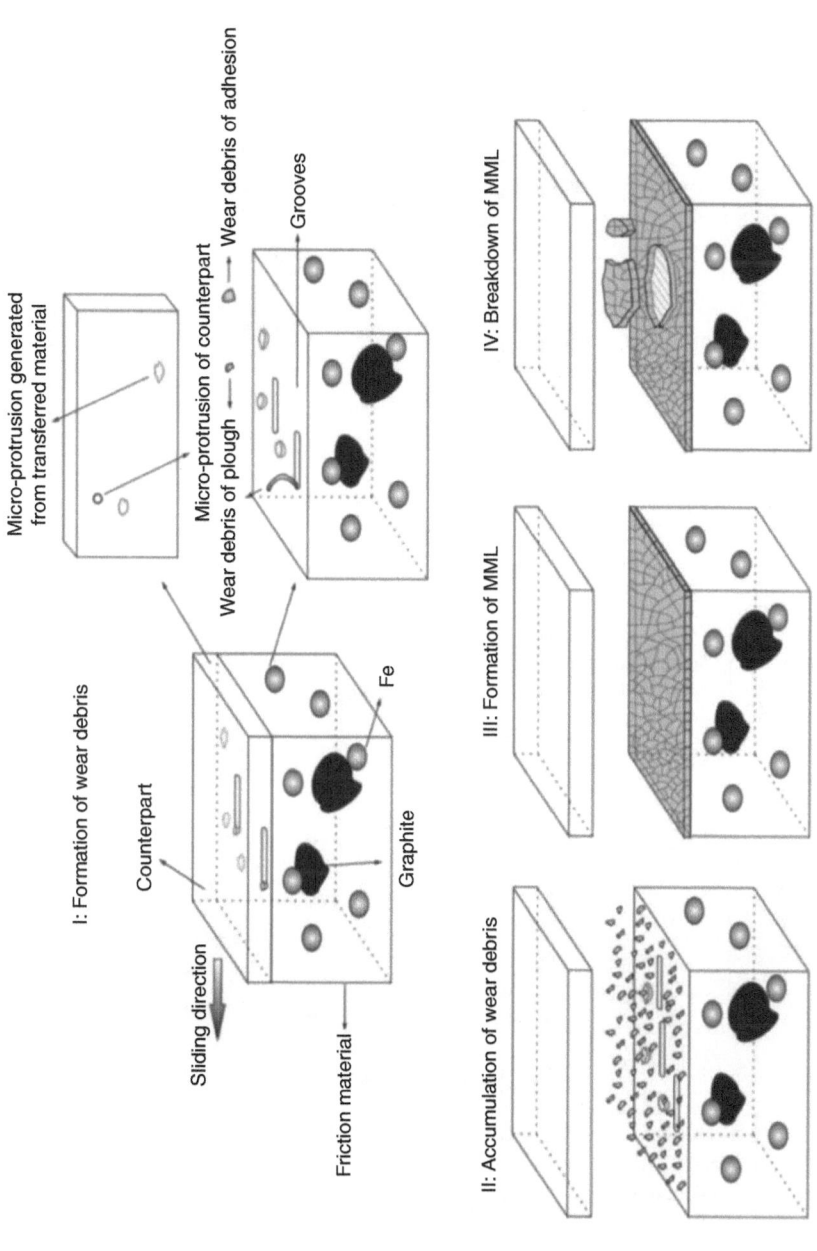

FIGURE 6.11 Oxidation-delamination.

Source: Zhou et al. (2018).

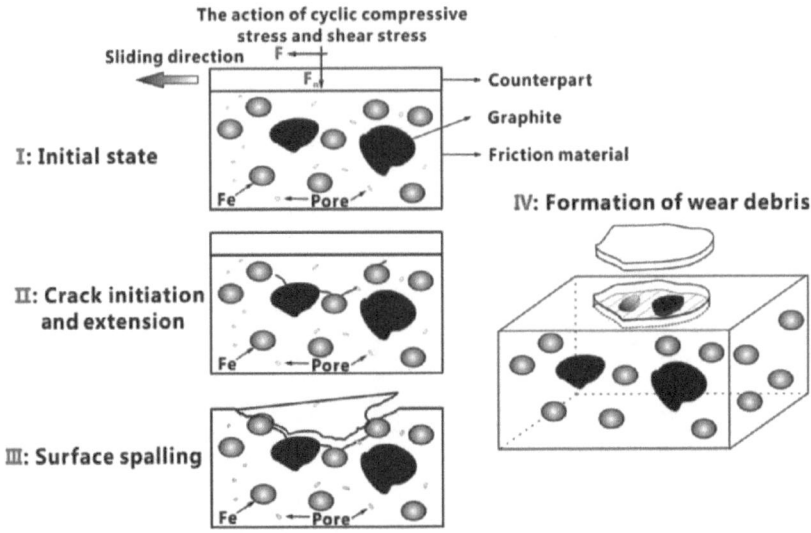

FIGURE 6.12 Delamination wear.

Source: Zhou et al. (2018).

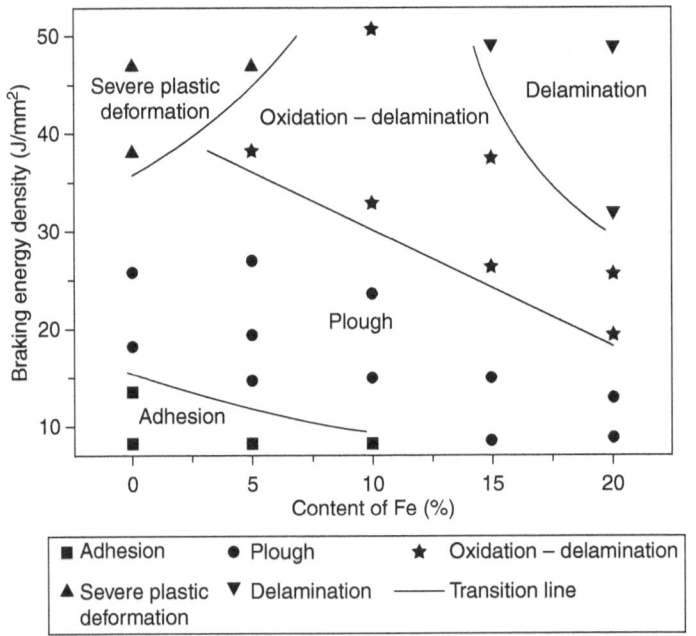

FIGURE 6.13 Wear map of Cu-Fe MMC.

Source: Zhou et al. (2018).

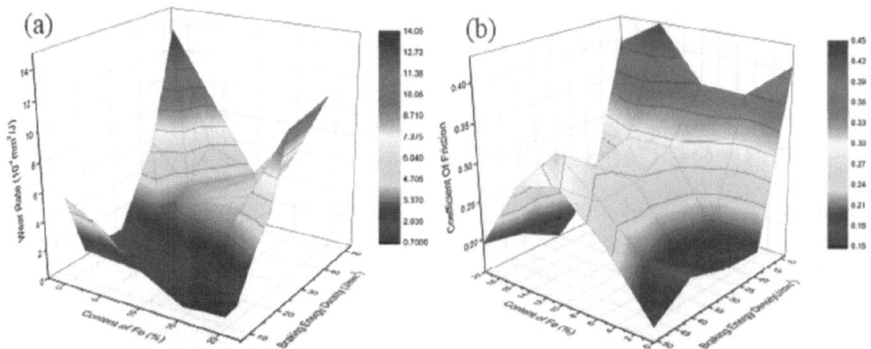

FIGURE 6.14 For Cu-Fe MMC, 3-D maps in terms of Fe content and braking energy density of (a) wear rate and (b) coefficient of friction.

Source: Zhou et al. (2018).

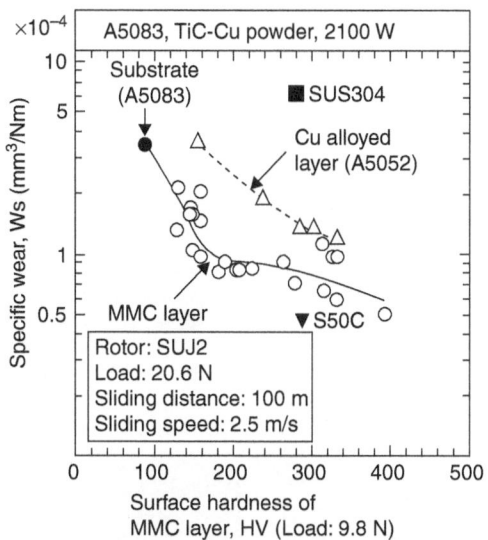

FIGURE 6.15 Relationship between the surface hardness of MMC layer and specific wear.

Source: Tomida et al. (2001).

A thick, hardened metal matrix composite (MMC) layer was formed to improve the wear resistance of a commercial Al–Mg alloy (A5083) plate by using laser surface alloying with Cu-coated TiC powders in sizes between 20 and 40 μm. In the MMC layer, TiC particles were not dissolved and were uniformly distributed in the molten matrix. The hardness of the MMC layer increased with increasing Cu content and the volume fraction of TiC particle and reached a maximum Vickers hardness value of approximately HV600. The wear rate decreased to one-sixth that of the substrate. The wear resistance of the MMC layer was much better than the Cu-alloyed layer without TiC powder and austenite stainless steel SUS304.

6.5 SEM CHARACTERIZATION TECHNIQUES USED IN TRIBOLOGICAL RESEARCH ON MMCS

Scanning electron microscope (SEM) characterization techniques used in tribological research on MMCs are reviewed in this section. Prabhu et al. (2014) investigated the effect of size and silicon carbide volume fraction on the wear of Fe-SiC-graphite hybrid composites under high-speed dry sliding conditions. The SEM and macrographs of wear tracks of iron-based hybrid composites are shown in Figure 6.16. Pull-out (delamination), an oxide layer (oxidative wear), interfacial crack, and grooves (abrasive wear) are evident from these micrographs and macrographs. Researchers confirm the wear mechanisms controlling the wear and friction behavior using this SEM characterization technique.

Bhingole et al. (2014) investigated the properties of dispersed alloy $MgO-Al_2O_3$-$MgAl_2O_4$ composites processed ultrasonically. Figure 6.17 shows the SEM images of magnesium composite specimens tested for a sliding distance of 2 km at 1 m/s speed under a load of 14.7 N. Worn surfaces are analyzed by SEM for understanding

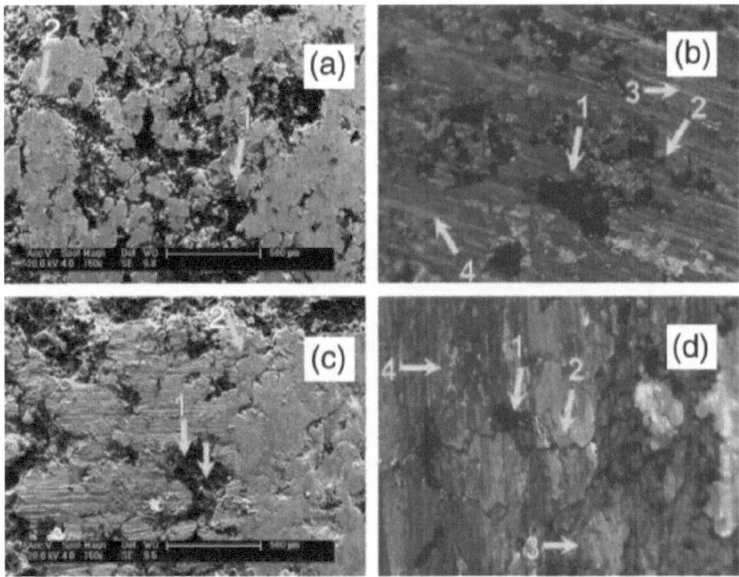

FIGURE 6.16 (a) SEM of wear tracks and (b) macrograph of Fe composites (Fe – 20% SiC (150–180 μm) – 18% Gr). (c) SEM of wear tracks and (d) macrograph of Fe composites (Fe – 20% SiC (1–30 μm) – 18% Gr). 1: Particulate pull-out; 2: interface crack; 3: oxide layer; 4: grooves.

Source: Prabhu et al. (2014).

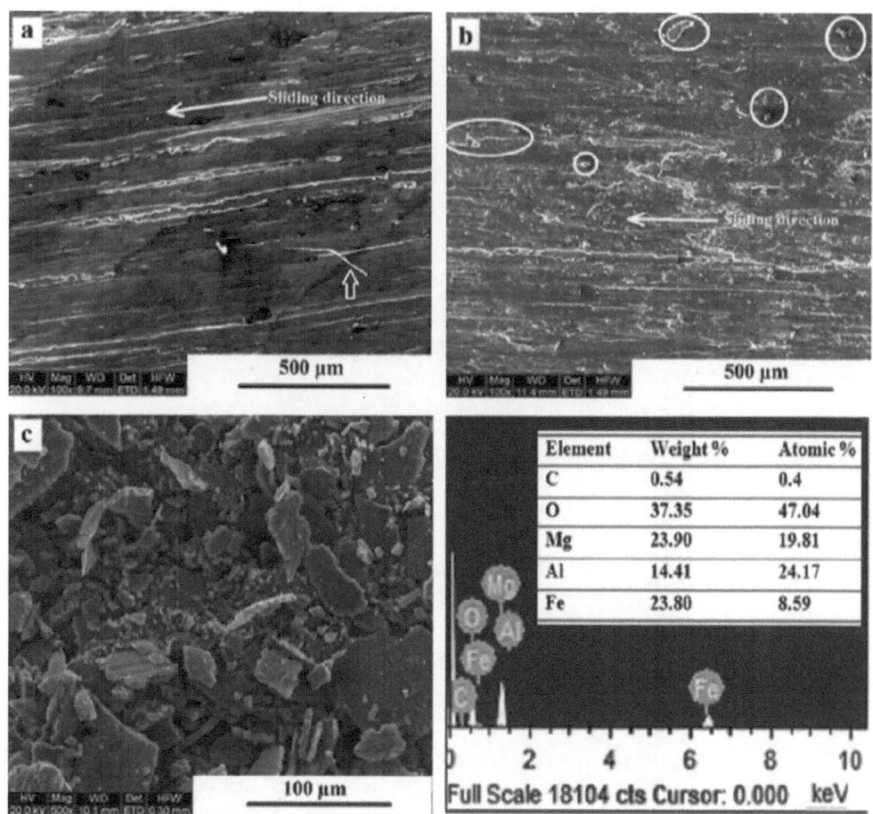

FIGURE 6.17 SEM of (a) AZ91, (b) AZ91-6.5-UST, (c) wear debris, and (d) EDS analysis.

Source: Bhingole et al. (2014).

friction and wear behavior and wear mechanisms. Ribbon-like strips and small fragments are removed by hard asperities of the meshing surface or by wear debris caught between rubbing surfaces as shown in Figure 6.17a. AZ91 is a soft alloy subjected to micro-welding during sliding. When micro-welds break, wear debris is formed, gets oxidized, and ploughs deeper grooves. Figure 6.17b shows the worn surface of AZ91-1.5-UTS, which has the highest wear resistance. Harder MMCs can resist abrasion. Shallower grooves are seen accompanied by plastic deformation. Figure 6.17c shows the SEM of wear debris, and the energy dispersive spectroscopy (EDS) analysis confirms that oxides of Mg, Al, and Fe were present as evidenced by Figure 6.17d.

Prakash et al. (2018) characterized the Ti-6Al-4V/B_4C MMC. Figure 6.18 displays the SEM of surfaces tested using a load of 10 N, at 1 m/s speed for a sliding distance of 1 km. Delamination, plastic deformation, abrasion, and cracks are found to be the reasons for material removal. The presence of crushed wear particles on the wear track indicates delamination wear. Figure 6.18a shows severe wear due to delamination, wear debris, and micro-cracks. Delamination is low and micro-cracks are not found in Figure 6.18b because the addition of 5% B_4C makes the composite stronger.

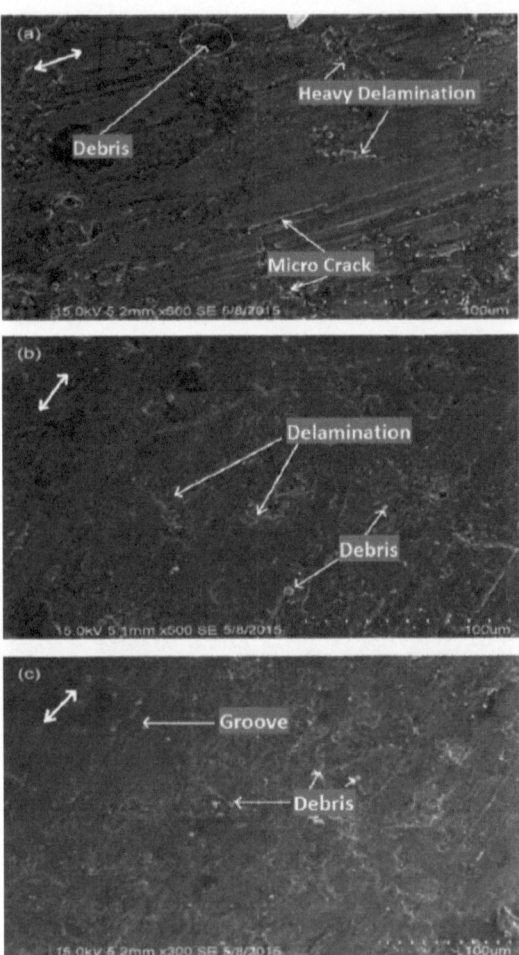

FIGURE 6.18 SEM of worn surfaces: (a) Ti-6Al-4V, (b) Ti-6Al-4V + 5% B₄C, and (c) Ti-6Al-4V + 10% B₄C.

Source: Prakash et al. (2018).

Scratches parallel to sliding direction and shallow grooves are seen in Figure 6.18c and the absence of delamination and micro-cracks indicate that the resistance to wear and hardness is high for this composite reinforced with 10% B_4C.

6.6 THEORETICAL APPROACHES USED FOR PREDICTING TRIBOLOGICAL PROPERTIES OF MMCS

Petre et al. (2017) conducted a phenomenological analysis on the friction of aluminium matrix graphite/cast iron. They have developed a theoretical model for predicting the friction coefficient influenced by adhesion coefficient, material deformation, and the

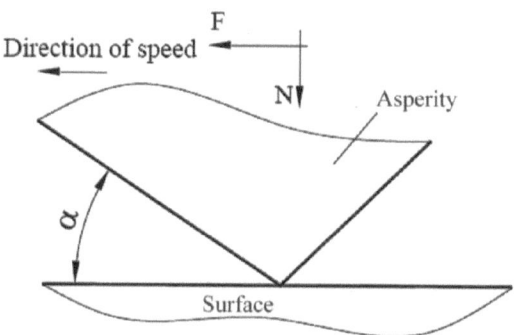

FIGURE 6.19 Displacement of asperity.

Source: Petre et al. (2017).

inclination angle of asperity. Figure 6.19 shows the displacement of asperity and also α, the asperity angle, direction of speed, direction of normal force, and direction of friction force acting on the asperity. Theoretical results are verified by experimental results.

$$\mu_a = \frac{A.\sin\alpha + \cos(\arccos f - \alpha)}{A.\cos\alpha + \sin(\arccos f - \alpha)} \tag{6.1}$$

$$A = 1 + \frac{\pi}{2} + \arccos(f) - 2\alpha - 2\arcsin\frac{\sin(\alpha)}{\sqrt{1-f}} \tag{6.2}$$

$$f = \frac{F_t}{F_n} = \frac{\tau.A_r}{p_r.A_r} = \frac{\tau_o}{p_r} + \beta \tag{6.3}$$

where μ_a = friction coefficient
α = the asperity angle
A = constant whose value is given by equation 6.2
f = molecular adhesion coefficient
τ = material shear strength
τ_0 = third-body layer shear strength
p_r = Mean real pressure
β = coefficient characteristic of the third body

Megahed et al. (2019) modeled wear characteristics of Al-Si/Al$_2$O$_3$ metal matrix compounds using multiple regression, ANOVA (analysis of variance), and the ANN (artificial neural network) approach. The sliding distance had more influence on the wear rate compared to alumina weight fraction and applied load. A wear model relating the sliding distance, weight fractions, and load has been developed as shown in Figure 6.20.

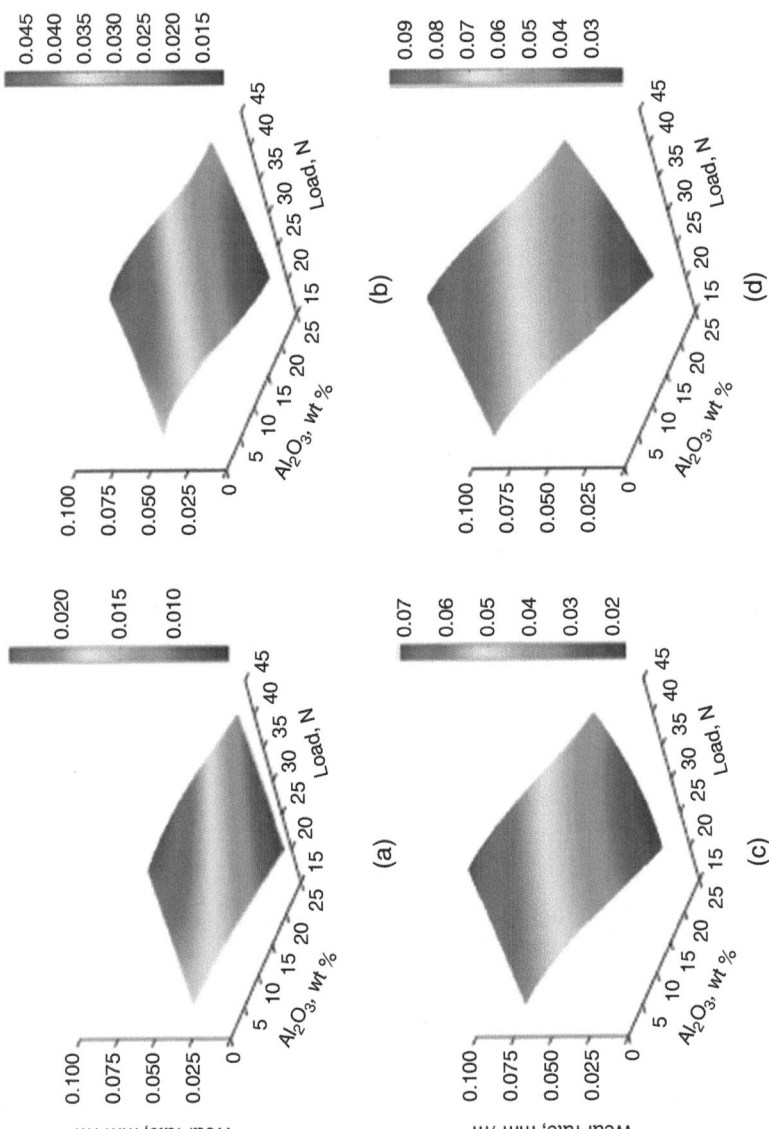

FIGURE 6.20 Relationship between wear rates, load, and the weight fraction of Al_2O_3 for different sliding distances: (a) 310, (b) 620, (c) 930 and (d) 1240 m.

Source: Megahed et al. (2019).

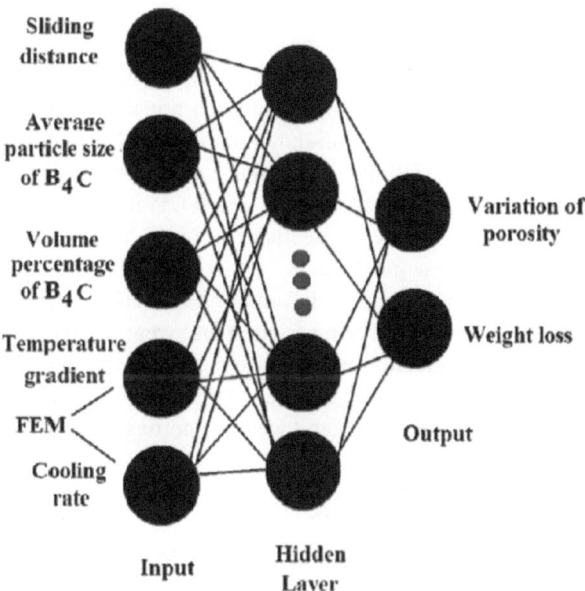

FIGURE 6.21   Schematic of the artificial neural network (ANN) used to predict wear of a composite.

Source: Shabani and Mazahery (2011).

Shabani and Mazahery (2011) predicted the wear characteristics of A356 composite reinforced by boron carbide particles using the finite element method (FEM) and artificial neural network (ANN). For discretization and for calculating the transient temperature field of quenching, FEM was used, and ANN along with FEM was applied to predict the wear behavior of the composite. Figure 6.21 is a schematic of the artificial neural network architecture. The number of input and output parameters decides the number of neurons in the input and output layers. The optimal structure was obtained by minimizing the error. The cooling rate and temperature gradient were obtained through FEM. Sliding distance, the particle size of boron carbide, and volume percentage of boron carbide are other input variables.

Weight loss and variation of porosity are output variables. The ANN was trained using 33 samples: 25 data sets were used directly for training and 8 data sets were used for validation. The predicted values using this model agreed with the results of the experiments.

6.7   SUMMARY

Classifications of MMC and their applications in different engineering fields are presented in the chapter. The tribological behavior of aluminium MMC, magnesium MMC, titanium MMC, and copper matrix composites are discussed in detail. Industrial applications of metal matrix composites are explained with suitable examples. Commonly used reinforcements are titanium carbide, silicon carbide, aluminium oxide, carbon nanotube, boron carbide, magnesium nitrate, magnesium boride, zirconium oxide, and fibers. Wear maps of composites showing wear mechanisms in different regions of operating conditions are illustrated. Schematic diagrams of adhesive, abrasive, oxidation-delamination, and delamination wear are provided with step-wise descriptions. When the volume fraction of reinforcements is less than 0.05%, adhesive wear takes place, and when it is more than 0.1%, abrasive wear takes place. The major wear mechanism for Al-MMC is delamination and adhesive wear; the major wear mechanism for Cu-Fe MMC is plastic deformation and oxidation-delamination; and plastic deformation takes place in Cu-SiO$_2$ composites. The usage of 3-D maps in analyzing friction and wear properties of composites are demonstrated. Scanning electron microscope characterization techniques used in tribological research on MMCs are explained with suitable examples. Theoretical approaches used for predicting friction and wear properties of MMCs are described.

6.8   SCOPE FOR FUTURE WORK

Applications in aircraft and high-speed trains require MMCs with tribological properties to withstand high temperature and high pressure. More research is required in the area of ceramic matrix composites. Theoretical modeling, simulations using finite element analysis, and thermo-mechanical analysis should be carried out to predict the performance of MMCs under extreme operating conditions. Transmission electron microscopy studies are needed to understand the behavior of the friction layer at the interface of tribo-contacts and the wear mechanisms operating under operating conditions. Applications of surface treatments, coatings, and films on MMCs for tribological applications are to be studied in depth for improving the lives of the parts and for refurbishing worn parts.

REFERENCES

Adebisi, A., Maleque, M., and Rahman, M. 2011. Metal matrix composite brake rotor: Historical development and product life cycle analysis. *Int. J. Automot. Mech. Eng.* 4:471–480. doi:10.15282/ijame.4.2011.8.0038

Ashwanth Kumar, M., Dhanasekaran, R., Reddy, S., and Vijayakumar, B. 2018. Review on wear characteristics of aluminum silicon carbide reinforced with basalt fiber. *Mater. Today: Proc.* 5:26948–26954. https://doi.org/10.1016/j.matpr.2018.08.183

Bhingole, P. P., Chaudhari, G. P., and Nath, S. K. 2014. Processing, microstructure, and properties of ultrasonically processed in situ MgO–Al$_2$O$_3$–MgAl$_2$O$_4$ dispersed magnesium alloy composites. *Composites Part A* 66:209–217.

Emamian, A., Alimardani, M., and Khajepour, A. 2014. Effect of cooling rate and laser process parameters on additive manufactured Fe-Ti-C metal matrix composites microstructure and carbide morphology. *J. Manuf. Process* 16:511–517.

Guo, W., Wang, D., Fu, Y., Zhang, L., and Wang, Q. 2016. Dry sliding wear properties of AZ31-Mg2Si magnesium matrix composites. *J. Mater. Eng. Perform.* 25:4109–4114. https://doi.org/10.1007/s11665-016-2263-5

Hayat, M. D., Singh, H., He, Z., and Cao, P. 2019. Titanium metal matrix composite: An overview. *Composites Part A* 121:418–438.

Hu, Y., and Cong, W. 2018. A review on laser deposition-additive manufacturing of ceramics and ceramic reinforced metal matrix composites. *Ceram. Int.* 44:20599–20612.

Iwai, Y., Honda, T., Miyajima, T., Iwasaki, Y., Surappa, M. K., and Xu, J. F. 2000. Dry sliding wear behavior of Al_2O_3 fiber reinforced aluminum composites. *Compos. Sci. Technol.* 60:1781–1789. https://doi.org/10.1016/s0266-3538(00)00068-3

Jiang, Q. C., Li, X. L., and Wang, H. Y. 2003. Fabrication of TiC particulate reinforced magnesium matrix composites. *Scripta Materialia* 48:713–717. https://doi.org/10.1016/s1359-6462(02)00551-1

Jiang, Q. C., Wang, H. Y., Ma, B. X., Wang, Y., and Zhao, F. 2005. Fabrication of B_4C particulate reinforced magnesium matrix composite by powder metallurgy. *J. Alloys Compd.* 386:177–181. https://doi.org/10.1016/j.jallcom.2004.06.015

Koczak, M. J., and Premkumar, M. K. 1993. Emerging technologies for the in-situ production of MMCs. *JOM* 45:44–48.

Kumar, N., Vannan, S. E., Kiran, M. D, Dsouza, V. L., and Nagaral, M. 2020. Dry sliding wear behavior of short carbon fiber reinforced aluminium matrix composites. *J. Crit. Rev.* 7:1110. ISSN-2394-5125.

Li, J., and Wang, H. M. 2010. Microstructure and mechanical properties of rapid direction-ally solidified Ni-base superalloy Rene'41 by laser melting deposition manufacturing. *Mater. Sci. Eng. A* 527:4823–4829.

Lim, C., Lim, S., and Gupta, M. 2003. Wear behavior of SiCp-reinforced magnesium matrix composites. *Wear* 255:629–637. https://doi.org/10.1016/S0043-1648(03)00121-2

Liu, B., Liu, Y., He, X. Y., et al. 2007. Preparation and mechanical properties of particulate-reinforced powder metallurgy titanium matrix composites. *Metall. Mater. Trans. A* 38:2825–2831.

Lu, D., Jiang, Y., and Zhou, R. 2013. Wear performance of nano-Al_2O_3 particles and CNTs re-inforced magnesium matrix composites by friction stir processing. *Wear* 305:286–290. https://doi.org/10.1016/j.wear.2012.11.079

Megahed, M., Saber, D., and Agwa, M. A. 2019. Modeling of wear behavior of Al-Si/Al_2O_3 metal matrix composites. *Phys. Met. Metallogr.* 120:1072–1082. ISSN 0031-918X.

Mindivan, H., Efe, A., Kosatepe, A. H., and Kayali, E. S. 2014. Fabrication and character-ization of carbon nanotube reinforced magnesium matrix composites. *Appl. Surf. Sci.* 318:234–243. https://doi.org/10.1016/j.apsusc.2014.04.127

Mishra, S. K., Biswas, S., and Satapathy, A. 2014. A study on processing, characterization and erosion wear behavior of silicon carbide particle-filled ZA-27 metal matrix composites. *Mater. Des.* 55:958–965. https://doi.org/10.1016/j.matdes.2013.10.069

Niu, F., Wu, D., Ma, G., Wang, J., Guo, M., and Zhang, B. 2015. Nanosized microstructure of Al_2O_3-ZrO_2 (Y_2O_3) eutectics fabricated by laser engineered net shaping. *Scr. Mater.* 95:39–41.

Patil, A., Banapurmath, N. R., Hunashyal, A. M., and Hallad, S. 2020. Enhancement of mech-anical properties by the reinforcement of fly ash in aluminium metal matrix composites. *Mater. Today: Proc.* 24:1654–1659. https://doi.org/10.1016/j.matpr.2020.04.487

Petre, I. C., and Popescu, I. N. 2017. Friction from the theoretical and experimental point of view of Al-Al_2O_3 graphite composite / cast iron, pin-on-disc sliding system. *Int. J. Mechatronics Appl. Mech.* 2:40–47.

Prabhu, T. R., Varma, V. K., and Vedantam, S. 2014. Effect of SiC volume fraction and size on dry sliding wear of Fe/SiC/graphite hybrid composites for high sliding speed applica-tions. *Wear* 309:1–10. doi.org/10.1016/j.wear.2013.10.006

Prakash, K. S., Gopal, P. M., Anburose, D., and Kavimani, V. 2018. Mechanical, corrosion, and wear characteristics of powder metallurgy processed Ti-6Al-4V/B_4C metal matrix com-posites. *Ain Shams Eng. J.* 9:1489–1496.

Qi, Y., Zhang, Y.-S., and Hu, L.-T. 2012. High-temperature self-lubricated properties of Al_2O_3/Mo laminated composites. *Wear* 280–281:1–4. https://doi.org/10.1016/j.wear.2012.01.010

Rohatgi, P., Tabandeh-Khorshid, M., Omrani, E., Lovell, M., and Menezes, P. 2013. Tribology of metal matrix composites. In *Tribology for Scientists and Engineers*. Springer. https://doi.org/10.1007/978-1-4614-1945-7_8

Shabani, M. O., and Mazahery, A. 2011. Prediction of wear properties in A356 matrix composite reinforced with B_4C particulates. *Synth. Met.* 161:1226–1231.

Shang, J., Ma, W., and Lu, J. 2012. Formation of laminar structure under unlubricated friction of Cu–SiO2 composite. *Tribol. Lett.* 48:249–254. https://doi.org/10.1007/s11249-012-0020-y

Thotsaphon, T. J., Katsuyoshi, K., Junko, U., and Hisashi, I. 2008. Friction and wear behavior of titanium matrix composite reinforced with carbon nanotubes under dry conditions. *Trans. JWRI* 37(2):51–56.

Tomida, S., Nakata, K., Saji, S., and Kubo, T. 2001. Formation of metal matrix composite layer on aluminum alloy with TiC-Cu powder by laser surface alloying process. *Surf. Coat. Technol.* 142–144:585–589. https://doi.org/10.1016/s0257-8972(01)01172-0

Vencl, A., Rac, A., and Bobic, I. 2004. Tribological behavior of Al-based MMCs and their application in the automotive industry. *Tribol. Ind.* 26 (3–4).

Zhang, D.-Y., Zhang, P.-B., Lin, P., Dong, G.-N., and Zeng, Q.-F. 2013. Tribological properties of self-lubricating polymer–steel laminated composites. *Tribol. Trans.* 56:908–918. https://doi.org/10.1080/10402004.2013.805348

Zhang, Y., Guo, Y., Chen, Y., et al. 2019. Ultrasonic-assisted laser metal deposition of the Al 4047 alloy. *Metals* 9:1111.

Zhou, H., Yao, P., Xiao, Y., et al., 2018. Friction and wear maps of copper metal matrix composites with different iron volume content. *Tribol. Int.* 132:199–210. https://doi.org/10.1016/j.triboint.2018.11.027

7 Achieving Exceptional Mechanical and Tribological Properties of Metal Matrix Composites through Stir Casting Followed by Cryorolling

S. Vigneshwaran,[1] P. Seenuvasaperumal,[2a]
C. Chinthanai Selvan,[3b] R. Palanivel,[4c]
Mohammad Abdur Rasheed[4d]*

[1] Department of Mechanical Engineering, National Institute of Technology Puducherry, Karaikal, India.

[2] School of Mechanical Engineering, Vellore Institute of Technology, Vellore, India.

[3] Department of Production Engineering, National Institute of Technology, Tiruchirappalli, India.

[4] Department of Mechanical Engineering, College of Engineering, Al-Dawadmi Shaqra University, Saudi Arabia.

*Corresponding author: svigneshwaranmech2010@gmail.com

Other email IDs: [a]seenumeed@gmail.com, [b]chinthanaiselvan.rc@gmail.com, [c]rpalanivelme@gmail.com, [d]marasheed@su.edu.sa

CONTENTS

DOI: 10.1201/9781003109723-7

7.1   INTRODUCTION

7.1.1   PRODUCTION OF MMCs AND REQUIREMENTS

Composite materials are fabricated with a combination of two or more materials which are insoluble in one another, and the composites exhibit properties that are better than any of the component materials. The composite materials are categorized based on the matrix and reinforcement material used for the fabrication. They are further classified according to matrix material as metal matrix composites (MMCs), ceramic matrix composites (CMCs), polymer matrix composites (PMCs) and carbon matrix composites. MMCs have capacity to resist high temperature and moisture, and they exhibit improved mechanical properties compared with various other composites (Rajan et al., 1998). MMCs hold a blend of ductile matrix metal or alloy reinforced with metals other than matrix metal, non-metals, and organic and inorganic compounds. The various matrix and reinforcement combinations used in the fabrication of MMCs are shown in Figure 7.1. MMCs are fabricated to improve specific strength, stiffness, wear and corrosion resistance properties along with their low density (Arunachalam et al., 2019). Owing to the attractive properties of MMCs, their significance in aerospace, automobile, defence and marine sectors is rising, which results in more research on MMCs through various advanced manufacturing methods (Dinaharan et al., 2020).

The commonly used matrix materials for the fabrication of ceramic-reinforced MMCs are aluminium (Al) and magnesium (Mg). However, iron-, copper- and titanium-based ceramic-reinforced MMCs have also been produced successfully. Ceramic-reinforced Al-MMCs are preferred over fibre- or whisker-reinforced Al-MMCs owing to their low-cost production. Moreover, isotropic properties and microstructural non-homogeneity are more common in fibre- and whisker-based MMCs. Widely used ceramic reinforcements include SiC, B_4C, TiC, TiB_2, Al_2O_3 (refer to Figure 7.1 for the various types of reinforcement used). The production techniques

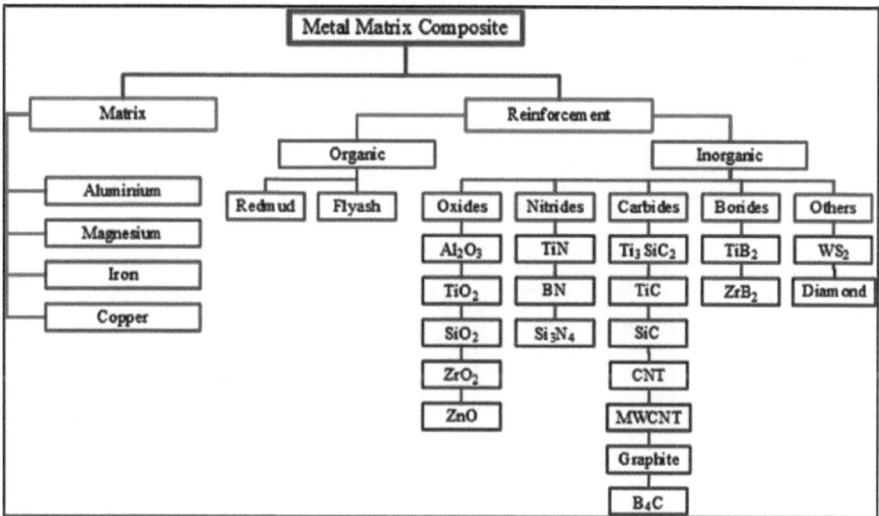

FIGURE 7.1   Matrix and reinforcement combinations used for the fabrication of MMCs.

Source: Arunachalam et al. (2019); reprinted with permission.

of ceramic-reinforced Al-MMCs are well associated with their structure and property correlation and better tribological and corrosion-resistance properties. As the particle content increases, improvement in the tensile strength with notable decrement in the elongation is observed.

7.1.2 VARIOUS PRODUCTION METHODS FOR AL-MMCs AND ASSOCIATED CHALLENGES

The fabrication method of Al-MMCs is based on the state of metal matrix as liquid, solid or other forms (including semi-solid and in-situ methods). Ceramic-reinforced MMCs have been fabricated by solid state methods, such as mechanical alloying and power metallurgy (Srinivasarao et al., 2009; Rahimian et al., 2009). Furthermore, liquid state methods such as stir casting (Kalaiselvan et al., 2011), squeeze casting (Venkatesan and Anthony Xavior, 2018), centrifugal casting (Kingsley and Suárez, 2011), infiltration method (Gecu et al., 2017), and various deposition methods (Ramkumar and Dinaharan, 2020) have also been adopted to produce ceramic-reinforced MMCs. MMCs made using solid state methods have certain challenges like size and shape limitations, reduced strength, high processing cost, density variation and dimensional distortion during sintering (Palanivel et al., 2020). The benefits of liquid state methods include improved binding of matrix to particle, easy processing and better microstructural control. In particular, the casting routes produce strongly bonded MMCs, which show notable mechanical and tribological properties (David Raja Selvam et al., 2013). The fabrication of MMCs through stir casting also results in a number of other benefits including low processing cost, simplicity, controllability of reinforcement (in matrix metal) and ability to produce large MMCs (Shau and Sahu, 2018).

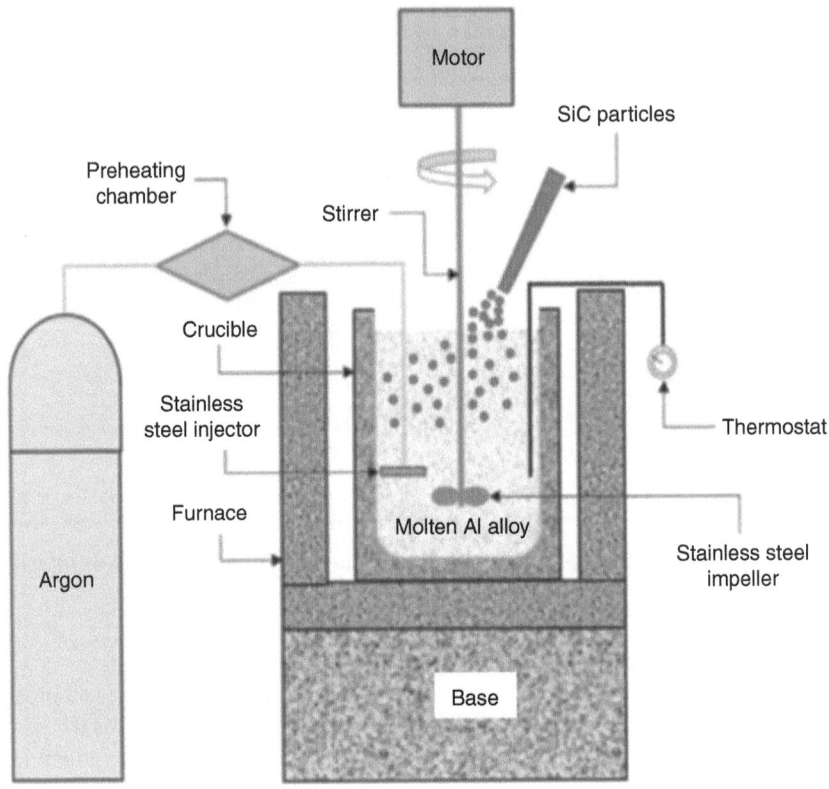

FIGURE 7.2   Stir casting unit.

Source: Idrisi and Mourad (2019); reprinted with permission.

7.1.3   STIR CASTING PROCESS

Stir casting is a liquid state manufacturing process often used for the production of MMCs. The typical simple stir casting setup is shown in Figure 7.2. The steps in the fabrication of MMCs through stir casting are listed in Figure 7.3. This involves melting of matrix metal and simultaneous stirring with the addition of reinforcement particles, which produces MMCs after solidification. For the successful production of MMCs in the stir casting process, the following process parameters need to be considered, based on the work of Arunachalam et al. (2019).

a. Particle size: The size of the reinforcement particle added should be small in order to attain better mechanical properties.
b. Temperature: Optimal melting temperature is desirable. When the temperature of melting is high, it improves the wettability but it also reduces the viscosity of the matrix metal. If the melting temperature is low, clustering of reinforcement particles occurs.

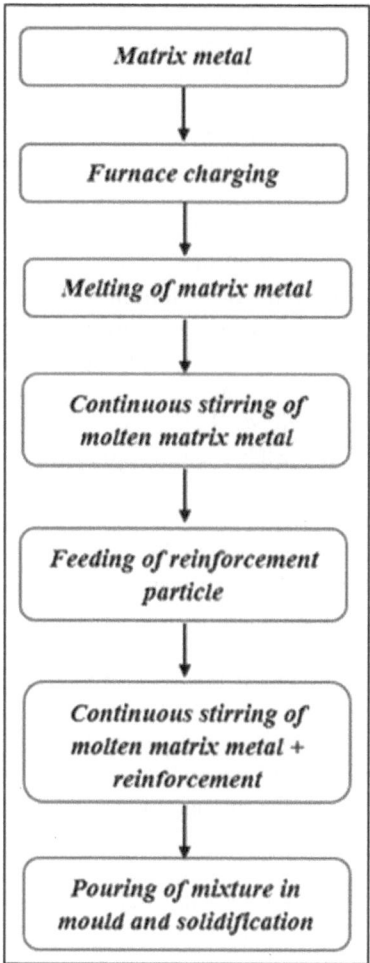

FIGURE 7.3 Steps involved in the fabrication of MMCs through stir casting.

    c. Stirring speed and duration: Stirring speed should not be too high, which can cause resistance to particle movement while stirring. If the speed is too low, the ability to suspend the particle decreases. Longer stir time is desirable as it produces uniform distribution.

    d. Impeller profile: The impeller profile plays a significant role in stirring speed and stirring duration. Furthermore, it creates a vortex by stirring action to attain proper matrix and reinforcement blend.

7.1.4 NEED FOR THE SECONDARY PROCESS AFTER STIR CASTING

Stir casting has been successfully used to prepare Al-MMCs with micron-sized particles, but it is challenging to produce Al-MMCs with nanoscale particles using the same method. This is attributed to poor wettability and agglomeration of ceramic particles (Arunachalam et al., 2019). It is especially challenging to fabricate Al-MMCs with nano-sized reinforcement particles because the surface area is greater than for the same volume fraction of micron-sized reinforcement particles (Harichandran and Selvakumar, 2016). The presence of a gas layer on the particle surface provides a gap between the molten matrix and reinforcement particles (Mazahery et al., 2009). Neither simultaneous stirring nor ultrasonic treatment show any capability to break down the cluster formation of nano-sized reinforcement particles in the molten Al matrix, and complete eradication of clusters of nano particles becomes difficult (Su et al., 2012). These challenges eventually deteriorate the mechanical properties of Al-MMCs. It is reported that post processing including extrusion and rolling can result in further improvement in properties needed for the use of Al-MMCs in the production of aerospace structural materials, automobile components, precision components and many other devices (Geng and Wu, 2018).

Recently, fabrication of Al-MMCs with ultra-fine-grained (UFG) microstructure has gained more attraction among research communities. The UFG-based Al-MMCs can be produced through methods like equal channel angular pressing (ECAP) (Bera et al., 2013), high pressure torsion (HPT) (Xue et al., 2015), friction stir processing (FSP) (Dinaharan et al., 2020) and accumulative roll bonding (ARB) (Ramkumar and Dinaharan, 2020). These methods can produce homogeneous distribution with more uniform grain structure; however, the size of the component is limited to laboratory scale. Though a large-scale product is possible through ARB, achieving good bond quality between layers in the fabrication of Al-MMCs is difficult. Processing through a powder metallurgy route produces non-uniform distribution of reinforcement particles, and producing a completely dense, compact material is also a challenging task (Yao et al., 2015). A hybrid manufacturing method that combines stir casting followed by cryorolling can potentially develop UFG Al-MMCs as it can suppress dynamic recovery (Deb et al., 2018; Vigneshwaran et al., 2018). Moreover, Al-MMCs for large-scale industrial components can be produced without any size constraints.

7.1.5 Cryorolling

Cryorolling can be defined as the rolling process that is carried out at cryogenic (sub-zero) temperatures. The process sequence of cryorolling is given in Figure 7.4. Before the initiation of cryorolling, the material to be rolled is immersed in a liquid nitrogen bath (-196°C) for around 30 to 45 minutes based on the thickness of the material. For each rolling pass, the rolling reduction is kept minimal (say, 0.1 mm) to restrict the influence of adiabatic heating. After cryorolling, the samples, especially Al alloys, are stored in a deep-freezer for protection from atmospheric influence. During the rolling process, a constant roll speed is set (Vigneshwaran et al., 2019). The cryorolling process is suitable for materials that have a very low temperature of transition from ductile to brittle (it is approximately -200°C in the case of Al alloys (Zhemchuzhnikova et al., 2013)).

Cryorolling results in the formation of UFG-structured materials (average grain size less than 1 μm) with relatively low plastic deformation and high density of dislocations. The cryorolled product can be a suitable material for the automotive and aerospace sectors (Valiev et al., 2000). The rolling of pure metals, alloys and

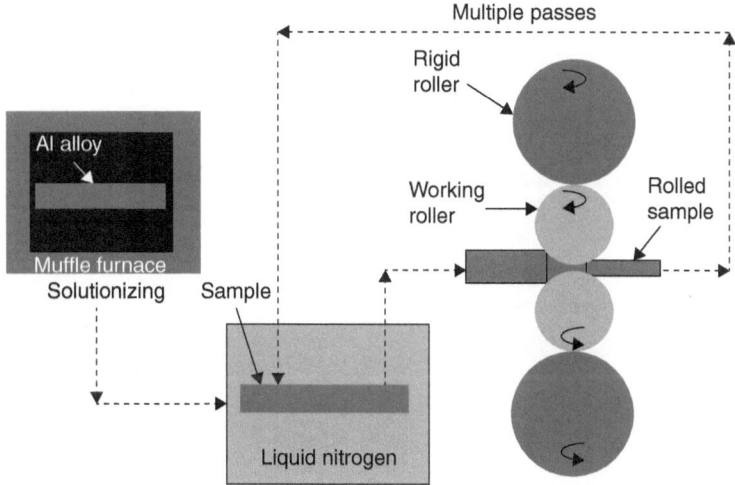

FIGURE 7.4 Schematic representation of process sequence involved in cryorolling of Al alloy.

Source: Kandarp et al. (2019); reprinted with permission.

composites at cryogenic temperature suppresses the dynamic recovery and the density of the accumulated dislocations thereby reaches a high steady level compared to the room-temperature-rolled counterparts. Cryorolling has been established as one of the leading routes to produce nanocrystalline/UFG pure metals, including Cu, Al and Ni, from their bulk forms. Cryorolling has been proved a better method to produce UFG microstructure than other severe plastic deformation (SPD) techniques because of its high scalability, low load requirement, simple processing procedure and the ability to produce long products. Wang et al., 2002, have produced UFG structure of commercial pure Cu by cryorolling at lower temperature compared to the SPD techniques that have been carried out at ambient temperature. The mechanical behaviour of AA2219 alloy subjected to cryorolling and post annealing treatment has been investigated by Shanmugasundaram et al., 2006, who observed a significant improvement in yield strength (YS) (485 MPa) and ultimate tensile strength (UTS) (540 MPa) in that alloy. Low temperature aging heat treatment on cryorolled age hardenable materials such as AA7075 and AA2024 alloy has been carried out by Zhao et al., 2005, and Cheng et al., 2007, who have obtained huge improvement of strength in both the materials. In the recent past, a 5xxx aluminium alloy with Sc (Al-Mg-Sc-Zr) has been introduced in the cryogenic environment and has gained attraction because of its improvement in strength, ductility and fracture toughness traits as reported by Zhemchuzhnikova et al., 2013. The cryorolled product is mostly in the form of sheet metal, which can be employed as a successful structural material in the avionic sectors, skins for automotive components, fabrication of storage tanks and so on.

The hybrid manufacturing route, which is the focus of this chapter, involves the production of Al-MMCs by the stir casting route followed by cyrorolling. This route is schematically shown in Figure 7.5. The microstructure, wear behaviour and the

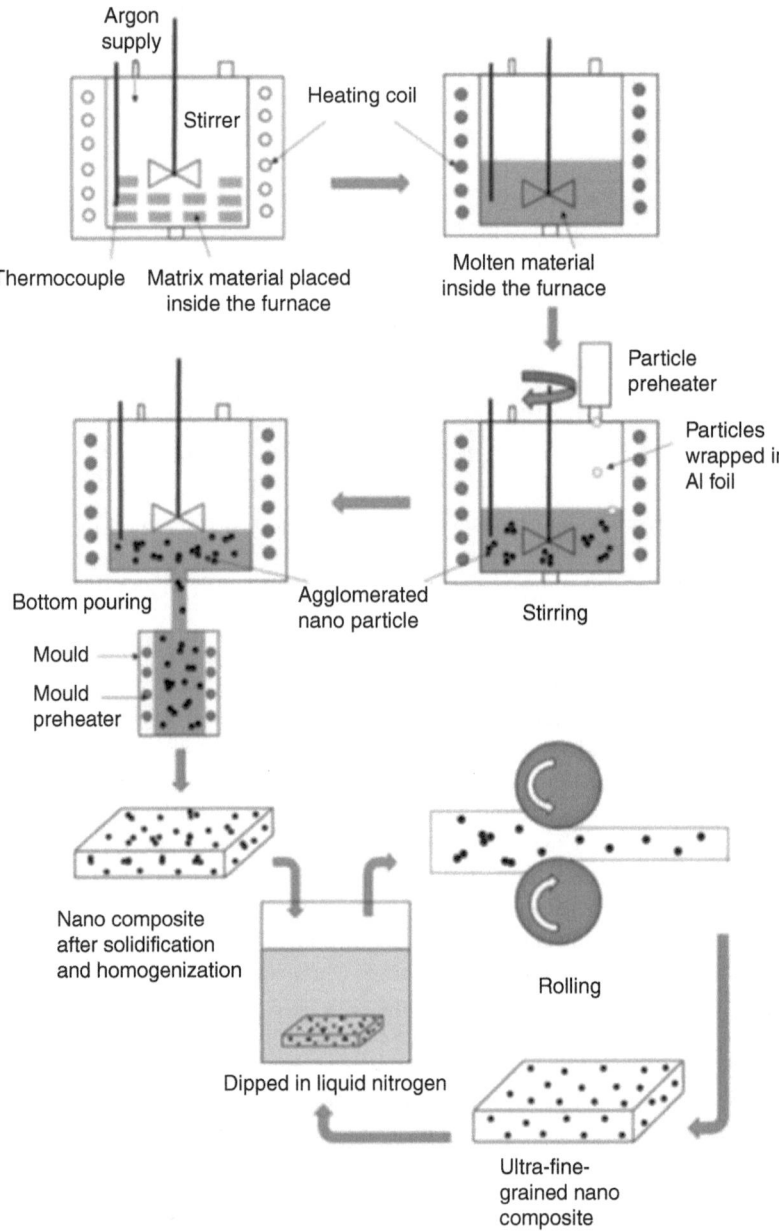

FIGURE 7.5 Process sequence involved in the production of Al-MMCs through stir casting followed by cryorolling (hybrid manufacturing route).

Source: Deb et al. (2018); reprinted with permission.

tensile properties of Al-MMCs produced by this hybrid manufacturing route are dealt with in the following sections.

7.2   MICROSTRUCTURAL ASPECTS OF AL-MMCS FABRICATED BY STIR CASTING FOLLOWED BY CRYOROLLING

Deb et al., 2018, fabricated bulk UFG AA1050 nano composite sheets through a hybrid manufacturing route. Initially, the Al–SiC MMC was prepared through stir casting, and cryorolling was carried out as a secondary process after stir casting. The authors studied the structure–property correlation of the fabricated nano composite sheets in two different stages, namely, (1) after performing the stir casting process and (2) after cryorolling the stir cast composite. They observed a gradual decrease in the average grain size (all were in micron levels) when the volume fraction of SiC increased (0, 0.5, 1, 2 wt.%) after stir casting. Furthermore, Figure 7.6 shows a significant number

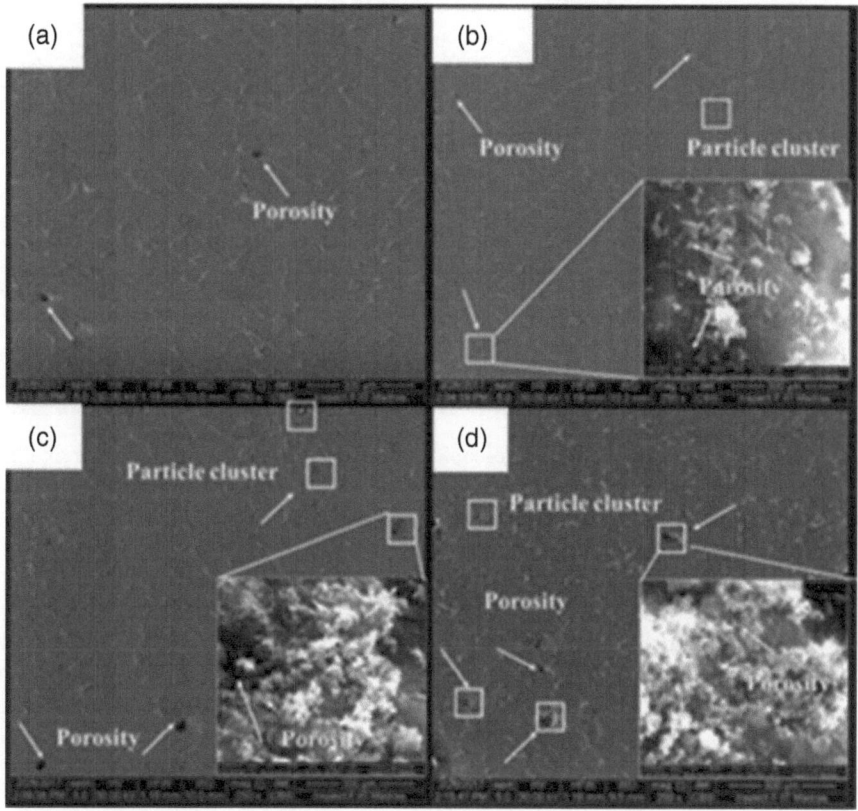

FIGURE 7.6   Microscopic images of stir cast AA1050–SiC MMC with various wt.% of reinforcement particles: (a) 0%, (b) 0.5%, (c) 1% and (d) 2%. Inset images show the particle clusters at greater magnification.

Source: Deb et al. (2018); reprinted with permission.

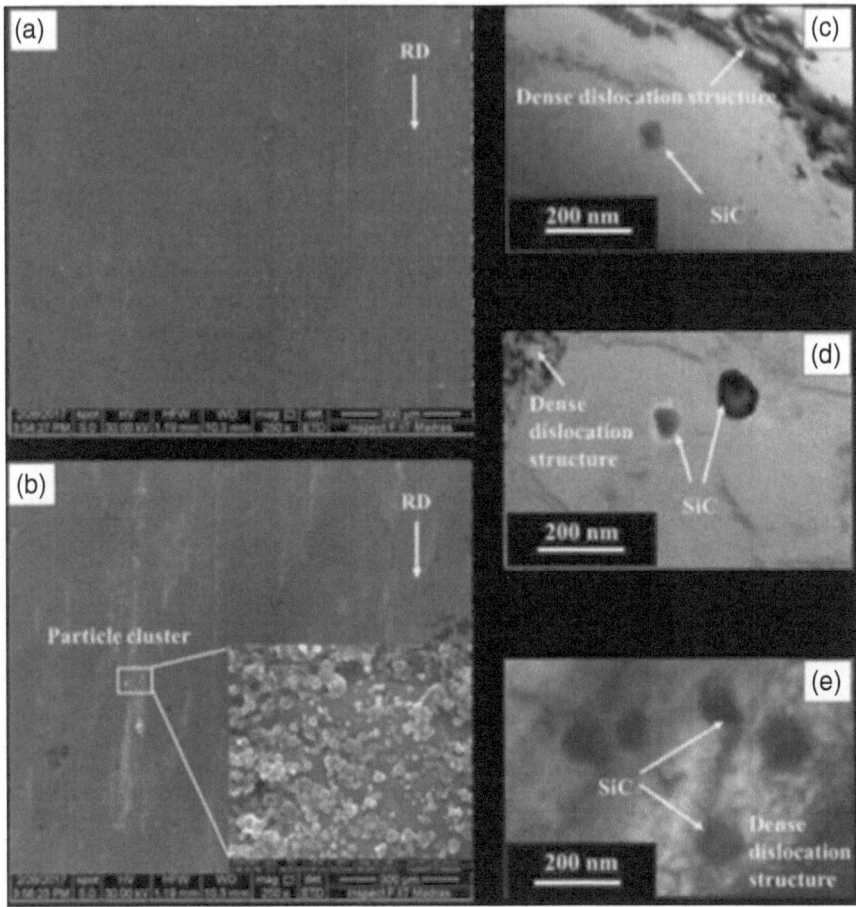

FIGURE 7.7 Scanning electron microscopic images of AA1050–SiC MMC after cryorolling with (a) 0.5 wt.% and (b) 2 wt.% reinforcements. Transmission electron microscopic (TEM) images of stir cast AA1050–SiC MMC after cyrorolling with (c) 0.5 wt.%, (d) 1 wt.%, (e) 2 wt.% reinforcement combinations.

Source: Deb et al. (2018), reprinted with permission.

of micro pores and particle clusters after stir casting. However, after cryorolling the average grain size reduced drastically to nanometric levels. In addition, the pores were completely closed and the particle clusters were minimal, as observed in Figure 7.7a–b. This is due to the influence of cryorolling, where the roll pressure extruded the Al matrix through the particle clusters and deformed the composite. Also, the cryorolling dispersed the clustered particles and improved the homogeneous distribution of SiC in the Al matrix (Reza et al., 2014). The restriction of dynamic recovery also results in the formation of a high density of dislocations as seen in Figure 7.7c–e.

Bembalge and Panigrahi, 2018, developed bulk UFG AA6063–SiC nano composite sheets through the hybrid manufacturing route (stir casting with cryorolling).

They varied the size of reinforcement particle in the Al matrix, namely, coarse SiC (12 μm), fine SiC (1 μm) and nano SiC (45 nm). The AA6063–SiC composite was prepared through stir casting for the above combination of particle sizes and the volume fraction of reinforcement particle was fixed at 4 wt.% for all the conditions. The electron backscatter diffraction (EBSD) image quality maps shown in Figure 7.8 reveals the microstructure with dendrites in all the conditions. However, the dendritic structure reduces in size as the particle size decreases from coarse to nanometric level (Figure 7.8b–d). This is because as the particle content increases, the particles tend to segregate in the grain boundaries, whichhinders grain growth. This leads to the refinement of grains as the particle size decreases. Furthermore, the formation of micro pores is visible in the stir cast microstructure. To eliminate the porosity, room temperature rolling was performed with 1 mm thickness reduction. Further, the solution heat treatment (530°C for 2 hrs) resulted in a more equiaxed microstructure for all the conditions. Solution heat treatment is the process where the alloys are heated to the required temperature and held there for a certain period to cause

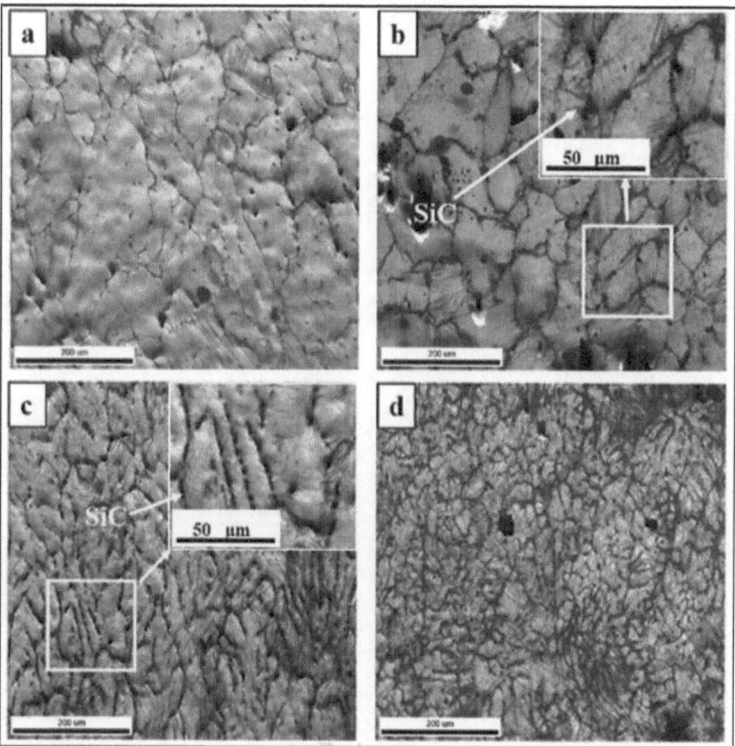

FIGURE 7.8   EBSD image quality maps of stir cast AA6063–SiC composites: (a) base alloy with no reinforcement, (b) AA6063 with coarse SiC, (c) AA6063 with fine SiC and (d) AA6063 with nano-sized SiC.

Source: Bembalge and Panigrahi (2018), reprinted with permission.

the phase constituents to enter into the solid solution before cooling in the solution. A better particle distribution was achieved after solution heat treatment. Additionally, cryorolling was executed on the solutionized composites and a 90% reduction in thickness was obtained (Bembalge and Panigrahi, 2018). After cryorolling, the base material with no reinforcement shows a high density of dislocation cell structures and UFGs (Figure 7.9a–b). In the case of composite sheet with coarse particles, the dislocation formation is noted in the matrix-particle interface zone along with the existence of UFGs (Figure 7.9c–d). Also, the Figure 7.9e–h micrographs show the microstructure of composites prepared by fine and nano-sized particles, which results

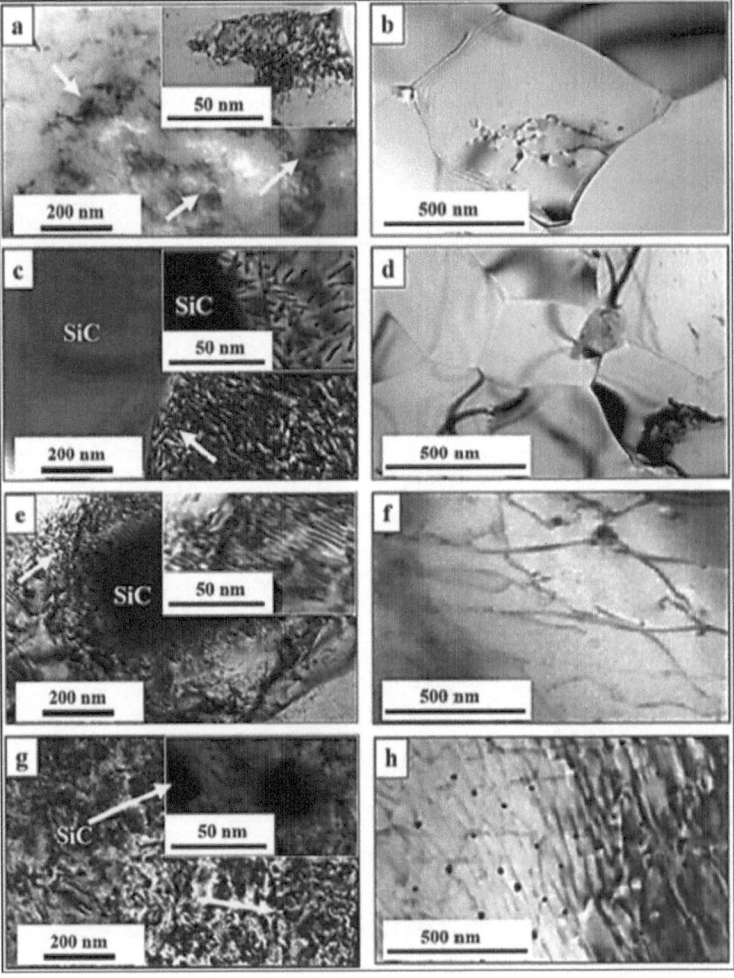

FIGURE 7.9 TEM images of (a, b) base alloy with no reinforcement, (c, d) composite sheet with coarse SiC, (e, f) composite sheet with fine SiC, (g, h) composite sheet with nano SiC.

Source: Bembalge and Panigrahi (2018), reprinted with permission.

in the formation of finer cell structure. This is attributed to the effect of cryorolling, which hinders the grains from elongating along the rolling direction because of the higher volume fraction of SiC particles compared to composites with coarse SiC particles. This resulted in highly refined grain structure in the composite, which had fine and nano-sized particles.

In another study by Deb et al., 2019, the recrystallization behaviour and thermal stability of AA1050–SiC MMC was evaluated, which was fabricated with the nano-sized particles (45 nm). The composite was prepared by the researchers through stir casting followed by cryorolling (90% reduction from its original thickness of 10 mm). Further, they conducted annealing ranging from 150°C to 300°C and studied the structure–property correlation. As seen in the EBSD images in Figure 7.10a–e, the cryorolled composite sheets possess a highly deformed grain structure and their orientation of grains changes after the annealing process. The cryorolled composite sheet annealed at 150°C shows a diffused microstructure similar to that of the unannealed cryorolled composite. As the annealing temperature increased to 200°C, a gradual growth in the sub-grains is observed (Figure 7.10c).

Moreover, the annealing carried out at 250°C produced a recrystallized microstructure; however, the average grain size was reported to be 507 nm, which is still in the UFG realm. In addition, a rapid grain coarsening was observed beyond 250°C, and annealing carried out at 300°C resulted in a fully recrystallized microstructure (average grain size ~6 μm). Further, the authors reported that the recrystallization initiated before the completion of the recovery phase (Deb et al., 2019). As Al is a high stacking fault energy material, it exhibits a tendency to recover at a faster rate even at low annealing temperatures. Also, the accumulated high density of dislocations enhanced the stored energy, which supported recrystallization. In contrast, the authors observed larger grain size on the same material without any reinforcement. It is clear that the presence of fine reinforcement (SiC) in the Al matrix pinned the grain boundaries and restricted the grain growth during annealing to some extent.

7.3 WEAR BEHAVIOUR OF AL-MMCS PRODUCED BY STIR CASTING FOLLOWED BY CRYOROLLING

Bembalge and Panigrahi, 2019a, fabricated AA6063–SiC composite through a hybrid manufacturing route (stir casting + solution heat treatment + cryorolling) and varied the particle size as coarse, fine and nano. They studied the influence of particle size in the composite on the wear behaviour. The hardness value increases as the particle size in the composite decreases, as reported in Figure 7.11a. The composite with nano-sized SiC particles exhibited the maximum hardness value of 200 HV when compared to the cryorolled base alloy and cryorolled composite with coarse and fine SiC particles. Pin-on-disc wear study was conducted for the various loads, for example, 10, 20, 30 and 50 N, and for each load, the sliding velocity was varied as 0.5, 1, 1.5 and 2 m/s. The applied load and the sliding velocity affected the specific wear rate of the cryorolled base alloy and cryorolled composite. A decrease in wear rate with increase in load at all sliding velocities was attained for all the conditions selected. When the load was increased to 50 N, the wear rate increased drastically

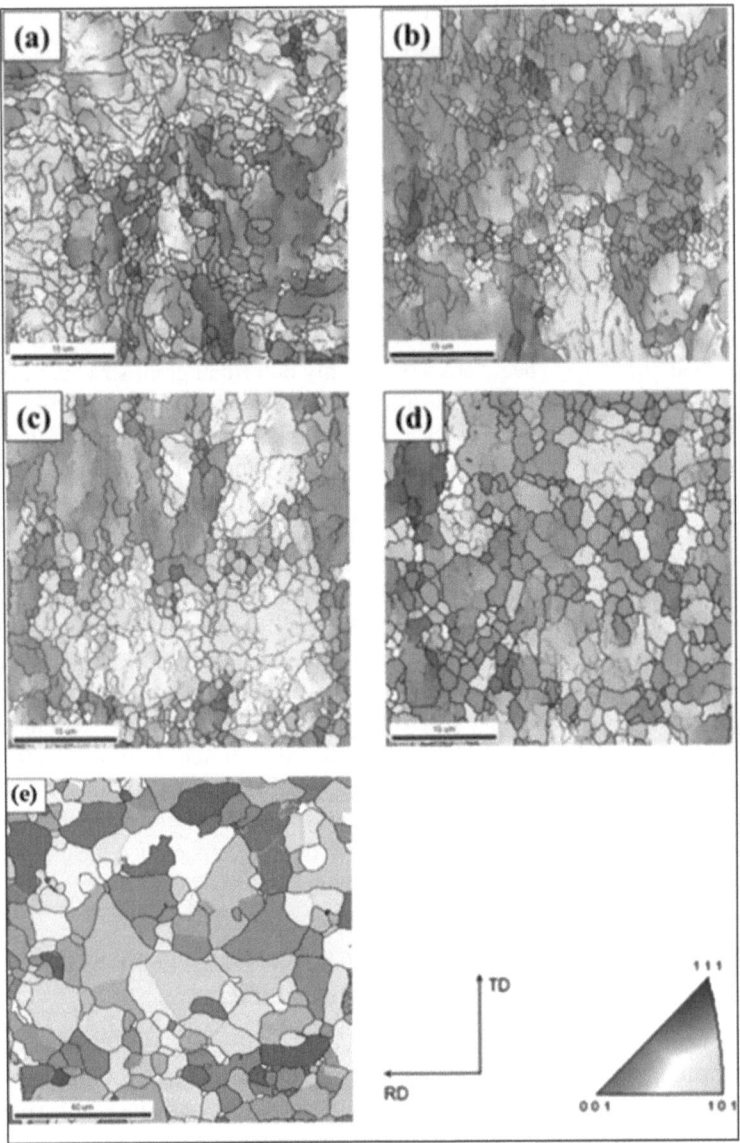

FIGURE 7.10 EBSD images of AA1050–SiC composite sheet after cryorolling and annealing: (a) cryorolled, (b) cryorolled and annealed at 150°C, (c) cryorolled and annealed at 200°C, (d) cryorolled and annealed at 250°C, and (e) cryorolled and annealed at 300°C.

Source: Deb et al. (2019), reprinted with permission.

and produced an adverse effect on the tested composites. An optimal wear rate was obtained for a load of 30 N at a sliding velocity of 2 m/s.

The wear rate for the various SiC particle sizes is shown in Figure 7.11b. Among the conditions selected, the cryorolled composite prepared with nano reinforcement

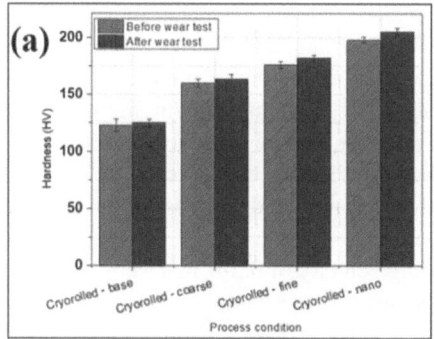
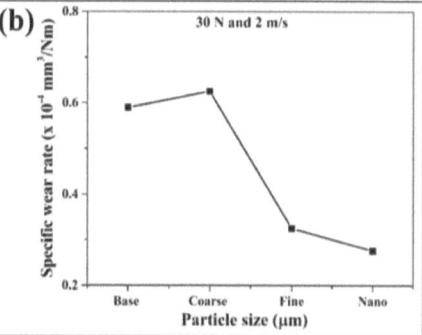

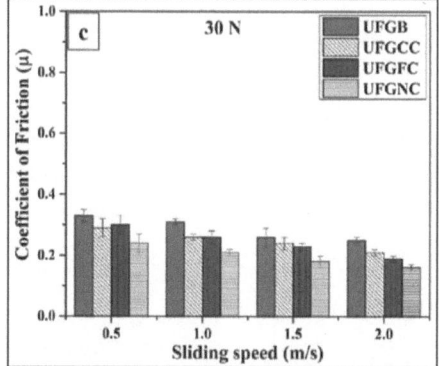

FIGURE 7.11 (a) Hardness value of AA6063–SiC MMC fabricated through a hybrid manufacturing route with varied particle sizes (wear test conducted at 30 N load with 2 m/s siding velocity). (b) Variation in specific wear rate with respect to varying SiC particle size. (c) Variation of coefficient of friction with respect to sliding speed and reinforcement particle size.

Source: Bembalge and Panigrahi (2019a). Part (a) redrawn based on their work, (b) and (c) reprinted with permission.

had the lowest wear rate. For the same volume fraction of reinforcement, as the particle size decreases, the number of SiC particles increases, which reduces the particle–particle distance in the Al matrix. This improves both the hardness and wear rate in the case of cryorolled composite with fine and nano-sized SiC particles. Further, the cryorolled composite with coarse SiC shows a higher wear rate than that of cryorolled base alloy. This is attributed to the broken SiC particles during cryorolling as reported by Bembalge and Panigrahi, 2019a. Apart from the particle size, the cryorolled microstructure (refer to Figure 7.9) shows higher dislocations due to the restriction of dynamic recovery. In addition, the variation in the thermal expansion coefficient between the matrix and reinforcement provoked the geometrically necessary dislocations. The accumulation of dislocations because of cryorolling and the dislocations due to unequal thermal expansion coefficients resulted in higher dislocation density in the case of cryorolled composites. Thus, the dislocation density is higher in the cryorolled composite than in cryorolled base alloy without any reinforcement. This leads to a fine-grained microstructure with higher dislocation density that supported additional hardness and wear resistance apart from the hardness attained by varying particle sizes. The coefficient of friction decreases with increase in sliding speed and with decrease in reinforcement particle size. Among the loads selected, the minimal coefficient of friction is noted at 30 N with a sliding speed of 2 m/s (refer to Figure 7.11c). This is attributed to the improved strength and hardness of composites produced with nano-sized reinforcement particles.

Lower loads, say 10–20 N with a low sliding velocity of 0.5–1 m/s, show abrasive marks with adhesive wear (Figure 7.12). The grooves are shallow in cryorolled base alloy (Figure 7.12a), and deep in the case of cryorolled composite with coarse SiC particles (Figure 7.12b). However, the grooves were average in depth for cryorolled composites with either fine or nano level SiC particles (Figure 7.12c–d).

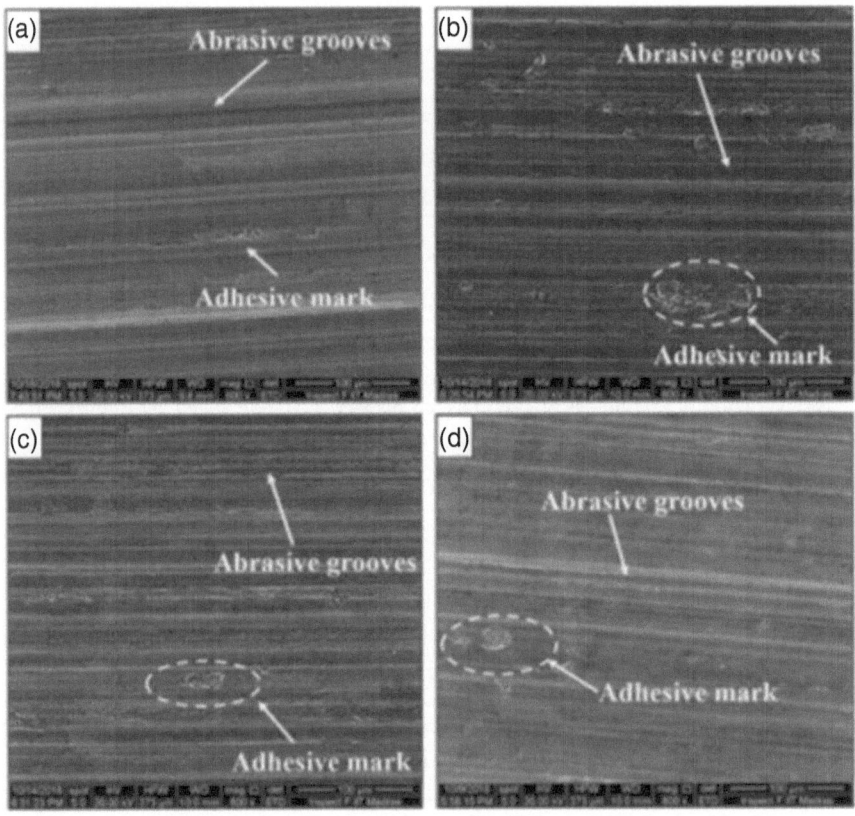

FIGURE 7.12 Surface morphology indicating abrasive and adhesive wear conducted at low load and at low sliding velocity (in AA1050–SiC MMC): (a) cryorolled – base alloy, (b) cryorolled – coarse, (c) cryorolled – fine and (d) cryorolled – nano.

Source: Bembalge and Panigrahi (2019a), reprinted with permission.

The greater wear in the cryorolled composite with coarse SiC is due to the large, weakly bonded particles which were pulled out from the Al matrix during the wear test. These pulled-out particles cause three-body wear that resulted in the formation of deep grooves with large abrasive and adhesive wear. In contrast, cryorolled composites with fine and nano particles show strong interfacial bonding with less tendency to pull out. This results in shallow grooves with fewer abrasive marks and less adhesive wear. The formation of oxide wear layers is observed in the wear surface morphology of the alloy and three types of composites discussed in this section (Figure 7.13). The oxide layers fill the micro voids and act as a protective shielding. The oxide layers are larger in the case of cryorolled composites (Figure 7.13b–d) than in cryorolled base alloy (Figure 7.13a). This shielding reduces the wear rate by minimizing the contact between the pin and test surface. Further, the pulled-out SiC particles may get fragmented and mix with the oxide layers, providing additional

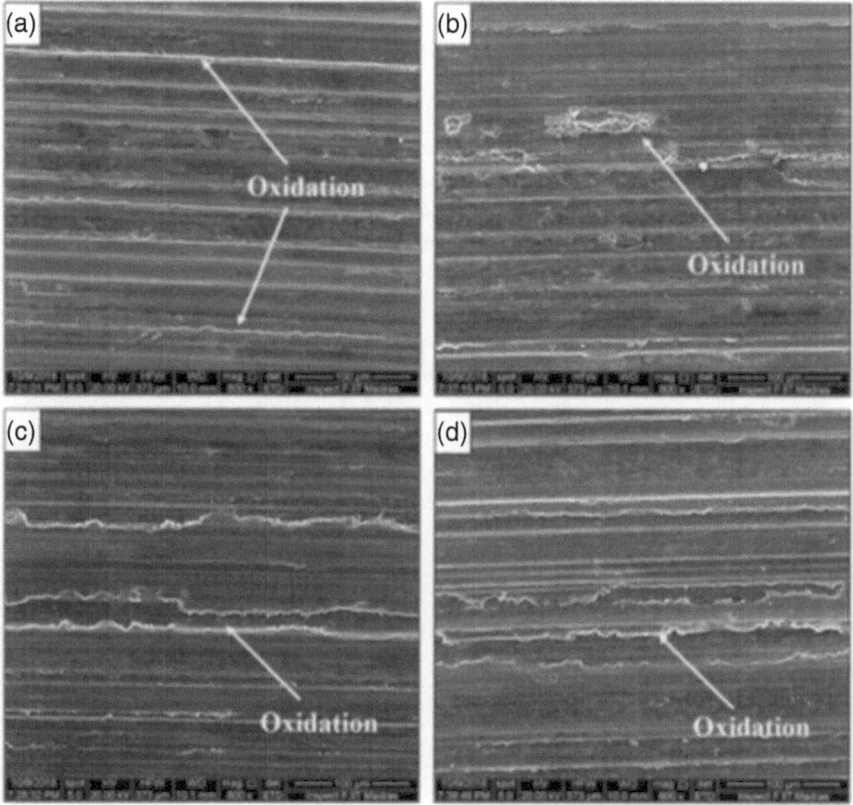

FIGURE 7.13 Surface morphology indicating the oxide wear layers in AA1050–SiC MMC tested at 30 N load and 2 m/s sliding velocity: (a) cryorolled – base alloy, (b) cryorolled – coarse, (c) cryorolled – fine and (d) cryorolled – nano.

Source: Bembalge and Panigrahi (2019a), reprinted with permission.

wear resistance in the case of cryorolled composites. Thus, favourable microstructure (refined microstructure with dislocations and geometrically necessary dislocations), better matrix–particle interfacial bonding and the formation of oxide wear layers enhanced the wear rate (specific) of the cryorolled composites with fine and nano-sized SiC particles.

7.4 TENSILE PROPERTY EVALUATION OF AL-MMCS PRODUCED BY STIR CASTING FOLLOWED BY CRYOROLLING

7.4.1 EFFECT OF VOLUME FRACTION OF REINFORCEMENT ON THE TENSILE BEHAVIOUR OF AL-MMC

Deb et al., 2018, produced Al-MMC from AA1050/SiC through the hybrid manufacturing route (stir casting + cryorolling), which resulted in a drastic increase in tensile strength with negligible variation in elongation, as seen in Table 7.1. The cryorolled composites exhibit greater strength when compared to the stir cast composite. The addition of nano-sized SiC particles and the increase in volume fraction of SiC particles in the Al matrix produced greater strength. However, addition of more reinforcement significantly reduced the ductility. The improved strength of composites compared to the unreinforced base alloy is because of grain refinement that happened as a result of SiC pinning, generation of geometrically necessary dislocations because of the thermal expansion coefficient between matrix and particles. Also, the nano particle clusters result in additional strength, which acts as a hindrance to dislocation motion. The geometrically necessary dislocations interact with the sessile dislocations during tensile deformation, improving strength (Mazahery et al., 2009). As the particle content increases (2 wt.%), the clustering of more nano-sized SiC particles with a higher volume fraction of micro pores between matrix and particle results in a reduction in strength as well as ductility. The clustering of a large number of particles provides a nucleation site for crack formation and reduces strength (Su et al., 2012). This results in the reduced strength values as the volume fraction of the particles reached a critical level. As discussed in section 7.1.4, the cryorolling process fragments the particle clusters and distributes the particles uniformly in the matrix. It also covers the micro pores, which are present after stir casting, between matrix and particle at their interface. The improvement in strength after cryorolling is due to the high density of dislocations and grain refinement. Also, the contribution of several strengthening mechanisms results in the improvement of cryorolled composites compared to the base alloy and stir cast composite (Amirkhanlou et al., 2015).

7.4.2 EFFECT OF POST ANNEALING TREATMENT ON THE TENSILE BEHAVIOUR OF AL-MMC

The tensile properties of the AA1050 base alloy and AA1050/SiC composite after the cryorolling process and post annealing process are shown in Table 7.2 based on the work of Deb et al., 2019. The strength of the material (base alloy and composite) increases after cryorolling. The strength of the material decreases with increases in annealing temperature as the material undergoes recovery and recrystallization. Beyond

250°C, the researchers observed a drastic improvement in the elongation property (both in base alloy and composite). This drop in strength with improvement in elongation is significant for annealing at 300°C. Further, a completely recrystallized microstructure is attained (refer to Figure 7.10e). The cryorolled material gained strength through dislocation strengthening. However, with increasing annealing temperature, the importance of grain boundary strengthening increased. The grain growth is visible in a microstructural examination (refer to Figure 7.10a–e) for annealing above 250°C. When compared to the cryorolled base alloy (unreinforced), the cryorolled composite showed greater strength due to smaller grain size. The smaller grain size is

TABLE 7.1
Effect of Variation in Reinforcement Addition on the Tensile Properties of Al–SiC MMC Fabricated by Stir Casting + Cryorolling

Al-MMCs

Base material	AA1050
Reinforcement	SiC
Particle size	45 nm
Volume fraction (vol.%)	0.5, 1, 2
Stir casting process	
Melting temperature of matrix	750°C and cooled to 650–670°C
Reinforcement	Preheated to 300°C for 2 hrs
Temperature and speed of stirring	720°C at 300 RPM for 10 min
Plate dimension L × W × T (mm³)	300 × 100 × 11
Homogenization	400°C for 2 hrs – water quenched
Machining operation	Machined to 300 × 100 × 10 (mm³)
Cryorolling process	
Cryo treatment	30 min dipped in liquid nitrogen (LN) before rolling and again dipped in LN for 5 min before each roll pass
Rolling reduction	90% from the initial 10 mm thickness
Reduction per pass	5%
Roll diameter and speed	110 mm and 8 RPM

Process condition	(before cryorolling)			(after cryorolling)		
Tensile properties	YS* (MPa)	UTS* (MPa)	Total elongation (%)	YS (MPa)	UTS (MPa)	Total elongation (%)
AA1050 – base alloy	34 ± 5	54 ± 5	15.4 ± 0.8	176 ± 4	188 ± 3	14 ± 0.75
AA1050/0.5 wt.% SiC	68 ± 6	109 ± 8	12.5 ± 0.6	250 ± 3	258 ± 4	13.7 ± 0.6
AA1050/1 wt.% SiC	95 ± 3	165 ± 7	6.5 ± 0.9	261 ± 3	269 ± 3	4.6 ± 0.5
AA1050/2 wt.% SiC	94 ± 4	134 ± 5	4 ± 0.5	224 ± 5	237 ± 3	3.9 ± 0.7

Source: Reproduced tensile property values with permission from Deb et al., 2018.

TABLE 7.2

Effect of Annealing Treatment on the Tensile roperties of Al–SiC MMC Fabricated by Stir Casting + Cryorolling + Annealing

Al-MMCs

Base material	AA1050
Reinforcement	SiC
Particle size	45 nm
Volume fraction (vol.%)	0.5
Stir casting process	
Melting temperature of matrix	750°C and cooled to 650–670°C
Reinforcement	Preheated to 300°C for 2 hrs
Temperature and speed of stirring	720°C at 300 RPM for 10 min
Plate dimension L × W × T (mm³)	300 × 100 × 11
Homogenization	400°C for 2 hrs – water quenched
Machining operation	Machined to 300 × 100 × 10 (mm³)
Cryorolling process	
Cryo treatment	30 min dipped in liquid nitrogen (LN) before rolling and again dipped in LN for 5 min before each roll pass
Rolling reduction	90% from the initial 10 mm thickness
Reduction per pass	5%
Roll diameter and speed	110 mm and 8 RPM
Post heat treatment	Annealed at 150, 200, 250 and 300°C for 1 hr

Process condition	Cryorolled and annealed without SiC reinforcement			Reinforced with 0.5 wt.% SiC + cryorolled + annealed			
Tensile properties	**YS (MPa)**	**UTS (MPa)**	**Total elongation (%)**		**YS (MPa)**	**UTS (MPa)**	**Total elongation (%)**
AA1050 – base alloy (not annealed)	175	188	14	AA1050/SiC (not annealed)	250	258	13.7
AA1050 – base alloy (150°C)	158	174	17.9	AA1050/SiC (150°C)	203	225	16.3
AA1050 – base alloy (200°C)	150	160	23	AA1050/SiC (200°C)	197	213	18.4
AA1050 – base alloy (250°C)	135	154	25.8	AA1050/SiC (250°C)	171	180	23
AA1050 – base alloy (300°C)	30	86	62	AA1050/SiC (300°C)	32	117	42.3

Source: Reproduced tensile property values with permission from Deb et al., 2019.

due to the pinning of reinforcement particles on the grain boundaries, which restricts grain growth.

7.4.3 Influence of Particle Size Variation on the Tensile Behaviour of Al-MMCs

The tensile properties of stir cast AA6063/SiC composite before and after cryorolling are given in Table 7.3, which is based on the report of Bembalge and Panigrahi, 2018. The SiC was added to the Al matrix in various sizes as coarse, fine and nano. The strength level increases with reduction in elongation as the particle size increases in the case of stir cast composite (Figure 7.14a). Moreover, the improvement in strength is significant in the cryorolled composites, and the composite with nano-sized SiC particles showed the maximum strength value (Figure 7.14b). This is due to the effect of cryorolling, which increases the dislocation density, eliminates the micro pores present in the stir cast composite and enhances the interfacial bond strength of matrix and particle. The lower strength in the stir cast composite is attributed to the presence of dendritic microstructure, which is brittle in nature. Also, the SiC particles present in these boundaries increase the stress concentration zones, cause early failure and lead to less strength. As evidence, the dendritic globules and dendritic zones are observed in the fracture surface morphology of stir cast composites (refer to Figure 7.15(a–d) for all combinations of SiC). However, the dendritic structures were eliminated by solution heat treatment and subsequent cryorolling process (refer to Table 7.3). The cryorolling process enhances the strength of the material as a result of dislocation strengthening, grain boundary strengthening and Orowan strengthening (dislocation loop occurs because of particle hindrance). Furthermore, the thermal coefficient difference between matrix and particle produces geometrically necessary dislocations, adding strength to the cryorolled composite. The fracture surface morphologies of failed tensile specimens show no traces of dendritic regions, only micro voids and dimples, which is an indication of the ductile fracture phenomenon (Figure 7.16a–d).

Another study was conducted by the same authors (Bembalge and Panigrahi, 2019b) by varying the particle size of SiC in AA6063 matrix. They also carried out cryorolling with post annealing treatments on the composites. The results show increased strength as the particle size decreases. Further, the cryorolled product after annealing showed a gradual decrease in strength with simultaneous restoration of ductility. This decrease in strength with nominal increase in ductility is significant after annealing at 300°C as a result of the presence of grain coarsening. However, the AA6063 with nano-sized SiC showed sufficient strength owing to better interfacial bonding as the particle size gets finer.

7.5 SUMMARY AND FUTURE SCOPE

This chapter gives an overview of the fabrication of Al-MMCs through stir casting followed by cryorolling, which is a hybrid manufacturing route. The microstructural, wear behaviour and tensile properties of different Al-MMCs subjected to cryorolling are explored. The influence of particle size, volume fraction, cryorolling and post

TABLE 7.3

Effect of Particle Size Variation on the Tensile Properties of Al–SiC MMC Fabricated by Stir Casting + Cryorolling

Al–MMCs

Base material	AA6063
Reinforcement	SiC
Particle size	12 µm, 1 µm, 45 nm
Volume fraction (vol.%)	4
Stir casting process	
Melting temperature of matrix	700°C
Reinforcement	Preheated to 600°C
Temperature and speed of stirring	700°C at 500 RPM for 10 min
Plate dimension L × W × T (mm³)	300 × 100 × 11
Room temperature rolling process	1 mm reduction from its original thickness of 11 mm
Cryorolling process	
Cryo treatment	30 min dipped in liquid nitrogen (LN) before rolling and again dipped in LN for 5 min before each roll pass
Rolling reduction	90% from the initial 10 mm thickness
Reduction per pass	5%
Roll diameter and speed	110 mm and 8 RPM

Process condition	*Stir casting with composite*			*Stir casting with composite + cryorolling*			
Tensile properties	YS (MPa)	UTS (MPa)	Total elongation (%)		YS (MPa)	UTS (MPa)	Total elongation (%)
AA6063 – base alloy	85	170	29	AA6063 – base alloy	253	275	15.2
AA6063/SiC (coarse)	117	174	14	AA6063/SiC (coarse)	289	321	8.9
AA6063/SiC (fine)	135	190	7.6	AA6063/SiC (fine)	320	353	7.3
AA6063/SiC (nano)	153	209	4.8	AA6063/SiC (nano)	371	409	2.2

Source: Tensile property values reproduced with permission from Bembalge and Panigrahi, 2018.

annealing treatment on the microstructure and its associated property changes are discussed. The composite produced by stir casting shows micro voids in the matrix–particle interface, which results in inferior strength. Further, the clustering of fine reinforcement particles, which act as potential nucleation sites for crack formation,

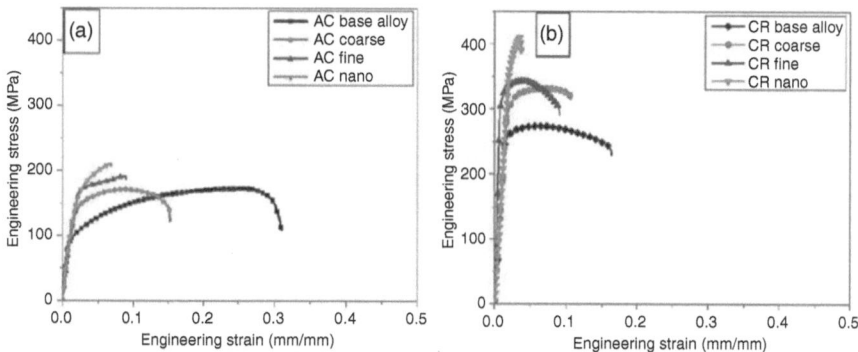

FIGURE 7.14 Engineering stress versus engineering strain plots of AA6063 reinforced with SiC with coarse, fine and nano particles (a) after stir casting and (b) after cryorolling.

Source: Bembalge and Panigrahi (2018), reprinted with permission.

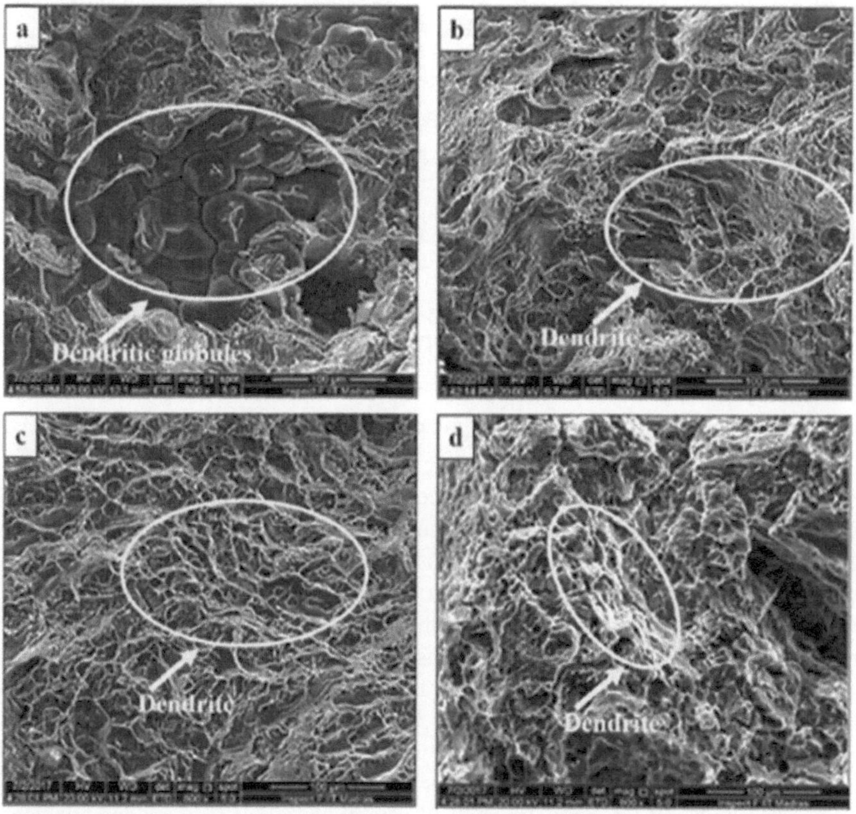

FIGURE 7.15 Fracture surface morphology of stir cast AA6063/SiC composite: (a) AA6063 base alloy, (b) AA6063 with coarse SiC, (c) AA6063 with fine SiC, (d) AA6063 with nano SiC.

Source: Bembalge and Panigrahi (2018), reprinted with permission.

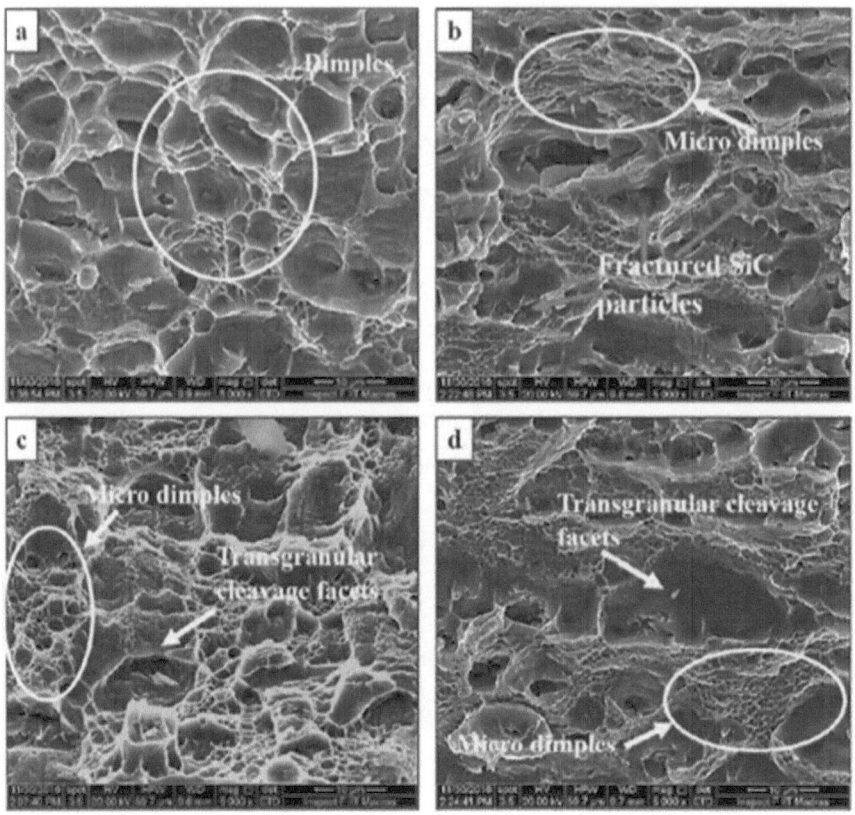

FIGURE 7.16 Fracture surface morphology of cryorolled AA6063/SiC composite: (a) AA6063 base alloy, (b) AA6063 with coarse SiC, (c) AA6063 with fine SiC, (d) AA6063 with nano SiC.

Source: Bembalge and Panigrahi (2018), reprinted with permission.

leads to poor strength in the stir cast composites. To overcome this challenge, cryorolling is used to close micro pores and also fragment the clustered particles. This results in a better distribution of reinforcement particles in the Al matrix and also exhibits better interfacial bonding. Cryorolling also suppresses dynamic recovery, leading to an accumulation of dislocations. Additionally, the matrix and particle create geometrically necessary dislocations owing to their differing thermal expansion coefficients. These dislocations interact with the sessile dislocations during a tensile test and exhibit additional strength. As the particle size gets finer, the strength increases. Also, the presence of fine particles in the Al matrix pinned the grain boundaries, which restricted the grain coarsening during annealing. Hence, the cryorolled composite which had fine particles showed a combination of strength and considerable ductility.

Most research on this hybrid manufacturing route has been conducted on Al alloys with ceramic reinforcements, particularly with SiC particles. Its general use may be significantly expanded if more research is conducted on other high strength materials and composites. However, there is also no detailed study on the influence of other ceramic particles through this route. Further, this hybrid manufacturing route can be extended to metallic reinforcements like Fe, Ni, stainless steel (SS), Cu particles and so on. Studies on the addition of metallic reinforcements might overcome challenges like particle breakdown that occur with coarse ceramic particles. The formation of brittle phases, which deteriorate the mechanical properties, can also be avoided. Moreover, production of high entropy alloy-based MMCs has gained attraction among researchers as it establishes improved mechanical properties. Hence, this hybrid manufacturing route can also be successfully employed on high entropy alloys. It is anticipated that the production of Al-MMCs with metallic reinforcement through stir casting followed by cryorolling could result in better structure–property relations, extending this route for wider applications.

REFERENCES

Amirkhanlou, S., M. Ketabchi, N. Parvin, A. Orozco-Caballero and F. Carreño. 2015. Homogeneous and ultrafine-grained metal matrix nanocomposite achieved by accumulative press bonding as a novel severe plastic deformation process. *Scripta Materialia* 100 (April): 40–43. https://doi.org/10.1016/j.scriptamat.2014.12.007.

Arunachalam, R., P.K. Krishnan and R. Muraliraja. 2019. A review on the production of metal matrix composites through stir casting: Furnace design, properties, challenges, and research opportunities. *Journal of Manufacturing Processes* 42 (June): 213–245. https://doi.org/10.1016/j.jmapro.2019.04.017.

Bembalge, O.B., and S.K. Panigrahi. 2018. Development and strengthening mechanisms of bulk ultrafine grained AA6063/SiC composite sheets with varying reinforcement size ranging from nano to micro domain. *Journal of Alloys and Compounds* 766 (October): 355–372. https://doi.org/10.1016/j.jallcom.2018.06.306.

Bembalge, O.B., and S.K. Panigrahi. 2019a. Influence of SiC ceramic reinforcement size in establishing wear mechanisms and wear maps of ultrafine grained AA6063 composites. *Ceramics International* 45, no. 16 (November): 20091–20104. https://doi.org/10.1016/j.ceramint.2019.06.274.

Bembalge, O.B., and S.K. Panigrahi. 2019b. Thermal stability, grain growth kinetics, and mechanical properties of bulk ultrafine-grained AA6063/SiC composites with varying reinforcement sizes. *Metallurgical and Materials Transactions A* 50A (July): 4288–4306. https://doi.org/10.1007/s11661-019-05342-6.

Bera, S., S.G. Chowdhury, Y. Estrin and I. Manna. 2013. Mechanical properties of Al7075 alloy with nanoceramic oxide dispersion synthesized by mechanical milling and consolidated by equal channel angular pressing. *Journal of Alloys and Compounds* 548 (January): 257–265. https://doi.org/10.1016/j.jallcom.2012.09.007.

Cheng, S., Y.H. Zhao, Y.T. Zhu and E. Ma. 2007. Optimizing the strength and ductility of fine structured 2024 Al alloy by nano-precipitation. *Acta Materialia* 55, no. 17 (October): 5822–5832. https://doi.org/10.1016/j.actamat.2007.06.043.

David Raja Selvam, J., D.S. Robinson Smart and I. Dinaharan. 2013. Microstructure and some mechanical properties of fly ash particulate reinforced AA6061 aluminum alloy composites prepared by compocasting. *Materials and Design* 49 (August): 28–34. https://doi.org/10.1016/j.matdes.2013.01.053.

Deb, S., S.K. Panigrahi and M. Weiss. 2018. Development of bulk ultrafine grained Al-SiC nano composite sheets by a SPD based hybrid process: Experimental and theoretical studies. *Materials Science and Engineering A* 738 (December): 323–324. https://doi. org/10.1016/j.msea.2018.09.101.

Deb, S., S.K. Panigrahi and M. Weiss. 2019. The effect of annealing treatment on the evolution of the microstructure, the mechanical properties and the texture of nano SiC reinforced aluminium matrix alloys with ultrafine grained structure. *Materials Characterization* 154 (August): 80–93. https://doi.org/10.1016/j.matchar.2019.05.023.

Dinaharan, I., N. Murugan and E.T. Akinlabi. 2020. Friction stir processing route for metallic matrix composite production. In *Encyclopedia of Materials: Composites*, ed. D. Brabazon, Amsterdam: Elsevier.

Gecu, R., Ş.H. Atapek and A. Karaaslan. 2017. Influence of preform preheating on dry sliding wear behavior of 304 stainless steel reinforced A356 aluminum matrix composite produced by melt infiltration casting. *Tribology International* 115 (November): 608–618. https://doi.org/10.1016/j.triboint.2017.06.040.

Geng, L., and K. Wu. Metal matrix composites. In *Composite Materials Engineering*, ed. X.S. Yi, S. Du and L. Zhang, Singapore: Springer.

Harichandran, R., and N. Selvakumar. 2016. Effect of nano/micro B_4C particles on the mechanical properties of aluminium metal matrix composites fabricated by ultrasonic cavitation-assisted solidification process. *Archives of Civil and Mechanical Engineering* 16, no. 1 (January): 147–158. https://doi.org/10.1016/j.acme.2015.07.001.

Idrisi, A.H., and A.H.I. Mourad. 2019. Conventional stir casting versus ultrasonic assisted stir casting process: Mechanical and physical characteristics of AMCs. *Journal of Alloys and Compounds* 805 (October): 502–508. https://doi.org/10.1016/j.jallcom.2019.07.076.

Kalaiselvan, K., N. Murugan and S. Parameswaran. 2011. Production and characterization of AA6061–B4C stir cast composite. *Material and Design* 32, no. 7 (August): 4004–4009. https://doi.org/10.1016/j.matdes.2011.03.018.

Kandarp, C., K. Hariharan and D. Ravi Kumar. 2019. Development of combined groove pressing and rolling to produce ultra-fine grained Al alloys and comparison with cryorolling. *Materials Science and Engineering A* 760, (July): 7–18. https://doi.org/10.1016/j.msea.2019.05.088.

Kingsley, T., and O.M. Suárez. 2011. Study of boride-reinforced aluminum matrix composites produced via centrifugal casting. *Material and Manufacturing Process*, 26, no. 2 (March): 338 – 345. https://doi.org/10.1080/10426910903124829.

Mazahery, A., H. Abdizadeh and H.R. Baharvandi. 2009. Development of high-performance A356/nano-Al_2O_3 composites. *Materials Science and Engineering A* 518, no. 1–2 (August): 61–64. https://doi.org/10.1016/j.msea.2009.04.014.

Palanivel, R., I. Dinaharan, and R.F. Laubscher. 2020. Casting routes for production of metallic based composite parts. In *Encyclopedia of Materials: Composites*, ed. D. Brabazon, Amsterdam: Elsevier.

Rahimian, M., N. Ehsani, N. Parvin and H.R. Baharvandi. 2009. The effect of particle size, sintering temperature and sintering time on the properties of Al–Al_2O_3 composites, made by powder metallurgy. *Journal of Materials Processing Technology* 209, no. 14 (July): 5387–5393. https://doi.org/10.1016/j.jmatprotec.2009.04.007.

Rajan, T.P.D., R.M. Pillai and B.C. Pai. 1998. Reinforcement coatings and interfaces in aluminium metal matrix composites. *Journal of Materials Science* 33, (July): 3491–3503. https://doi.org/10.1023/A:1004674822751.

Ramkumar, K.R., I. Dinaharan. 2020. Accumulative roll bonding route for composite materials production. In: *Encyclopedia of Materials: Composites*, ed. D. Brabazon, Amsterdam: Elsevier.

Reza, M., K. Ardakani, S. Khorsand and S. Amirkhanlou. 2014. Application of compocasting and cross accumulative roll bonding processes for manufacturing high-strength, highly uniform and ultra-fine structured Al / SiCp nanocomposite. *Materials Science and Engineering A* 592, (January): 121–127. https://doi.org/10.1016/j.msea.2013.11.006.

Shanmugasundaram, T., B.S. Murty and V. SubramanyaSarma. 2006. Development of ultrafine grained high strength Al–Cu alloy by cryorolling. *Scripta Materialia* 54, no. 12 (June): 2013–2017. https://doi.org/10.1016/j.scriptamat.2006.03.012.

Shau, M.K., and R.K. Sahu. 2018. Fabrication of aluminum matrix composites by stir casting techniques and stirring process parameter optimization. In: *Advanced Casting Technologies*, ed. T.R. Vijayaram, London: IntechOpen. http://dx.doi.org/10.5772/int echopen.73485.

Srinivasarao, B., C. Suryanarayana, K. Oh-ishi and K. Hono. 2009. Microstructure and mechanical properties of Al–Zr nanocomposite materials. *Materials Science and Engineering A* 518, no. 1–2 (August): 100–107. https://doi.org/10.1016/j.msea.2009.04.032.

Su, H., W. Gao, Z. Feng and Z. Lu. 2012. Processing, microstructure and tensile properties of nano-sized Al_2O_3 particle reinforced aluminum matrix composites. *Materials and Design* 36 (April): 590–596. https://doi.org/10.1016/j.matdes.2011.11.064.

Valiev, R.Z., R.K. Islamgaliev and I.V. Alexandrov. 2000. Bulk nanostructured materials from severe plastic deformation. *Progress in Materials Science* 45 (January): 103–189. https://doi.org/10.1016/S0079-6425(99)00007-9.

Venkatesan, S., and M. Anthony Xavior. 2018. Tensile behavior of aluminum alloy (AA7050) metal matrix composite reinforced with graphene fabricated by stir and squeeze cast processes. *Science and Technology of Materials* 30, no. 2, (May–August): 74–85. https://doi.org/10.1016/j.stmat.2018.02.005.

Vigneshwaran, S., K. Sivaprasad, R. Narayanasamy and K. Venkateswarlu. 2018. Formability and fracture behaviour of cryorolled Al-3Mg-0.25Sc alloy. *Materials Science and Engineering A* 721 (April): 14–21. https://doi.org/10.1016/j.msea.2018.02.072.

Vigneshwaran, S., K. Sivaprasad, R. Narayanasamy and K. Venkateswarlu. 2019. Superior strength with enhanced fracture resistance of Al-Mg-Sc alloy through two-step cryo cross rolling. *Metallurgical and Materials Transactions A* 50A (May): 3265–3281. https://doi.org/10.1007/s11661-019-05253-6.

Wang, Y., M. Chen, F. Zhou and E. Ma. 2002. High tensile ductility in a nanostructured metal. *Nature* 419 (October): 912–915. https://doi.org/10.1038/nature01133.

Xue, Y., B. Jiang, L. Bourgeois, P. Dai, M. Mitome, C. Zhang, M. Yamaguchi, A. Matveev, C. Tang, Y. Bando, K. Tsuchiya and D. Golberg. 2015. Aluminum matrix composites reinforced with multi-walled boron nitride nanotubes fabricated by a high-pressure torsion technique. *Materials and Design* 88 (December): 451– 460. https://doi.org/10.1016/j.matdes.2015.08.162.

Yao, X., Y.F. Zheng, J.M. Liang and D.L. Zhang. 2015. Microstructures and tensile mechanical properties of an ultrafine grained AA6063–5vol%SiC metal matrix nanocomposite synthesized by powder metallurgy. *Materials Science and Engineering A* 648 (November): 225–234. https://doi.org/10.1016/j.msea.2015.09.059.

Zhao, Y.H., X.Z. Liao, Y.T. Zhu and R.Z. Valiev. 2005. Enhanced mechanical properties in ultrafine grained 7075 Al alloy. *Journal of Materials Research* 20 (March): 288–291. https://doi.org/ 10.1557/JMR.2005.0057.

Zhemchuzhnikova, D., A. Mogucheva and R. Kaibyshev. 2013. Mechanical properties and fracture behavior of an Al–Mg–Sc–Zr alloy at ambient and subzero temperatures. *Materials Science and Engineering A* 565 (March): 132–141. https://doi.org/10.1016/j.msea.2012.12.01.

8 Tribological Properties of Ceramic-Reinforced Metal Matrix Composite

V. Kavimani,[1]* P. M. Gopal,[1a] Titus Thankachan,[1b] K. Soorya Prakash[2c]

[1] Department of Mechanical Engineering, Karpagam Academy of Higher Education, Coimbatore, 641 021, India.

[2] Department of Mechanical Engineering, Anna University Regional Campus, Coimbatore, 642 047, India.

*Corresponding author: manikavi03@gmail.com

Other email IDs: [a]gopal33mech@gmail.com, [b]titusmech007@gmail.com, [c]k_soorya@yahoo.co.in

CONTENTS

8.1 INTRODUCTION

Technological advancements in modern industrial sectors have enabled material scientists worldwide to focus on revolutionary ideas in developing advanced materials that combine high performance with low density. Commercial non-ferrous materials such aluminium, magnesium, copper and titanium exhibit a lower density than commercial Fe-based material including cast iron, steel and others. Low-density materials, especially aluminium and magnesium, possess better strength-to-weight ratios, but these materials have inferior resistance to corrosion and wear, which limits a wider range of applications of these established and readily available materials (Thankachan et al. 2019, Kavimani et al. 2019, Soorya Prakash et al. 2017, Soorya Prakash et al. 2018). However, some basic and certain other functional and/or essential behaviours of these material types can be improved by various practices such as surface treatment, coating, alloy development, composite fabrication and others. Amongst these, composite

DOI: 10.1201/9781003109723-8

fabrication has been adopted by numerous material scientists as a recognized tool to improve the required properties of the base material; the composite is made by adding suitable hard ceramic materials such as SiC, TiC, WC, and B_4C. Composite materials are usually classified into three different types – polymer matrix composite (PMC), metal matrix composite (MMC) and ceramic matrix composite (CMC) – mainly based on the type of matrix material. Usually, MMCs show better values for specific strength and elastic modulus compared with the base matrix material. Depending on the nature of reinforcements, MMCs also display higher thermal stability, damping capacity, excellent wear and corrosion resistance (Kavimani et al. 2020, Zhou et al. 2020). Recent studies reveal that addition of nano-sized reinforcement particles improves the modulus and strength of base matrix materials without affecting their bulk properties. Nevertheless, attaining uniform distribution of nano reinforcement in matrix material was quite difficult owing to its clustering tendency and strong van der Waals interactions. Recently, extensive research has been done on MMCs to overcome technical challenges such as in material design, reducing the process cost, controlling interfacial bonding strength between base material and reinforcements and composite characterization. In general, MMC design strives to improve strength by selecting suitable reinforcement. For example, aluminium-based materials are ductile with better formability and toughness, and the addition of reinforcement in aluminium base matrix increases the hardness and thermal stability. The increased resistance to corrosion and wear further decreases the thermal expansion. The improved properties of these MMCs have gained major attention in defence, building and transport industries (in rail transport mainly) for fabricating mobility components that include brake drum, engine components, compressor piston unit, gears, cylinder liner and others. Better understanding of the wear behaviour of MMCs triggered the researchers to concentrate on further progress in compositional design, wear parameters and reinforcement selections (the main criteria of MMCs) based on their promising commercial applications (Ghandvar et al. 2016).

The tribological behaviour of MMCs plays a significant role in their application in wear-based environments. Tribology is the study of wear that takes place during the relative motion of two surfaces under various adverse service conditions. Based on the wear mechanism, wear conditions have been classified as adhesive wear, abrasive wear, fretting wear, delamination wear, corrosive wear, erosive wear and fatigue wear. In general, at higher loads and sliding velocity, delamination is the dominant mechanism, whereas at lower loads and sliding velocity abrasion wear dominates. Wear loss of MMC depends on uniform distribution of reinforcement, percentage of reinforcement, size of reinforcement, interfacial attraction between reinforcement and matrix material, sliding velocity, applied load and sliding distance. Depending on the wear mechanism, the wear resistance of MMC may be improved by reinforcement with ceramic particles that include SiC, TiC, WC and B_4C and with solid lubricants and carbon-based material such as graphite, graphene, CNT, BN and MoS_2. Wear resistance of MMCs significantly increases for any incremental value of hardness of base material that shields the composite surface from ploughing during sliding wear conditions (Soorya Prakash et al. 2016). The addition of hard ceramic particles in the base matrix leads to increases in the coefficient of friction (COF) as these ceramic particles detach from matrix material at higher wear conditions, spread over the wear track and initiate third-body abrasion that results in severe damage over the composite surface. Addition of solid lubricants in matrix material results in the formation of a tribolayer or solid lubricating layer over the composite surface that decreases the COF of MMC and increases the wear resistance. In the next section, the tribological behaviour of MMC containing various classifications of reinforcements is discussed systematically. The section focuses on lightweight composites, namely, Al MMC, Mg MMC and other composites including copper- and titanium-based MMCs (Carvalho et al. 2015; Raghav et al. 2018).

8.2 INFLUENCE OF CERAMIC PARTICLES ON THE TRIBOLOGICAL BEHAVIOUR OF MMC

8.2.1 EFFECT OF CERAMIC PARTICLES ON AL MMC

Aluminium and its alloys have a good strength-to-weight ratio; however, incorporation of ceramics like SiC particles (10–25 wt.%) into Al matrix drastically improves wear resistance (Table 8.1). These hard ceramic particles uniformly dispersed over the matrix surface help increase the hardness of matrix material. The addition of SiC is a good example of incremental trends in wear resistance and COF of composite. There are two major reasons for improved wear resistance: (1) Presence of SiC particles in the surface of a pin in a tribometer are significantly harder than the disc surface. (2) SiC particles do not deform plastically near the pin surface; such deformation would result in formation of scratches over the disc surface which might get oxidized during wear conditions, thus increasing the heat near the contact surface. These facts result in oxidation of the material near the pin surface, tending to form a mechanically mixed layer between disc and pin that further reduces the contact surface between them. The material also has the capacity to withstand higher frictional heat without localized adhesion, thus slowing wear rate reductions (Rao and Das 2011).

Likewise, aluminium metal matrix composite reinforced with hybrid SiC-graphene oxide processed by powder metallurgy exhibits better tribological properties compared with basic Al-Si alloy because the combination of hard ceramic particles and solid lubricants decreases friction coefficient values. This developed hybrid composite (10 wt.% SiC–0.3 wt.% reduced graphene oxide (RGO)) exhibits a 15.9% decrement in COF value and a 32.7% improvement in hardness compared to Al MMC reinforced with 10 wt.% SiC (Zeng et al. 2018). Al355 alloy matrix reinforced with SiC nanoparticles (5 vol.%) that are encapsulated by graphene (1 vol.%) shows better tribological properties. Here the wear rate and COF of developed composite decrease 98.0% and 35.9%, respectively, compared to base Al matrix. Graphene-encapsulated SiC nanoparticles from the composite smeared on the wear track and formed a micro tribolayer over the pin surface during the sliding wear, thus increasing wear resistance (Zhang et al. 2019). Addition of SiC (up to 15%) in B_4C–Al6061 matrix increased the hardness and wear resistance up to 44%. Addition of the hybrid form of SiC (3 wt.%) and B_4C (9 wt.%) in Al6061 matrix undergoes various wear mechanisms depending on load applied and sliding distance (Halil et al. 2019). Better improvement in abrasive wear resistance was evidenced in LM13 and ADC12 alloy while incorporating 7 wt.% SiC and 7 wt.% TiB_2. These hard particles inhibit the formation of flank wear and crater at the time of sliding over an abrasive surface (Yadav et al. 2020). Addition

TABLE 8.1
Hardness and Wear Behaviour of Ceramic-Reinforced Aluminium Composite

Reinforcement Percentage	Composites	Hardness	Wear behaviour	COF
Nano zirconia: 10 wt.%	Al	Al: ~38 HRC	WI = 19.01 mg	–
Graphite: 2–8 wt.%	Al + 10ZrO$_2$	Al + 10ZrO$_2$: ~53 HRC	WI = 7.29 mg	
	Al + 10ZrO$_2$+ 8Gr	Al + 10ZrO$_2$ +6 Gr: ~46 HRC	WI = 3.14 mg	
		Al + 10ZrO$_2$ + 8Gr: ~41 HRC	WI = 4.57 mg	
B$_4$C: 4–12 wt.%	Al7075		WI = 0.07	–
MoS$_2$: 3 wt.%	Al7075 + 12B$_4$C + 3MoS$_2$		WI = 0.03	
B$_4$C: 5–20 vol.%	Al7075	Al7075: ~120 BHN	Wr = 0.30 mg/m	
	Al7075 + 20B$_4$C	Al7075 + 20B$_4$C: ~240 BHN	Wr = 0.01 mg/m	
SiC: 5–20 vol.%	Al	Al: 51 HV	Wr = 0.04 mm^3/m	1.62
MoS$_2$: 2–6 vol.%	Al + 10SiC	Al + 10SiC: 54 HV	Wr = 0.017 mm^3/m	0.43
	Al + 20SiC	Al + 20SiC: 53 HV	Wr = 0.010 mm^3/m	0.72
	Al + 10SiC + 6MoS$_2$	Al + 10SiC + 6MoS$_2$; 42 HV	Wr = 0.003 mm^3/m	0.6
SiC: 10 wt.%	LM30		Wr = 9.07 × 10^{-11} mm^3/m	
	LM30 + 10SiC (Abrasive wear)		Wr = 3.22 × 10^{-11} mm^3/m	
Si$_3$N$_4$: 4–12 wt.%	Al7075	Al7075: 161 HV	WI = 97 microns	0.49
	Al7075 + 12Si$_3$N$_4$	Al7075 + 12Si$_3$N$_4$: 203 HV	WI = 62 microns	0.43
TiB$_2$ (100%): 18.2 vol.%	Al6082		Wr = 23.75 × 10^{-5} mm^3/Nm	
TiB$_2$ (50%) + BN (50%): 18.2 vol.%	Al6082 + 18.2TiB$_2$		Wr = 15.50 × 10^{-5} mm^3/Nm	
	Al6082 +18.2(TiB$_2$ + BN)		Wr = 13.00 × 10^{-5} mm^3/Nm	
SiC: 10–25 wt.%	Al7010		Wr = 2.20 ×10^{-12} m^3/m	0.450
	Al7009		Wr = 2.36 × 10^{-12} m^3/m	0.451
	Al2024		Wr = 2.92 × 10^{-12} m^3/m	0.458
	Al7010 + 25 wt.%		Wr = 1.1 × 10^{-12} m^3/m	0.78
	Al7009 + 25 wt.%		Wr = 1.35 × 10^{-12} m^3/m	0.74
	Al2024 + 25 wt.%		Wr = 1.63 × 10^{-12} m^3/m	0.71
Al$_2$O$_3$: 2–4 vol.%	Al6061 –	Al6061: 104 HV	Wr = 0.052 × mm^3/m	–
SiC: 4–8 vol.%	Al6061 + 4Al$_2$O$_4$ + 8SiC	Al6061 + 4Al$_2$O$_3$ + 8SiC: 132 HV	Wr = 0.002 × mm^3/m	
Nano TiC: 5 wt.%	Al7050	Al7050: 136 HV	Wr = 3.89 × mm^3/km	0.87
Gr: 2–10 wt.%	Al7050 + 5TiC	Al7050 + 5TiC: 181 HV	Wr = 1.11 × mm^3/km	0.59
	Al7050+5 TiC + 5 Gr	Al 7050 + 5TiC + 5Gr: 165 HV	Wr = 0.52 × mm^3/km	0.52
ZrC: 5–15 wt.%	Al-12Si		Wr = 6.13 × 10^{-6} mm^3/Nm	0.60
	Al-12Si + 15ZrC		Wr = 1.20 × 10^{-6} mm^3/Nm	0.54

Note: BHN, Brinell hardness number; HRC, hardness on the Rockwell C scale; HV, hardness on the Vickers scale; WI, wear loss; Wr, wear rate.

of hybrid reinforcement of B$_4$C (10 wt.%) and graphite (5 wt.%) in Al6061 and Al7075 matrix improve the wear resistance of the base matrix material (Baradeswaran et al. 2014). Usage of these hybrid combinations of reinforcement improves the hardness of matrix material up to 8% for Al 6061 matrix and ~10% improvement in Al7075. This fact signals a positive trend in the wear resistance of developed composite. Presence of this hard ceramic reinforcement enhances the plastic-forming capability of matrix material, and graphite acts as an active insulating layer that in turn decreases the contact angle between the pin and disc surface, thus decreasing the wear rate. Further, such reinforcement helps decrease the tendency for deeper grooves to form on the wear surface of composites by means of the oxidative wear mechanism. Other researchers incorporated B$_4$C in Al359 alloy matrix to produce a homogeneous mixture suitable for wear resistance application. B$_4$C was found to increase the micro hardness of homogeneous composite (i.e., composite with uniform properties throughout) up to 32.6% and heterogeneous composite up to 37.9%. Increasing the percentage of B$_4$C increases the wear resistance; heterogeneous composite undergoes abrasive and delamination wear mechanisms, and homogeneous composite predominantly undergoes fretting mechanism (Radhika et al. 2020). Figure 8.1 depicts various wear mechanisms of hybrid B$_4$C-SiC composite. At a lower applied load, increase in sliding distance causes abrasion wear, which initiates the formation of deeper grooves over the worn-out surface. Hence increases in sliding distance can increase the temperature near the pin surface, a factor in abrasion wear.

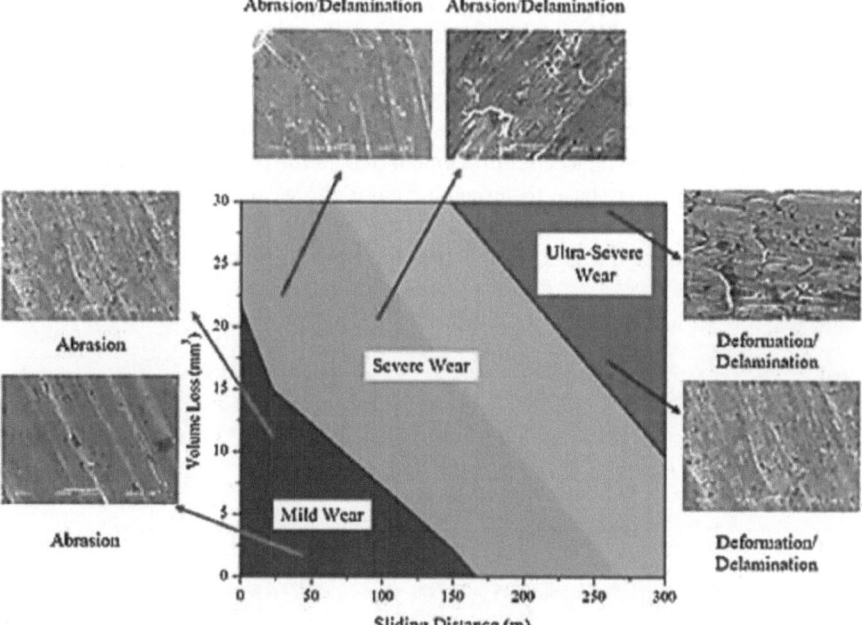

FIGURE 8.1   Wear mechanisms in ceramic matrix composite.

Source: Halil et al. (2019).

Increase in applied load results in the delamination mechanism of wear, which is evident from cracks and material removal or ploughings over the worn-out surface at the higher loading condition; matrix material undergoes a work-hardening effect that decreases the interfacial bonding between ceramic particles and Al matrix, resulting in crack formation. Addition of oxide-coated ceramic whisker in aluminium matrix material increases the abrasive wear resistance of Al MMC. Comparative wear analysis on Al MMC was done with copper oxide (CuO)–coated aluminium borate whiskers as the reinforcement for Al6061 matrix. Results revealed that CuO-coated whiskers exhibit 6% improvement in hardness and minimal abrasive mass loss compared with uncoated aluminium borate whiskers as reinforcement for Al MMC. These coated ceramic whiskers act as a load-bearing component that reduces the stress-strain gradient near the pin surface; furthermore, the oxide coating of ceramic particles enhances the interface wettability, thus increasing the wear resistance (Yue et al. 2017). A hybrid combination of mica and SiC assists in better improvement of wear resistance in Al356 matrix. Addition of mica (0–6 wt.%) along with SiC (10 wt.%) increases the hardness of Al MMC to about six times that of the base matrix material. The presence of mica initiates the formation of oxide films; oxide has a lower shear strength as well as lower ductility, which limits further oxide growth that tends to separate the contact between composite pin and counterface disc, thus decreasing the COF (Rajmohan et al. 2013). Addition of alumina up to 10 vol.% in Al6061 matrix acts as a barrier against deformation and abrasion; this fact prevents the hard asperities from causing abrasive wear. Incorporation of alumina initiates formation of metallic films over the pin surface that helps improve adhesive wear resistance by delaying the conversion from mild wear to severe wear. Alumina-reinforced Al MMC has valuable features for brake pad application in automobiles because oxygen presence during wear causes continuous re-oxidation (Pramanik 2016).

Figure 8.2a–b is a schematic representation of wear mechanisms that may occur in the composites and matrix. The reinforcement of ceramic particles of defined size and quantity into the matrix (base material) increases the wear resistance. The decrease in the probability of wear loss may well be attributable to the sustainment of oxide layer formation. In the unreinforced matrix, the creation of an oxide film may occur;

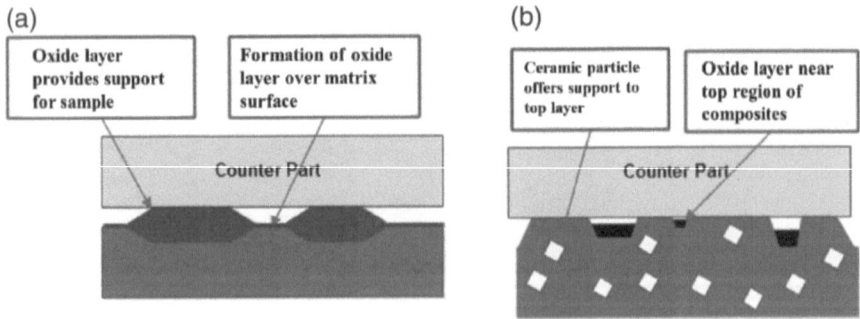

FIGURE 8.2 Wear mechanism in (a) composite, (b) matrix.

however, the support for the growth of the oxide layer is from the metal surface due to the surface hardening effect. Any such phenomenal facets will obstruct the layer formation of oxides, resulting in an enhanced material removal rate (MRR) across the pin surface. All the removed wear particles will get relocated over and above the disc track; eventually it will get increased appropriate to cyclic loading and further avoiding hard ceramics to plough out the bigger materials encountering metal debris accumulation. As far as composites are concerned, material oxidation commonly occurs nearer to the contact surface, which would increase oxide layer formation, increasing the resistance to wear.

Usage of nano alumina particles (5 wt.%) in Al matrix increases the hardness of the base material up to 94.2% and provides a 62.4% improvement in wear resistance. Al MMC processed with stir casting and varying addition of TiC (3–7 vol.%) in Al6061 matrix shows increased hardness and wear resistance. The wear rate of developed composite decreases linearly with respect to increasing volume percentage of TiC, and improved wear resistance was observed compared to the matrix material. Scratches and parallel grooves are formed over the worn surface as a result of abrasion actions. Addition of TiC reduces oxidation wear; however, increases in the sliding velocity result in adhesion wear (Gopalakrishnan and Murugan 2012). Usage of hybrid alumina (2–8 wt.%)–graphite (5 wt.%) reinforcement in Al7075 matrix displays a wear rate that is 36% lower than that of base material. Presence of alumina and graphite in base material forms a solid lubricant layer over the pin surface that decreases the COF up to 51%. This hybrid reinforcement results in occurrence of the delamination and abrasion wear mechanisms (Baradeswaran and Perumal 2014). Incorporation of silicon nitride particles (4–12 wt.%) in Al7075 matrix improves its tribological behaviour. Increasing the weight percentage of silicon nitride improves the hardness of the base material up to 26%; according to Archard's law, an increase in material hardness increases the wear resistance. The addition of silicon nitride increases the wear resistance by up to 37.17% and decreases the COF 11.03% compared with Al7075 matrix. During wear conditions, silicon nitride forms a tribo-chemical reaction layer (silicon oxide layer) over the pin surface that results in a lower COF (Mistry and Gohil 2019). Addition of TiB_2 (up to 10 wt.%) in Al–Si alloy improves the hardness of base material up to 45%. TiB_2 accelerates the grain refinement that restricts the growth of Si particles into needles of Si, thus increasing the hardness. Delamination wear was the dominant wear mechanism in TiB_2-reinforced Al MMC because these particles create higher stresses near the contact surface. This aids in removal of iron particles from the counterpart disc surface and converts the coarse grain wear debris to finer debris that forms a harder mechanical mixed layer between pin and disc, thus improving the wear resistance. Hence an increase of TiB_2 particles decreases the wear rate and increases the COF compared with that of base Al–Si alloy (Mandal et al. 2009).

However, reinforcing TiB_2 (up to 10 wt.%) in Al–4Cu alloy has shown better improvement of tribological properties. Addition of TiB_2 shows drastic improvement in hardness, up to 61%. During wear conditions, TiB_2 forms Al–Cu–Fe–Ti intermetallic bonds and also decreases α-Al cell size with respect to the eventual increase in the final weight percentage that helps in attaining better wear resistance (Mandal et al. 2007).

Powder metallurgy–processed ZrC (up to 15 wt.%) reinforcement of Al-12Si composite exhibits better tribological properties. Increase in addition of ZrC at adverse wear conditions, namely, greater applied load and sliding distance, results in lower COF and specific wear rate. Presence of ZrC decreases the chance of deformation and crack formation of Al matrix at higher loading conditions, problems that decrease the wear resistance of composite material (Lenin et al. 2018). ZrB_2 displays excellent wear resistance properties as reinforcement of Al6061 matrix. Addition of ZrB_2 up to 10% shows nonlinear decrement in wear rate, owing to the variation in thermal expansion between ZrB_2 and matrix that results in high density dislocation in the composite phase. Further, these ZrB_2 particles exhibit interfacial bonding with the base matrix, indicating a higher load-bearing capability and thereby reducing the chance for plastic deformation near the pin surface, which in turn increases the wear resistance (Dinaharan and Murugan 2012).

8.2.2 EFFECT OF CERAMIC PARTICLES ON MG MMC

Incorporation of hard ceramic particles in Mg-based material helps improve the wear resistance of the base material (Table 8.2). Addition of 5 wt.% of SiC particles in WE43 Mg matrix shows better anti-wear performance. Usage of SiC particles restricts the plastic deformation and increases the load-bearing capability of base materials under dry sliding condition. As a result of these superior properties, the wear rate decreases by as much as 81% and COF values decrease up to 56% compared with the base material. SiC particles constrain crack propagation formed by fatigue and inhibit the formation of a brittle delaminating lubrication layer, thus improving the wear resistance of the base material (Xie et al. 2020). Addition of SiC particles up to 15 vol.% in Mg matrix improves tribological performance at both room and elevated temperatures (up to 200°C). An increase in SiC percentage decreases the wear rate. When the wear conditions are observed at room temperature, oxidation wear is the dominant mechanism; any further increase in temperature results in plastic deformation and the adhesion wear mechanism. At higher temperature wear conditions, the presence of SiC initiates the formation of a work-hardened layer over the pin surface that acts as a tribo-layer, thus lowering the wear rate for Mg MMC compared with the base material (Labib et al. 2016). Incorporation of a hybrid of SiC (up to 10 wt.%) and graphite (up to 10 wt.%) in Mg matrix helps increase the wear resistance, making the composite a promising candidate for higher temperature applications (in brake pads and so on). A combination of 10 wt.% of SiC and 5 wt.% of graphite increases the hardness of Mg up to ~166% and further shows that SiC addition up to 10 wt.% increases the wear resistance; however, addition of graphite above 5 wt.% causes brittle fracture near the pin surface, thus increasing the wear rate. Any increase in graphite (up to 10 wt.%) decreases the COF value of Mg. For any composite materials, size of reinforcement acts as one of the key factors governing essential properties of developed composites (Soorya Prakash et al. 2016). Addition of varying particle sizes of SiC (5, 10, 20 μm) up to 20 wt.% improves the tribological behaviour of AZ91D Mg alloy. Smaller particle size assists in increasing the micro hardness of developed composite; addition of 15 vol.% of 5 μm SiC results in 103% improvement in hardness

TABLE 8.2
Hardness and Wear Behaviour of Ceramic-Reinforced Magnesium Composite

Reinforcement Percentage	Composites	Hardness	Wear Behaviour	COF
SiC: 5–10 vol.%	AZ91	—	Wr = .013 mm^3/m	0.23
	AZ91 + 5 vol.%		Wr = 0.02 mm^3/m	0.32
Nano Al$_2$O$_3$: 0.3 wt.%	AZ31	AZ31: 82 HV	Wl = 0.018 mg	0.48
CNT: 0.3 wt.%	AZ31 + 0.1Al$_2$O$_3$ + 0.2CNT	AZ31 + 0.1Al$_2$O$_3$ + 0.2CNT: 104 HV	Wl = 0.015 mg	0.45
TiB$_2$: 4–8 wt.%	RZ5	RZ5: 64 HV	VWl = 6.42 mm^3	—
	RZ5 + 8TiB$_2$	RZ5 + 8TiB$_2$: 86 HV	VWl = 5.37 mm^3	
B$_4$C: 1.5 wt.%	AZ91D	AZ91D: 20.5 BHN	Wl = 0.02 g	0.42
Gr: 1.5 wt.%	AZ91S + 1.5 B4C	AZ91 + 1.5 B$_4$C: 27.1 BHN	Wl = 0.012 g	0.45
	AZ91 + 1.5B4C + 1.5Gr	AZ91 + 1.5B$_4$C + 1.5Gr: 22.5 BHN	Wl = 0.019 g	0.43
TiC: 5–20 wt.%	Mg	Mg: 29 HV	Wl = 0.010 g	0.005
	Mg + 20TiC	Mg + 20TiC: 103 HV	Wl = 0.006 g	0.010
Nano WC: 1–2 wt.%	AZ31	AZ31: 69 HV	Wr = 27.31 × 10^{-12} kg/Nm	0.85
	AZ31 + 2WC	AZ31 + 2WC: 106 HV	Wr = 3.03 × 10^{-12} kg/Nm	0.83
ZrB$_2$: 5–20 wt.%	AZ31	AZ31: 41 BHN	—	0.35
	AZ31 + 20ZrB$_2$	AZ31 + 20ZrB$_2$: 95 BH		0.45

Note: BHN, Brinell hardness number; HRC, hardness on the Rockwell C scale; HV, hardness on the Vickers scale; VWl, volumetric wear loss; Wl, wear loss; Wr, wear rate.

of base material, and addition of 20 vol.% of 10 μm SiC increases the hardness 91%. This fact results in incremental wear resistance, and delamination wear was the dominant mechanism for wear in fabricated composite (Liu et al. 2020).

Addition of nano alumina (up to 5 wt.%) in Mg matrix improves the corrosion and wear resistance of Mg; a 50% decrement in wear rate was observed for developed Mg MMC. Hence reinforcement particles exhibit better interfacial bonding that hinders delamination and plastic deformation, thus reducing the wear rate; abrasive wear was the dominant mechanism for developed composite. From Archard's law, it could well be stated that better hardness results in better resistance to abrasion (Rahmani et al. 2020).

Usage of 1.5 wt.% of nano alumina in AZ31B matrix decreases its wear rate up to 20%. The composite undergoes predominantly adhesion and abrasion wear mechanisms (Srinivasan et al. 2012). Reinforcing hybrid TiC–TiB$_2$ in AZ91 Mg matrix has better improvement in both microstructural and tribological properties. Usage of these hybrid particles improves the resistance of material to wear from delamination, oxidation, melts and abrasive wear. Hybrid reinforcements have high thermal stability that decreases the chance of thermal softening effects, thus increasing the wear resistance compared to the particular base material and decreasing the wear rate by up to 120%. Formation of hard Mg17Al12 precipitates over the pin surface inhibits void formation and crack propagation near the contact surface and thus is likely to increase the wear resistance. Higher hardness of reinforcement generates high frictional heat near the pin surface that increases the development of an oxide layer between the metallic surfaces, in turn reducing the COF (Sahoo and Panigrahi 2019). An improvement of 44% in hardness has been attained for AZ91 Mg composite when reinforced with B$_4$C.

Mg-SiC composite undergoes abrasion and delamination wear mechanisms at lower sliding velocity, resulting in oxide layer generation over the sample composite that is highly dependent on the surface temperature near the pin surface (Figure 8.3).

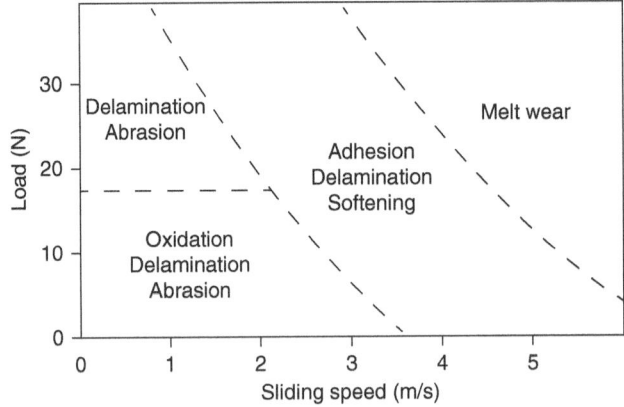

FIGURE 8.3 Variation of wear mechanism in ceramic composite with respect to speed.

Source: Lim and Gupta (2003).

A higher applied load removes the oxide layer, and at that time it is expected that oxidation wear would change to delamination wear, which leads to abrasion. At this juncture, the wear resistance of the sample composite decreases mainly because of bond weakening (mechanical) between the matrix material and reinforcements, initiating propagation of cracks. Further increments in sliding velocity result in a transition from delamination to adhesion wear; however, the greater hardness of the sample composite increases load resistance potentials, which yield better resistance towards adhesion. An even higher range in sliding velocity ends in thermal softening, which leads to composite failure. Addition of B_4C in Mg matrix decreases the grain size of the composite, thus increasing the hardness. Developed composite undergoes abrasive and severe adhesive wear mechanisms at lower sliding velocity, and oxidative and delamination wear are the major wear mechanisms that happen at higher sliding velocity (Patle et al. 2020). Addition of Ti (4 wt.%) particles in $Mg–B_4C$ composite increases the hardness and wear resistance of the base material, and the resulting composite undergoes delamination and adhesion wear mechanisms (Yao et al. 2015). Increasing the percentage of TiB_2 (up to 15%) in Mg matrix composite improves its hardness up to 156% and increases the wear resistance up to 41%. Loose and non-adherent wear debris formation in wear tracks are due to the higher hardness of TiB_2 and result in occurrence of the abrasive wear mechanism during wear condition (Stalin et al. 2020). Improvement in tribological behaviour was exhibited by AZ91D Mg metal matrix composite with the addition of TiC (up to 20 wt.%) particles. Increases in TiC increase the hardness of base material up to 19% and wear rate decreases up to 37.5%. Presence of TiC in Mg matrix increases the load-bearing capacity of the base material, thus increasing the hardness. The developed composite undergoes abrasion wear at lower loading conditions and experiences delamination and oxidation wear at higher loading conditions (Dash et al. 2020). Mg composite reinforced with SiC (up to 20 wt.%) and tungsten disulphide (10 wt.%) has better self-lubricating behaviour and better tribological properties compared with Mg and AZ31 alloy. The composite increases the hardness of the base material to ~110%, and at normal wear condition (i.e., room temperature) the wear rate of developed composite decreases up to 72% and the COF decreases to 37.5% of that of the base material; however, an 88% decrement in wear loss and a 54.5% decrement in COF at high temperature wear condition (110°C) are witnessed. These hybrid reinforcements significantly reduce the depth of plastic deformation near the pin surface and thus reduce the wear rate. The crystal structure of the reinforcements lowers the shear strength, which results in a minimal COF. Adhesion wear is one of the dominant mechanisms that occurs in MMC at higher temperature wear conditions (Zhu et al. 2020). Addition of nano WC (2 wt.%) particles to AZ31 alloy matrix increases the abrasive wear resistance of the base material up to 56%, and a 53% improvement in microhardness was reported (Banerjee et al. 2020). Tribological behaviour of AZ31 alloys could be improved by adding hard ceramic particles such as B_4C, WC and nano-ZrO_2 in both mono and hybrid forms.

The hybrid forms $B_4C + ZrO_2$ and $WC + ZrO_2$ assist in increasing the hardness and wear resistance of AZ31 base material. In mono form, addition of B_4C, WC or ZrO_2 increases the hardness of base matrix material up to 112%, 100% or 120%, respectively. Addition of hybrid reinforcement with $B_4C + ZrO_2$ or $WC + ZrO_2$ shows drastic

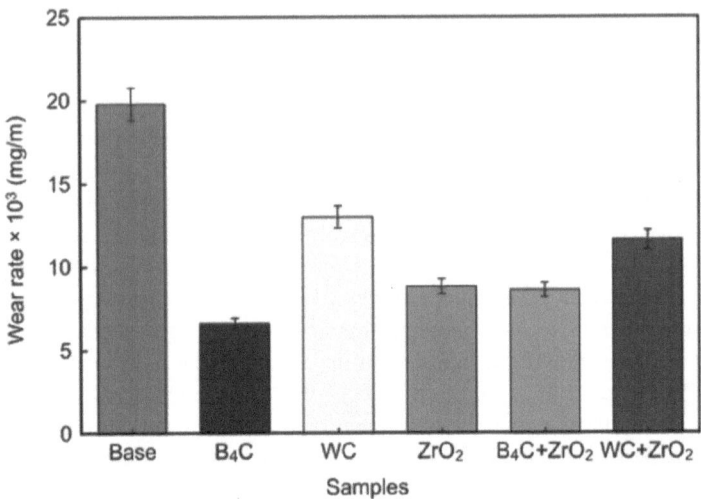

FIGURE 8.4   Wear behaviour of magnesium (AZ31) composite with respect to reinforcement.

Source: Jalilvand and Mazaheri (2020).

improvement in micro hardness of 128% or 114%, respectively. The result is a decremented wear rate compared with the base material: the addition of mono B_4C, WC or ZrO_2 results in a wear rate decrease of 70%, 30% or 55%, respectively (Figure 8.4). Likewise for hybrid reinforcement, 63% and 40% decrements in wear rate were noted for composites reinforced with $B_4C + ZrO_2$ and WC+ ZrO_2, respectively. COF values of base material decrease with respect to addition of reinforcements on the order of 22%, 38%, 62%, 46% and 52% for B_4C, WC, ZrO_2, $B_4C + ZrO_2$ and WC+ ZrO_2, respectively. In this research abrasive wear was the dominant wear mechanism for developed composite (Jalilvand and Mazaheri 2020).

8.2.3   Effect of Ceramic Particles on Cu- and Ti-Based MMCs

Cu-based composite displays improved resistance to sliding wear with respect to nature of reinforcement (Table 8.3). Usage of ZrO_2 particles in Cu matrix composite improves its tribological performance and maintains homogeneity of the Cu matrix with stronger bonding strength, which in turn increases the hardness and provides a high COF. Wear performance experimentation was conducted based on varying the braking energy density for Cu matrix composite reinforced with ZrO_2 in two phases, namely, cubic and monoclinic. At severe braking conditions monoclinic ZrO_2 undergoes continuous deformation and forms unstable, thicker, plastically deformed layers. These layers have low hardness and could be crushed under a continuous cyclic load that initiates crack formation.

Addition of hard ZrC in Cu matrix produces greater hardness to facilitate resistance to wear (Figure 8.5). During hard wear, ZrC spreads all through the pin surface, acting as a stable lubricant in layers that help decrease the wear rate. Good interfacial

TABLE 8.3

Effect of Ceramics on Copper and Titanium Matrix Composite

Reinforcement Percentage	Composites	Hardness	Wear Behaviour	COF
TiC: 5–15 vol.%	Cu	Cu: 65 HV	$Wr = 6.91 \times 10^{-4}$ mm^3/m	–
Gr: 5–10 vol.%	Cu-15TiC + 5Gr	Cu-15TiC + 5Gr: 98.8 HV	$Wr = 1.16 \times 10^{-4}$ mm^3/m	0.48
	Cu-15TiC + 10Gr	Cu-15TiC + 10 Gr: 89.2 HV	$Wr = 0.40 \times 10^{-4}$ mm^3/m	0.85
SiC: 4 wt.%	Cu	Cu: 94 HV	$Wr = 6.25 \times 10^{-4}$ mm^3/m	0.46
CNT: 1–4 wt.%	Cu + 4SiC	Cu + 4SiC: 131 HV	$Wr = 4.37 \times 10^{-4}$ mm^3/m	0.78
	Cu + 4SiC + 3CNT	Cu + 4SiC + 3CNT: 175 HV	$Wr = 2.4 \times 10^{-4}$ mm^3/m	0.69
SiC: 10–20 wt.%	Cu	Cu: 84.5 HV	$Wl = 1.65$ mg	0.82
	Cu + 10SiC	Cu + 10SiC: 122.3 HV	$Wl = 0.98$ mg	
	Cu + 20SiC	Cu + 20SiC: 138.6 HV	$Wl = 0.68$ mg	
TiB$_2$: 6–18 vol.%	Cu	Cu: 64 HV	$Wr = 24 \times 10^{-5}$ mm^3/m	0.82
	Cu + 18TiB$_2$	Cu + 18TiB$_2$: 128 HV	$Wr = 14 \times 10^{-5}$ mm^3/m	0.60
AlN 25 wt.% + BN 75 wt.%: 5–15 vol.%	Cu	Cu: 60 HV	$Wr = 0.85 \times 10^{-5}$ mm^3/Nm	0.73
	Cu + 15(AlN + BN)	Cu + 15(AlN + BN): 77 HV	$Wr = 0.59 \times 10^{-5}$ mm^3/Nm	0.64
B$_4$C: 4–8 wt.%	Cu–Cr	Cu–Cr: 69 HV	$Wl = 0.012$ g	0.70
	Cu–Cr-8B$_4$C	Cu–Cr-8B$_4$C: 109 HV	$Wl = 0.007$ g	0.45
B$_4$C: 0.94–3.74 wt%	Ti	Ti: 134 HRB	$Wl = 0.58$ mm^3	0.28
	Ti + 3.74B$_4$C	Ti + 3.74B4C: 306 HRB	$Wl = 0.13$mm^3	0.61
TiC: 3–30 vol.%	Ti	Ti: 4.7 GPa	$K = 0.0013$	0.78
	Ti + 30TiC	Ti + 30TiC: 14.1GPa	$K = 0.002$	0.70
Nano ZrO$_2$: 3–10 wt.%	Ti	Ti: 300 HV	$Swr = 3 \times 10^{-6}$ g/m	30.1
	Ti +10ZrO$_2$	Ti +10ZrO$_2$: 570 HV	$Swr = 1.1 \times 10^{-6}$ g/m	5.60
TiC: 2.5 – 40 vol.%	Ti	Ti: 297 HV	$Ml = 41$ mg	9.30
TiB$_2$: 5–40 vol.%	Ti + 40TiC	Ti + 40TiC: 413 HV	$Ml = 7.60$ mg	2.80
Si$_3$N$_4$: 2.5–20 vol.%	Ti + 20TiB$_2$	Ti + 20TiB$_2$: 610 HV	$Ml = 12.61$ mg	
	Ti + 20Si$_3$N$_4$	Ti + 20Si$_3$N$_4$: 1199 HV	$Ml = 3.7$ mg	

Note: HRB, hardness on the Rockwell B scale; HV, hardness on the Vickers scale; K, wear coefficient; Ml, mass loss; Swr, specific wear rate; Wl, wear loss; Wr, wear rate.

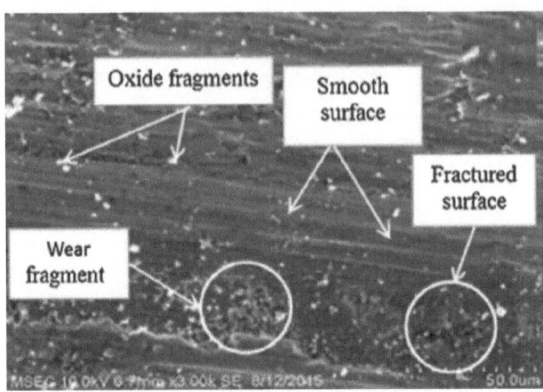

FIGURE 8.5   Influence of ceramics on copper matrix composite.

Source: Selvakumar and Singh (2016).

attractions amongst nano ZrC particles in matrix constantly protect the pin surface, resulting in a smoother surface that demonstrates the improved wear resistance and a lower COF of sample composite (Selvakumar and Singh 2016). Cubic ZrO_2–reinforced Cu matrix material forms three different layers over the material surface, namely, the plastic deformation layer, the matrix layer and the mechanical mixture layer. These multilayers have a high capability to withstand more shear and compressive stresses, thus resisting any plastic deformations near the material surface. Oxidation-induced delamination wear is the major mechanism of wear in ZrO_2-based composite. Incorporation of SiC (up to 4 wt.%)–graphite (up to 4 wt.%) hybrid reinforcement in Cu matrix shows better wear resistance. Here an increase in percentage of graphite decreases the hardness of composite but increases the wear resistance owing to its solid lubricating properties. Uniform dispersion of reinforcement prevents plastic deformation of the wear surface, and addition of reinforcements up to 8% lowers the wear rate. A hybrid combination of nano alumina (up to 4 wt.%) with graphite (up to 6 wt.%) reinforcement in Cu matrix displays better improvements in wear, thermal and corrosion properties. Uniform distribution of nano Al_2O_3 in the base matrix allows the dislocation movements that prevent the occurrence of plastic deformation during induction process. Increment in graphite content also increases the hardness of base material; increases of up to 4 wt.% in graphite improve the micro hardness equal to the combination containing nano Al_2O_3. A combination of 4 wt.% of alumina and 2 wt.% of graphite increases the hardness of copper up to 69.9 HV and has a lower wear rate compared with other combinations of composite samples (Venkatesh and Rao 2018). Addition of nano ZrO_2 particle up to 10 vol.% in Ti matrix increases the hardness of the composite up to 96%. Uniformly dispersed ZrO_2 particles act as protuberances over the pin surface and restrict the plastic deformation near the contact surface, thus tending to increase the wear resistance. During wear conditions ZrO_2 particles pull out from Ti and slip near the contact surface between disc and pin in the form of a lubricating film that decreases COF (Abd-Elwahed et al. 2020). Addition of hard B_4C (10 wt.%) particles in Ti alloys results in drastic improvement in hardness and wear properties compared with the base material (Soorya Prakash et al. 2018).

8.3 CONCLUSION

Composite material undergoes various wear mechanisms with respect to loading condition. It is important to investigate the tribological behaviour of matrix material with different reinforcement particles under the same wear condition in order to gather more evidence for selecting a suitable composite for a specific application or developing composite material for particular mobility components. Extensive studies of the tribological behaviour of Al- and Mg-based composites have been carried out by researchers; however, only a few works are focused on Cu- and Ti-based ceramic metal matrix composites. Hence studies of these composites will definitely help improve the possible range of applications for components with enhanced wear resistance. Addition of ceramic particles increases the wear resistance needed for major commercial application; however, efforts have to be made to decrease the cost of reinforcement in conjunction with low-cost fabrication techniques. Some domestic and agriculture wastes that are major sources of ceramic particles, like rice husk and rock dust, can be used to reduce the reinforcement cost and also provide an eco-friendly product. Numerical approaches for wear behaviour of ceramic-reinforced composites that hold diverse shapes and can be produced at the desired scale would be used for ultimate research to save research time and costs. Additionally, the mechanisms of effects caused by the interfacial bonding between reinforcements and matrix have not been fully explicated and so they are to be focused on. Further, efficient models for predicting tribological behaviour of composite should also be developed in order to select appropriate reinforcement for a specified application.

REFERENCES

Abd-Elwahed, M. S., A. F. Ibrahim, and M. M. Reda. 2020. "Effects of ZrO2 Nanoparticle Content on Microstructure and Wear Behavior of Titanium Matrix Composite." *Journal of Materials Research and Technology* 9 (4). Elsevier: 8528–34.

Banerjee, Sudip, Suswagata Poria, Goutam Sutradhar, and Prasanta Sahoo. 2020. "Abrasive Wear Behavior of WC Nanoparticle Reinforced Magnesium Metal Matrix Composites." *Surface Topography: Metrology and Properties* 8 (2). IOP Publishing: 25001.

Baradeswaran, A., and A. Elaya Perumal. 2014. "Study on Mechanical and Wear Properties of Al 7075/Al2O3/Graphite Hybrid Composites." *Composites Part B: Engineering* 56. Elsevier: 464–71.

Baradeswaran, A., S. C. Vettivel, A. Elaya Perumal, N. Selvakumar, and R. Franklin Issac. 2014. "Experimental Investigation on Mechanical Behaviour, Modelling and Optimization of Wear Parameters of B 4 C and Graphite Reinforced Aluminium Hybrid Composites." *Materials & Design* 63. Elsevier: 620–32. https://doi.org/10.1016/j.matdes.2014.06.054.

Carvalho, M. Buciumeanu, S. Madeira, D. Soares, F. S. Silva, G. Miranda. 2015. "Dry Sliding Wear Behaviour of AlSi-CNTs-SiCp Hybrid Composites." *Tribology International* 90. Elsevier: 148–56. https://doi.org/10.1016/j.triboint.2015.04.031.

Dash, Dharmeswar, Ram Singh, Sutanu Samanta, and Ram Naresh Rai. 2020. "Influence of TiC on Microstructure, Mechanical and Wear Properties of Magnesium Alloy (AZ91D) Matrix Composites." NISCAIR-CSIR, India.

Dinaharan, I., and N. Murugan. 2012. "Dry Sliding Wear Behavior of AA6061/ZrB2 in-Situ Composite." *Transactions of Nonferrous Metals Society of China* 22 (4). Elsevier: 810–18.

Ghandvar, Hamidreza, Saeed Farahany, and Mohd Hasbullah Idris. 2016. "Effect of Wettability Enhancement of SiC Particles on Impact Toughness and Dry Sliding Wear Behavior of

Compocasted A356/20SiCp Composites." *Tribology Transactions* 1–2. https://doi.org/
10.1080/10402004.2016.1275902.

Gopalakrishnan, S., and N. Murugan. 2012. "Production and Wear Characterisation of AA
6061 Matrix Titanium Carbide Particulate Reinforced Composite by Enhanced Stir
Casting Method." *Composites Part B: Engineering* 43 (2). Elsevier: 302–8.

Halil, Karakoç, Ovalı İsmail, Dündar Sibel, and Çıtak Ramazan. 2019. "Wear and Mechanical
Properties of Al6061/SiC/B4C Hybrid Composites Produced with Powder Metallurgy."
Journal of Materials Research and Technology 8 (6). Elsevier: 5348–61.

Jalilvand, Mohammad Mahdi, and Yousef Mazaheri. 2020. "Effect of Mono and Hybrid Ceramic
Reinforcement Particles on the Tribological Behavior of the AZ31 Matrix Surface
Composites Developed by Friction Stir Processing." *Ceramics International*. Elsevier.

Kavimani, V., K. Soorya Prakash, and Titus Thankachan. 2019. "Experimental Investigations
on Wear and Friction Behaviour of SiC@r-GO Reinforced Mg Matrix Composites
Produced through Solvent-Based Powder Metallurgy." *Composites Part B: Engineering*
162 (April): 508–21. https://doi.org/10.1016/j.compositesb.2019.01.009.

Kavimani, V., K. Soorya Prakash, Titus Thankachan, and R. Udayakumar. 2020. "Synergistic
Improvement of Epoxy Derived Polymer Composites Reinforced with Graphene
Oxide (GO) plus Titanium Di Oxide(TiO2)." *Composites Part B: Engineering* 191.
Elsevier: 107911. https://doi.org/10.1016/j.compositesb.2020.107911.

Labib, F., H. M. Ghasemi, and R. Mahmudi. 2016. "Dry Tribological Behavior of Mg/SiCp
Composites at Room and Elevated Temperatures." *Wear* 348–49. Elsevier: 69–79.
https://doi.org/10.1016/j.wear.2015.11.021.

Lenin, A. Haiter, S. C. Vettivel, T. Raja, Lulseged Belay, and S. Christopher Ezhil Singh. 2018. "A
Statistical Prediction on Wear and Friction Behavior of ZrC Nano Particles Reinforced with
AlSi Composites Using Full Factorial Design." *Surfaces and Interfaces* 10. Elsevier: 149–61.

Lim, C. Y. H., S. C. Lim, and M. Gupta. 2003. "Wear Behaviour of SiCp-Reinforced
Magnesium Matrix Composites." *Wear* 255 (1–6): 629–37. https://doi.org/10.1016/
S0043-1648(03)00121-2.

Liu, Huan, Nannan Lu, Xiaojun Wang, Xiaoshi Hu, and Deqiang Chen. 2020. "Different
Tribological Behaviors of SiC p/AZ91 Composites Induced by Tailoring the Distribution
of SiC Particles." *Metals and Materials International* 3. Springer: 1–14.

Mandal, A., M. Chakraborty, and B. S. Murty. 2007. "Effect of TiB2 Particles on Sliding Wear
Behaviour of Al–4Cu Alloy." *Wear* 262 (1–2). Elsevier: 160–66.

Mandal, A., B. S. Murty, and M. Chakraborty. 2009. "Wear Behaviour of near Eutectic Al–Si
Alloy Reinforced with in-Situ TiB2 Particles." *Materials Science and Engineering: A*
506 (1–2). Elsevier: 27–33.

Mistry, Jitendra M., and Piyush P. Gohil. 2019. "Experimental Investigations on Wear and
Friction Behaviour of Si3N4p Reinforced Heat-Treated Aluminium Matrix Composites
Produced Using Electromagnetic Stir Casting Process." *Composites Part B: Engineering*
161. Elsevier: 190–204.

Patle, Hemendra, B. Ratna Sunil, and Ravikumar Dumpala. 2020. "Sliding Wear Behavior
of AZ91/B4C Surface Composites Produced by Friction Stir Processing." *Materials
Research Express* 7 (1). IOP Publishing: 16586.

Pramanik, A. 2016. "Effects of Reinforcement on Wear Resistance of Aluminum Matrix
Composites." *Transactions of Nonferrous Metals Society of China* 26 (2). Nonferrous
Metals Society of China: 348–58. https://doi.org/10.1016/S1003-6326(16)64125-0.

Radhika, N., J. Sasikumar, J. L. Sylesh, and R. Kishore. 2020. "Dry Reciprocating Wear
and Frictional Behaviour of B4C Reinforced Functionally Graded and Homogenous
Aluminium Matrix Composites." *Journal of Materials Research and Technology* 9 (2).
Elsevier: 1578–92.

Raghav, G. R., A. N. Balaji, D. Muthukrishnan, V. Sruthi, and E. Sajith. 2018. "An Experimental Investigation on Wear and Corrosion Characteristics of Mg-Co Nanocomposites." *Materials Research Express* 5 (6). IOP Publishing.

Rahmani, K., A. Sadooghi, and S. J. Hashemi. 2020. "The Effect of Al2O3 Content on Tribology and Corrosion Properties of Mg-Al2O3 Nanocomposites Produced by Single and Double-Action Press." *Materials Chemistry and Physics*. Elsevier, 123058.

Rajmohan, T., K. Palanikumar, and S. Ranganathan. 2013. "Evaluation of Mechanical and Wear Properties of Hybrid Aluminium Matrix Composites." *Transactions of Nonferrous Metals Society of China* 23 (9). Elsevier: 2509–17.

Rao, R. N., and S. Das. 2011. "Effect of SiC Content and Sliding Speed on the Wear Behaviour of Aluminium Matrix Composites." *Materials & Design* 32 (2). Elsevier: 1066–71.

Sahoo, B. N., and S. K. Panigrahi. 2019. "Development of Wear Maps of In-Situ TiC+ TiB2 Reinforced AZ91 Mg Matrix Composite with Varying Microstructural Conditions." *Tribology International* 135. Elsevier: 463–77.

Selvakumar, Natarajan, and Sreedharan Christopher Ezhil Singh. 2016. "Influence of Nano ZrC Content on Tribological Analysis, Microstructure and Mechanical Properties of Cu–4Cr Matrix Composites Produced by Hot Extrusion." *Archives of Civil and Mechanical Engineering* 16 (3). Elsevier: 537–52.

Soorya Prakash, K., P. Balasundar, S. Nagaraja, P. M. Gopal, and V. Kavimani. 2016. "Mechanical and Wear Behaviour of Mg–SiC–Gr Hybrid Composites." *Journal of Magnesium and Alloys* 4 (3). Elsevier B.V.: 197–206. https://doi.org/10.1016/j.jma.2016.08.001.

Soorya Prakash, K., P. M. Gopal, and V. Kavimani. 2017. "Effect of Rock Dust, Cenosphere and E-Waste Glass Addition on Mechanical, Wear and Machinability Behaviour of Al 6061 Hybrid Composites." *Indian Journal of Engineering and Materials Sciences* 24 (4). NISCAIR-CSIR, India: 270–82.

Soorya Prakash, K., P. M. Gopal, D. Anburose, and V. Kavimani. 2018. "Mechanical, Corrosion and Wear Characteristics of Powder Metallurgy Processed Ti-6Al-4V/B4C Metal Matrix Composites." *Ain Shams Engineering Journal* 9 (4). Elsevier: 1489–96. https://doi.org/10.1016/j.asej.2016.11.003.

Srinivasan, M., C. Loganathan, M. Kamaraj, Q. B. Nguyen, M. Gupta, and R. Narayanasamy. 2012. "Sliding Wear Behaviour of AZ31B Magnesium Alloy and Nano-Composite." *Transactions of Nonferrous Metals Society of China* (English Edition) 22 (1): 60–65. https://doi.org/10.1016/S1003-6326(11)61140-0.

Stalin, B., M. Ravichandran, V. Mohanavel, and P. L. Raj. 2020. "Investigations into Microstructure and Mechanical Properties of Mg-5wt.% Cu-TiB2 Composites Produced via Powder Metallurgy Route." *Journal of Mining and Metallurgy B: Metallurgy* 56 (1): 99–108.

Thankachan, Titus, K. Soorya Prakash, and V. Kavimani. 2019. "Investigating the Effects of Hybrid Reinforcement Particles on the Microstructural, Mechanical and Tribological Properties of Friction Stir Processed Copper Surface Composites." *Composites Part B: Engineering* 174. Elsevier: 107057. https://doi.org/10.1016/j.compositesb.2019.107057.

Venkatesh, R., and Vaddi Seshagiri Rao. 2018. "Thermal, Corrosion and Wear Analysis of Copper Based Metal Matrix Composites Reinforced with Alumina and Graphite." *Defence Technology* 14 (4). Elsevier: 346–55.

Xie, Zhiwen, Feng Guo, Xinfang Huang, Kangsen Li, Qiang Chen, Yongjun Chen, and Feng Gong. 2020. "Understanding the Anti-Wear Mechanism of SiCp/WE43 Magnesium Matrix Composite." *Vacuum* 172. Elsevier: 109049.

Yadav, Pradeep Kumar, Gajendra Dixit, Basil Kuriachen, Manoj Kumar Verma, Surendra Kumar Patel, and Raj Kumar Singh. 2020. "Effect of Reinforcements and Abrasive

Size on High-Stress Tribological Behaviour of Aluminium Piston Matrix Composites." *Journal of Bio-and Tribo-Corrosion* 6 (1). Springer: 1–14.

Yao, Yan-tao, Lan Jiang, Gao-feng Fu, and Li-qing Chen. 2015. "Wear Behavior and Mechanism of B4C Reinforced Mg-Matrix Composites Fabricated by Metal-Assisted Pressureless Infiltration Technique." *Transactions of Nonferrous Metals Society of China* 25 (8). Elsevier: 2543–48.

Yue, H. Y., B. Wang, X. Gao, S. L. Zhang, X. Y. Lin, L. H. Yao, and E. J. Guo. 2017. "Effect of Interfacial Modifying on the Microstructures, Mechanical Properties and Abrasive Wear Properties of Aluminum Borate Whiskers Reinforced 6061Al Composite." *Journal of Alloys and Compounds* 692. Elsevier: 395–402.

Zeng, Xiang, Jingang Yu, Dingfa Fu, Hui Zhang, and Jie Teng. 2018. "Wear Characteristics of Hybrid Aluminum-Matrix Composites Reinforced with Well-Dispersed Reduced Graphene Oxide Nanosheets and Silicon Carbide Particulates." *Vacuum* 155. Elsevier: 364–75.

Zhang, Jiangshan, Shufeng Yang, Zhixin Chen, Hui Wu, Jingwei Zhao, and Zhengyi Jiang. 2019. "Graphene Encapsulated SiC Nanoparticles as Tribology-Favoured Nanofillers in Aluminium Composite." *Composites Part B: Engineering* 162. Elsevier: 445–53.

Zhou, M. Y., L. B. Ren, L. L. Fan, Y. W. X. Zhang, T. H. Lu, G. F. Quan, and M. Gupta. 2020. "Progress in Research on Hybrid Metal Matrix Composites." *Journal of Alloys and Compounds*. Elsevier: 155274.

Zhu, Juanjuan, Jiahui Qi, Dikai Guan, Le Ma, and Rob Dwyer-Joyce. 2020. "Tribological Behaviour of Self-Lubricating Mg Matrix Composites Reinforced with Silicon Carbide and Tungsten Disulfide." *Tribology International* 146. Elsevier: 106253.

9 Tensile and Wear Behaviour of MMCs Reinforced with Metallic Particles by Solid-State Technique

S. Vigneshwaran,[1]* R. Palanivel,[2a] A. Alblawi,[2b] R. F. Laubscher[3c]

[1]Department of Mechanical Engineering, National Institute of Technology Puducherry, Karaikal, India.

[2]Department of Mechanical Engineering, College of Engineering, Al-Dawadmi Shaqra University, Saudi Arabia.

[3]Department of Mechanical Engineering Science, University of Johannesburg, Auckland Park Kingsway Campus, Johannesburg 2006, South Africa.

*Corresponding author: svigneshwaranmech2010@gmail.com

Other email IDs: [a]rpalanivelme@gmail.com, [b]aalblawi@su.edu.sa, [c]rflaubscher@uj.ac.za

CONTENTS

DOI: 10.1201/9781003109723-9

9.1 INTRODUCTION

Among the many potential classes of engineering materials, metal matrix composites (MMCs) have developed rapidly to become a significant player in the modern era. Their use and significance in the aerospace, marine, automotive, and defence sectors is forever increasing owing to attractive properties including low density, high specific strength, high-temperature thermal stability, creep resistance and their fatigue and wear resistance (Kaczmar *et al.*, 2000; Miracle, 2005). Several industries are actively replacing their conventional alloy-based products with MMCs owing to these benefits. Furthermore, the various combinations of matrix material and reinforcement, along with the development of more advanced manufacturing technologies, have led to a significant increase in research related to MMCs (Dinaharan *et al.*, 2020).

MMCs are produced by combining a matrix with reinforcements by utilizing various processing methods, as shown in Figure 9.1. Typically the matrix is either a metal, ceramic, or polymer, whereas whiskers, fibres, and particulates are widely used as reinforcements (Taha, 2001). Liquid-state processes (casting/pressure infiltration/spray deposits/ in-situ) are most commonly used to manufacture MMCs because of their simplicity, mass producibility, and cost effectiveness (Rozak *et al.*, 1992, Muralidharan *et al.*, 2018, David Raja Selvam *et al.*, 2018). Apart from the aforementioned simplicity and affordability, MMCs produced by the liquid-state process do, however, display significant disadvantages. These include poor wettability of matrix and reinforcement, debonding of matrix and particles, settling or floating of particles due to density variation, agglomeration of particles, and the formation of undesirable intermetallic particles (in-situ process). This typically leads to brittle interfacial elements that cause poor ductility (Yadav and Bauri, 2011; Rajan *et al.*, 2016).

The production of Al-based MMCs by various techniques along with their most significant advantages and disadvantages are listed in Table 9.1. The primary difficulties associated with the production of Al-based MMCs by liquid and powder metallurgy routes are listed in the following subsections (Bauri and Yadav, 2018).

9.1.1 CASTING

Generally, in the casting routes, poor wettability and clustering of reinforcement particles are the main causes for the deterioration of the mechanical properties of

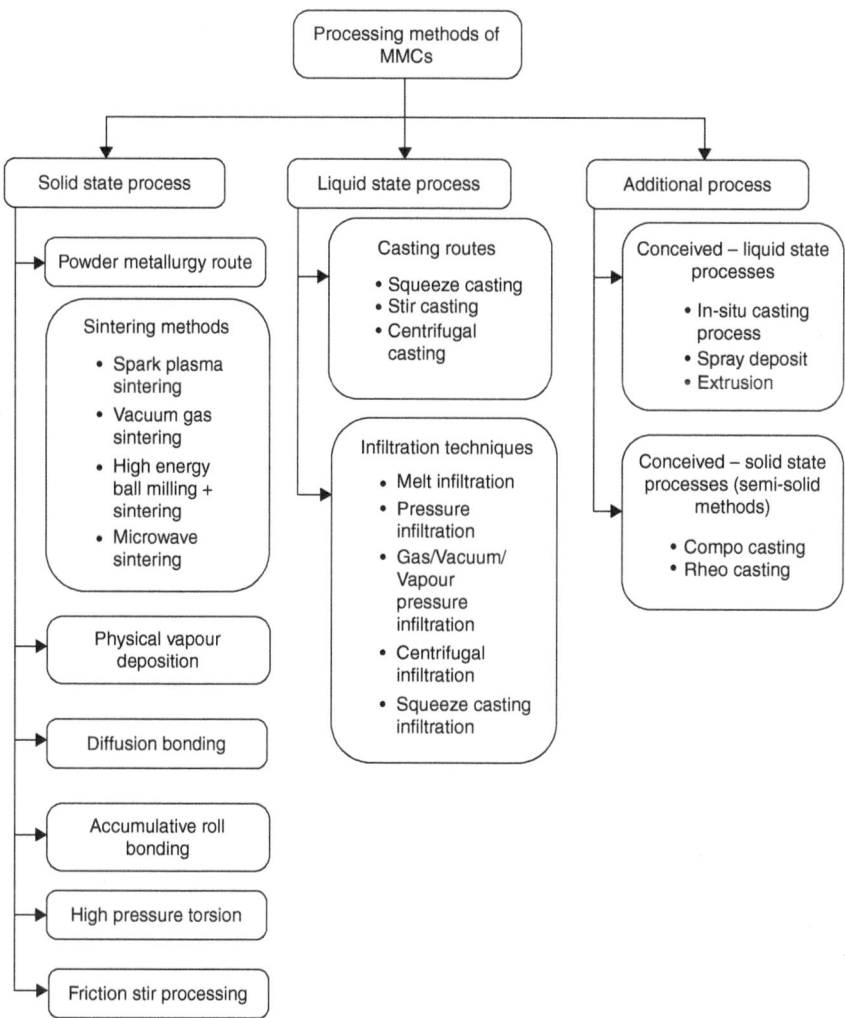

FIGURE 9.1 Various methods of processing MMCs.

Source: Modified from Arunachalam et al., 2019.

MMCs. Furthermore, inadequate degassing results in entrapped gases that leads to significant porosity. Prolonged liquid matrix and reinforcement particle interaction can develop by-products as a result of the interfacial reaction, which may affect the properties of MMCs (Venkatesan and Anthony Xavior, 2018; Arunachalam *et al.*, 2019).

9.1.2 IN-SITU PROCESS

Controlling the morphology (size and shape) of the reinforcement particles is difficult in this process. It is only suitable for a select few material systems (Liu *et al.*, 2017).

TABLE 9.1

Advantages and Disadvantages of Various Production Methods for Al-Based MMCs

Method	Process	Advantages	Disadvantages
Liquid-state process – Castingroutes	*Stir casting* (Arunachalam *et al.*, 2019)	Simple process Mass producibility Mechanized completely Cost affordability	Slow process Post heat treatment is essential.
	Squeeze casting (Venkatesan and Anthony Xavior, 2018)	Uniform distribution of reinforcement particle	The cost of production is high owing to more manufacturing setup.
	Vacuum casting (Singh *et al.*, 2015)	The dispersion of reinforcement is uniform.	Additional equipment such as vacuum pumps and accessories incurs high costs.
	Centrifugal casting (Kingsley and Suárez, 2011)	Free from pores Mould filling is better Controllable wall thickness Good mechanical strength	The quality of the inner surface of the casting is affected.
	In-situ casting (Liu *et al.*, 2017)	Matrix-particle bond is good. Fine particle size with homogeneous distribution. Good strength at elevated temperature applications	Reduction in ductility Limited to certain materials
Liquid-state process – Infiltration techniques	*Melt infiltration* (Gecu *et al.*, 2017)	Wear resistance properties of MMCs are good. Dense MMCs with uniform microstructure MMCs can be achieved at nearly net shape.	Temperature range is limited. Blockage of infiltration
	Pressure infiltration (Blucher, 1992; W. Yang *et al.*, 2018)	Better tribological behaviour of MMCs	Not suitable for large scale MMCs Porosity is unavoidable
	Gas, vacuum and vapour infiltration (Arunachalam *et al.*, 2019)	Any combination of MMCs with large scale sizes can be produced. The porosity of MMCs is low except in vapour infiltration. Residual stress is minimal. Complex shapes can be made. Mechanical properties are good compared to the stir casting method.	Solidification consumes more time. The need for inert gases Slower production Wettability is low Formation of cracks
Solid-state process – Deposition	*Physical vapour deposition, spray atomization*	Flexibility in fabricating MMCs	Porosity will be present. Setup cost is high.

TABLE 9.1 (Continued)
Advantages and Disadvantages of Various Production Methods for Al-Based MMCs

Method	Process	Advantages	Disadvantages
	(Ramkumar and Dinaharan, 2020; Kaur and Pandey, 2010)	Matrix-reinforcement bond is excellent.	Optimized deposition parameters essential to achieve adhesion quality are challenging.
Solid-state process	*Diffusion bonding* (Bauri and Yadav, 2018)	High joint integrity of matrix and reinforcement in MMCs	Non-uniform distribution of fibres Low volume fraction of fibres Formation of interfacial elements in matrix-fibre-matrix regime.
	Accumulative roll bonding (Ramkumar and Dinaharan, 2020)	Clustering of the reinforcement particle is minimal. Improved uniform distribution of reinforcement produces good strength.	Reinforcement particles used should be of minimal size (mostly nano size), and the volume of reinforcement used is also lower.
	High pressure torsion (Seenuvasaperumal et al., 2018)	Improved distribution of particles with better wear-resistance properties	Size of the sample is less and usage at large scale will involve high cost.
Solid-state process-Powder metallurgy	*Spark plasma sintering* (Dash et al., 2013; Razavi et al., 2017)	One-stage process of compaction and sintering in a uniform manner	Expensive owing to setup costs Shape constraint for MMCs (only symmetrical objects can be made)
	High energy ball milling + sintering (Kong et al., 2002)	Mixing is uniform.	Welding of particles occurs. Time-consuming
	Microwave sintering (Reddy et al., 2017)	Improved mechanical properties	Not suitable for nitrides, alumina (owing to dielectric properties)
Solid-state process	*Friction stir processing* (Bauri et al., 2011)	Microstructural modification and refinement of grains are possible. The dispersion of reinforcement is uniform in the matrix along with a change in initial microstructure due to three-dimensional material flow. High mechanical properties with good ductility.	Multiple passes are needed to create sound MMCs.

Note This table involves studies based on the fabrication of Al-based MMCs with various types of reinforcement particles (ceramics, oxides, etc.). Only the process feasibility is discussed, which may vary for MMCs with other types of matrix (ceramic, polymer, or other matrix metals).

9.1.3 INFILTRATION

The infiltration route of MMC production involves self-supported reinforcement assembling. This may cause damage to fibres or clustering of reinforcement during the infiltration process that may degrade mechanical properties owing to the non-homogeneity of the MMCs produced (Gecu *et al.*, 2017; W. Yang *et al.*, 2018).

9.1.4 POWDER METALLURGY

Achieving uniform reinforcement spacing is complex. Using an extrusion-based powder metallurgy route potentially eliminates porosity but results in texturing of reinforcement particles along the extrusion axis (Surappa, 2003).

Solid-state processes may therefore be preferred for the production of MMCs as no melting is involved (Clyne and Jones, 2006; Ramkumar and Dinaharan, 2020). A selection of solid-state processes used for the fabrication of MMCs is shown in Figure 9.2. A brief description of these include:

i.   *Diffusion bonding (DB):* Interdiffusion of atomic particles produced between two layers of the metal matrix, which contains aligned fibres. The application of an external load induces the bond between metal and fibres (Bhanu Prasad *et al.*, 1991).

ii. *Vapour deposition (VD):* Fibres are passed into a chamber, which is filled with vapours of the metal matrix that were produced by electron beams. The MMCs are produced by gathering the coated fibres as bundles or arrays. These bundles or arrays are consolidated by hot pressing (Clyne and Jones, 2006).

iii. *Accumulative roll bonding (ARB):* Reinforcement particles are added in between two matrix sheets. Subsequent stacking and rolling produce atom-to-atom bonds between the sheets and the reinforcement particles are bonded to the sheets, forming MMCs (Ramkumar and Dinaharan, 2020).

iv. *High pressure torsion (HPT):* Metallic powders and reinforcement particles are consolidated in ambient temperature and densely packed MMCs are produced between two anvils, which rotate with high compressive force leading to torsional strain on the samples (Seenuvasaperumal *et al.*, 2018).

v.   *Friction stir processing (FSP):* FSP is a successor to friction stir welding (FSW), which uses frictional heat and mechanical action produced by the rotating FSP tool. FSP is a thermomechanical process developed for microstructural modification and refinement of grains. As the material flow is three-dimensional, the dispersion of reinforcement is uniform in the matrix along with a change in initial microstructure. Hence, FSP has become an effectively utilized tool in the fabrication of surface and bulk MMCs (Mishra *et al.*, 2003; Rathee *et al.*, 2018). Fabrication of MMCs using metallic reinforcement by FSP is the main focus of this chapter, and the metallurgical, tensile, and wear behaviour of MMCs manufactured by FSP are discussed in detail.

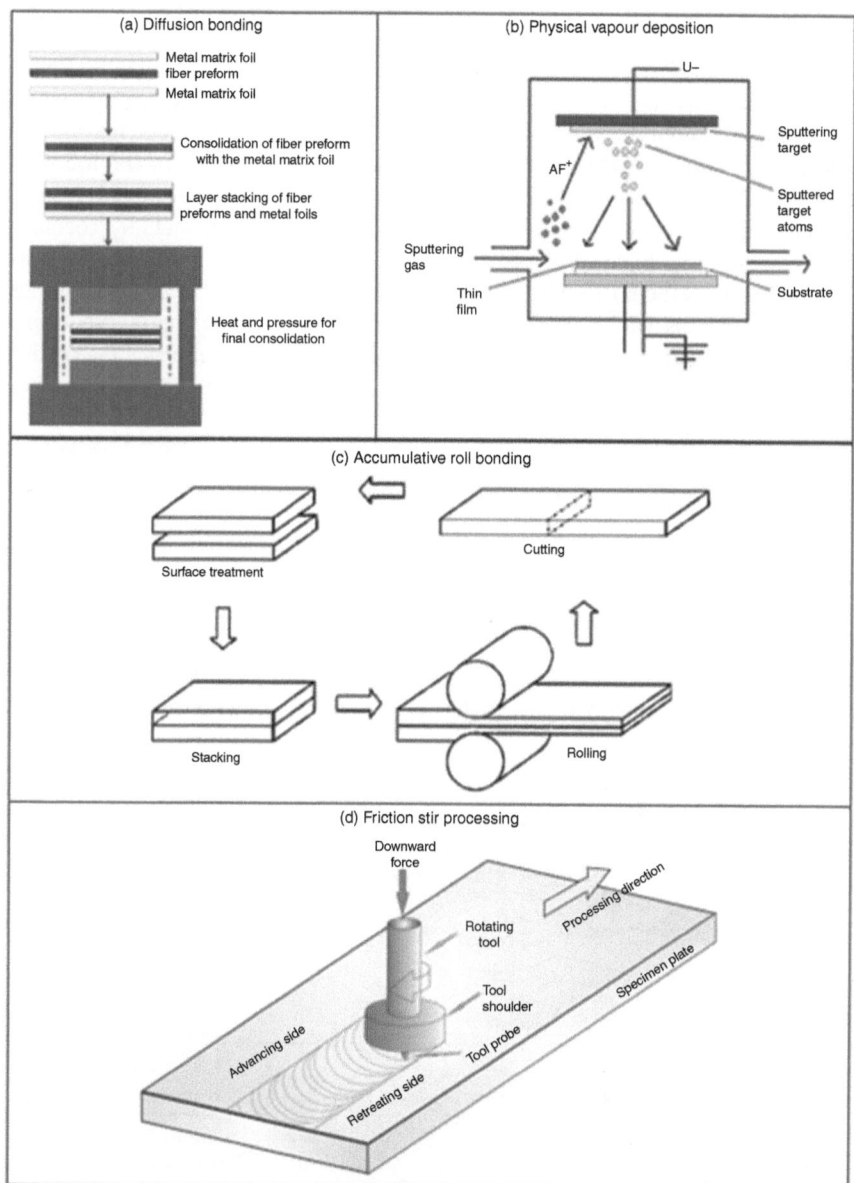

FIGURE 9.2 Schematic representation of four procedures followed for the fabrication of Al-based MMCs by solid-state process.

Sources: (a) Shirvanimoghaddam *et al.,* 2017; (b) Kalra *et al.,* 2018; (c) Shaarbaf and Toroghinejad, 2008; (d) Sharma *et al.,* 2015, all reprinted with permission.

9.2   FRICTION STIR PROCESSING OF AL-BASED MMCS

Mishra *et al.*, 2003, fabricated Al-SiC MMCs by FSP. Various research projects following this approach were initiated by researchers using different matrix and reinforcement combinations. These included both surface and bulk metal matrix composites. Fabrication of MMCs by FSP is presented and discussed in the following subsections. Various factors influence the microstructure, material flow, and temperature distribution during the manufacture of MMCs by FSP. These include processing parameters (tool rotational speed (RPM), tool traverse speed (mm/min), the direction of rotation, plunge depth (mm), etc.), the geometry of the tool (pin type, pin and shoulder dimensions, shoulder profile, etc.) and the type of work material used. Apart from these factors other significant aspects are also at play and influence the production quality of the MMC by FSP. These are presented in Figure 9.3 in the part of the process where they come into play.

9.2.1   Microstructure Formation

FSP locally softens and plasticizes the work material by frictional heat and strain. This heat and strain transform the initial microstructure into several zones on both sides (retreating and advancing). These include the stir zone (SZ), thermomechanically affected zone (TMAZ), and heat-affected zone (HAZ) (Figure 9.4). The SZ, also denoted as the nugget, is the zone that attains maximum frictional heat and strain. Grain refinement is typically higher in the SZ compared to other zones. The TMAZ is

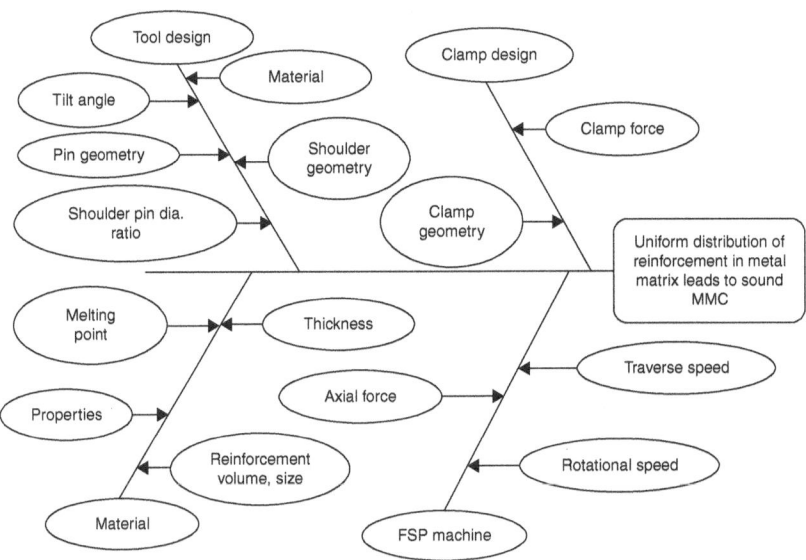

FIGURE 9.3   Diagram of causes and final effect indicating the various factors responsible for the production of MMCs using FSP.

Source: Modified from Jayaraman *et al.,* 2009.

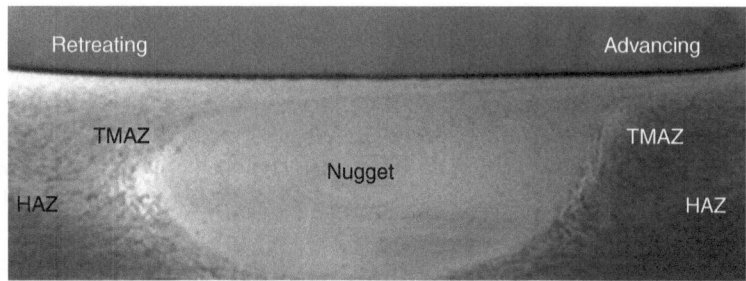

FIGURE 9.4 Cross section of the various microstructural zones of an Al alloy processed by FSP.

Source: Mishra and Ma, 2005, reprinted with permission.

observed in close proximity to the SZ. Its width is largely material based. This zone is less influenced by heat and strain compared to the SZ, and the grains are slightly elongated. Generally, Al-based alloys display a wider TMAZ. The HAZ is located next to the TMAZ and just before the base metal. The HAZ involves no strain; however, frictional heat influence causes grain growth, which depends on the thermal conductivity of the material. Owing to the high thermal conductivity of Al and its alloys, the grain size in the HAZ is significantly bigger than the base metal microstructure (Bauri and Yadav, 2018).

9.2.2 PROCESS PARAMETERS

Frictional heat plasticizes the work material. This heat conducts towards the work material at a rate that is a function of the tool rotational speed and tool traverse speed. The amount of heat produced (adiabatically) is directly proportional to tool rotational speed. In contrast, high tool traverse speeds imply less contact time with the work material and produce less heat. Therefore, maximum frictional heat is generated at high tool rotational speed and low traverse speed. Sahraeinejad *et al.*, 2015, conducted FSP on 5xxx Al alloy with Al_2O_3 and SiC particles at various tool rotational speeds (56–1800 RPM) and tool traverse speeds (11–2000 mm/min). They determined that as the tool rotational speed increased, the particle distribution was homogeneous, which they attributed to the higher heat input. Furthermore, they stated that a decrease in tool traverse speed improved the distribution of particles in MMCs. Faraji and Asadi, 2011, also suggested that the distribution of particles is more uniform when the tool traverse speed is 40 mm/min rather than 80 mm/min at 900 RPM. Besides these studies, a number of other studies on FSP revealed that more than one pass is required for the improved dispersion of reinforcement particles in the matrix (Dolatkhah *et al.*, 2012; Ni *et al.*, 2014; Guo *et al.*, 2014; Girish and Anandakrishnan, 2019). Considering the direction of tool rotation, Sahraeinejad *et al.*, 2015, recommended a change in the direction of tool rotation between each FSP pass, noting a more balanced material flow as the result. This results in a perceived improvement of particle dispersion in both the advancing and retreating sides.

9.2.3 TOOL GEOMETRY

The stirring and material flow direction are greatly influenced by the geometry of the tool. The tool shoulder generates 80% of the total heat during FSW with the pin responsible for the rest. In addition, the pin profile has a significant effect on material flow behaviour. The primary function of the non-consumable rotating tool pin is to stir the plasticized material and transport it rearwards to form a solid joint. The pin profile plays a crucial role in material flow and, in turn, regulates the welding speed of the FSW process (Oosterkamp et al., 2004). Padmanaban and Balasubramanian, 2009, used different types of tool pin profiles that included cylindrical, tapered cylindrical, square, and triangular, and reported that the triangular and square pin profiles (Figure 9.5) produce a pulsating stirring action in the flow due to their flat faces. This pulsating action in the stir zone of the triangular and square pin profiled tools produces fine-grained microstructure with uniform distribution. Bahrami *et al.*, 2014, reported that a tapered tool with threads exhibited the most uniform distribution of SiC particles in Al matrix when FSP was performed with five different types of tool pin profiles (thread with taper, straight triangular, straight square, four fluted square, and four fluted cylinder). The uniform distribution using a tapered tool with threads is due to the improvement in material flow in the downward direction along the threads. Moreover, threaded tools typically display higher heat generation for superior material flow behaviour of the reinforcement particles along the tool surface (Sahraeinejad *et al.*, 2015).

Palanivel *et al.*, 2018, conducted research utilizing different types of tool shoulder profiles including a partial impeller (PI), full impeller (FI), and flat grooved (FS) shoulder profile (Figure 9.6). The function of the shoulder profile is to produce heat and to act as an escape route for the material displaced by the pin, thereby enhancing the extrusion and consolidation of the material during welding. The shoulder

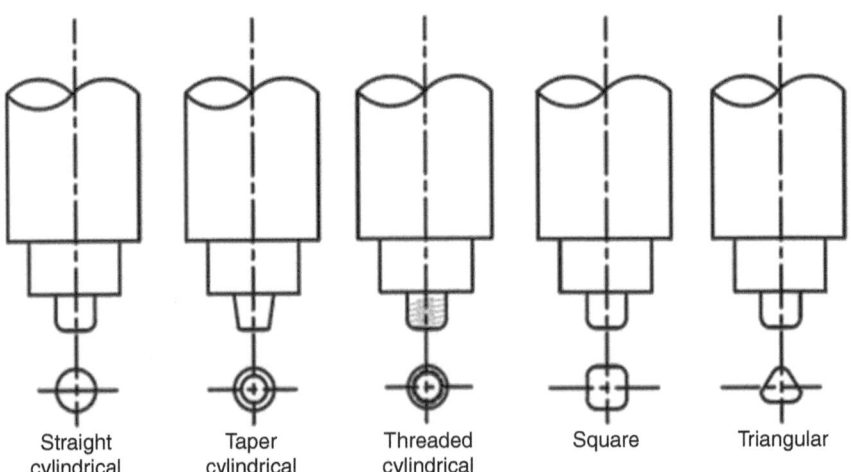

FIGURE 9.5 Various tool pin profiles used for friction stir welding (FSW) and FSP.

Source: Padmanaban and Balasubramanian, 2009, reprinted with permission.

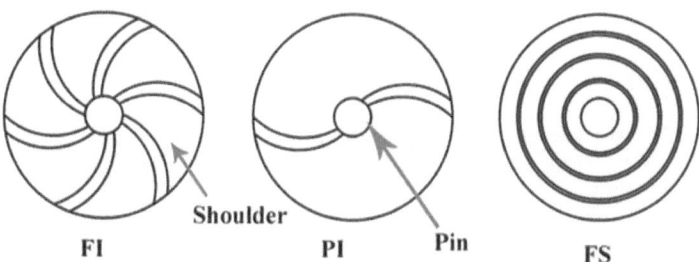

FIGURE 9.6 Examples of three shoulder profiles: partial impeller (PI), full impeller (FI), and flat grooved (FS).

promotes the transport of material at the top surface, from the retreating to the advancing side, and displaces it down within the pin diameter. A full impeller profile (FI) is sufficient, and owing to its multiple impeller region, it does not allow the material to escape, thereby preventing the formation of defects, unlike the PI and FS tool shoulder profiles. However, the FI shoulder profile also struggles to generate optimum heat to ensure appropriate plasticization of the material, leaving the end of the pin too far from the anvil at the lowest and highest ranges of welding speed and rotational speed.

9.2.4 Processing Methods

The different methods for introducing the reinforcement into the matrix metal during FSP are categorized as (i) groove filling (Mishra *et al.*, 2003), (ii) hole filling (M. Yang *et al.*, 2010), (iii) sandwich-type (Mertens *et al.*, 2012), (iv) direct filling–FSP (Y. Huang *et al.*, 2014), and (v) surface coating/surface modification–FSP (Kurt *et al.*, 2011). Mishra *et al.*, 2003, used the groove-filling method (Figure 9.7a and b) on AA 5083 plate with SiC reinforcement and FSP to create a composite. They attained a uniform distribution of metal and reinforcement after overlapping passes. Most of the MMCs produced by FSP follow the groove-filling method (Yadav and Bauri, 2011; Rejil *et al.*, 2012; Sahraeinejad *et al.*, 2015; Kumar *et al.*, 2019). Mostly, after filling the groove with reinforcement particles, the groove is closed using a pinless FSP tool (Figure 9.7b). The FSP tool is then used with either a single pass or multiple passes to fabricate the MMC. This method produces a uniform distribution of particles, and the distribution is more homogeneous as the number of passes increases (C. Lee et al., 2006; Bauri and Yadav, 2018).

Similarly to the groove-filling method, in the hole-filling method a sequence of drilled holes are filled with reinforcement particles. FSP then follows to close the drilled holes. A pictorial representation of the hole filling and closing method by FSP is shown in Figure 9.7c–d). Dixit *et al.*, 2007, fabricated Al–NiTi MMC using the hole-filling method. The drilled holes were filled with NiTi particles and friction stir–processed with a pinless tool. Hole diameters were smaller than the FSP tool shoulder diameter. A uniform distribution of the NiTi particles was subsequently obtained. Similarly, two studies using the hole-filling method of FSP for the production of MMCs on AA 6061 were conducted, one with Al_2O_3 particles (M. Yang *et al.*, 2010)

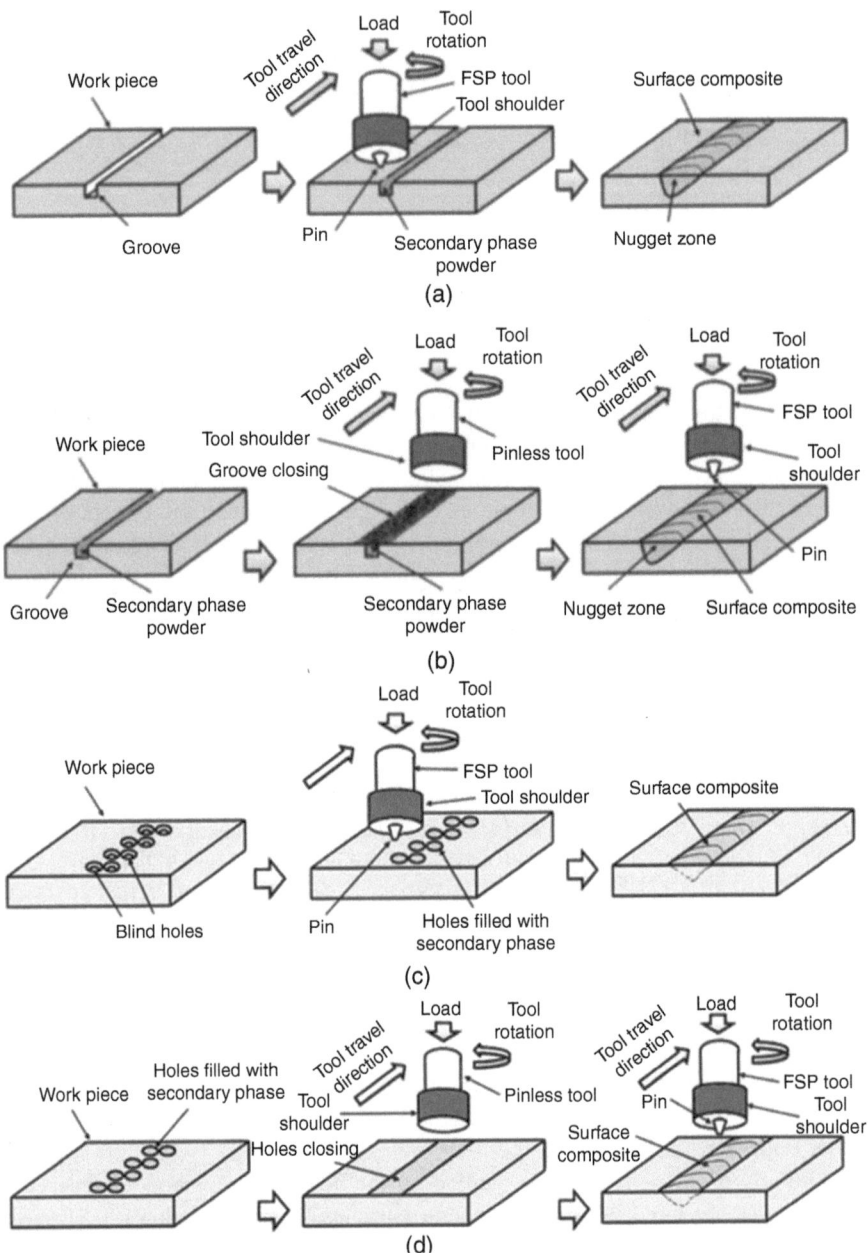

FIGURE 9.7 FSP processing methods: (a) groove filling with reinforcement, followed by (b) groove closing using FSP tool; and (c) hole filling with reinforcement, followed by (d) hole closing using FSP tool.

Source: Ratna Sunil, 2016, reprinted with permission.

and one with NiTi particles (Ni *et al.*, 2014). In line with Dixit *et al.*, 2007, above, they also attained a uniform distribution of particles in the Al matrix.

Among the aforementioned processing methods, the groove- and hole-filling techniques accompanied by subsequent FSP displayed uniform dispersion of reinforcement in the matrix metal. The other methods of reinforcement filling and FSP processing displayed promising results, but further strategies and research are needed to improve performance (Ratna Sunil, 2016).

9.3 TENSILE PROPERTY EVALUATION OF AL-BASED MMCS PRODUCED BY FSP

Several types of organic (fly ash, red mud, rice husk ash) and inorganic (oxides, carbides, nitrides) materials as well as fibres have been used as reinforcement particles for the fabrication of Al-based MMCs by FSP. FSP is versatile enough to fabricate most combinations of matrix and reinforcement regardless of physical and chemical properties, which is a challenging task in other routes. Sufficient literature has been reported by various researchers on homogeneous distribution and improved microstructural, mechanical, and tribological properties of MMCs fabricated by FSP (Bauri and Yadav, 2018; Arunachalam *et al.*, 2019; Dinaharan *et al.*, 2020). Apart from the aforementioned reinforcement particles, Al-based MMCs produced by FSP with metallic reinforcements have also shown competitive properties as far as the combination of strength and ductility is concerned (Table 9.2).

9.3.1 Effect of Matrix-Particle Interaction on Tensile Strength

The MMCs manufactured from commercially pure aluminium (CP-Al) with Ni particles as reinforcement show a drastic improvement in strength while retaining

TABLE 9.2
Comparison of Tensile Properties of Al-Based MMCs Fabricated by FSP Using Inorganic and Metallic Reinforcements

Matrix Metal	Reinforcement	Vol.%	UTS (MPa)	Elongation (%)
AA 5083 (G. Huang *et al.*, 2018a)	WC (inorganic)	11	425	7
AA 5083 (Bauri *et al.*, 2014)	W (metal)	14	404 ± 23	30 ± 1
AA 5083 (Huang and Shen, 2017)	Ti (metal)	30	383 ± 4	19 ± 1
AA 6082 (Thangarasu *et al.*, 2014)	TiC (inorganic)	18	320	8
AA 6082 (Selvakumar *et al.*, 2017a)	Mo (metal)	18	305	15
AA 6082 (Selvakumar *et al.*, 2017b)	SS (metal)	18	293	18

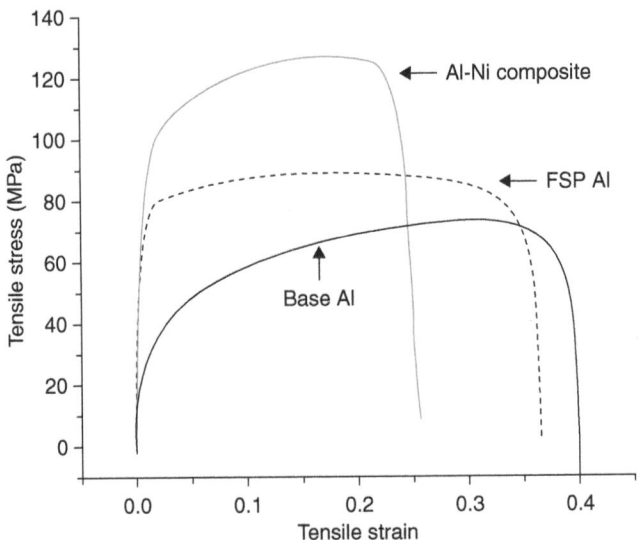

FIGURE 9.8 Stress-strain plot of CP-Al, friction stir processed (FSP) Al, and FSP CP-Al with Ni reinforcement.

Source: Yadav and Bauri, 2011, reprinted with permission.

ductility, as seen in Figure 9.8. The strength improvement is possible owing to a well-developed Al-Ni bond, which is capable of transferring load effectively from the matrix to the reinforcement via the interface. Furthermore, the dislocation density present in the matrix, as well as the Ni reinforcements, hinders the dislocation motion that leads to higher strength. Moreover, the FSP also refines the initial grain size and contributes additional strength. The MMCs retain ductility owing to the absence of brittle interfaces in the MMCs processed by FSP, the ductile phenomenon of Ni particles, and well-developed interfacial bonds. Hence, a combination of strength and ductility can be achieved in the MMCs fabricated through FSP (Yadav and Bauri, 2011).

9.3.2 Effect of Reinforcement Particle Addition on Tensile Strength

MMCs produced by FSP exhibit improved tensile strength compared to the parent material. The enhancement in strength is attributed to the collaborative effect of grain refinement by FSP and strengthening due to reinforcement. The strength of AA 6082 improved notably with the addition of metallic reinforcements (Mo and stainless steel, SS) and further improved with an increase in reinforcement volume (Selvakumar *et al.*, 2017a; Selvakumar *et al.*, 2017b). Furthermore, the total elongation of the MMCs reduced as the volume percentage of reinforcement increases. The tensile strength (UTS) and total elongation with respect to an increase in the volume percentage of Mo and SS reinforcements are shown in Figure 9.9. Similarly, improvement in tensile strength has been reported for varieties of MMCs fabricated by FSP

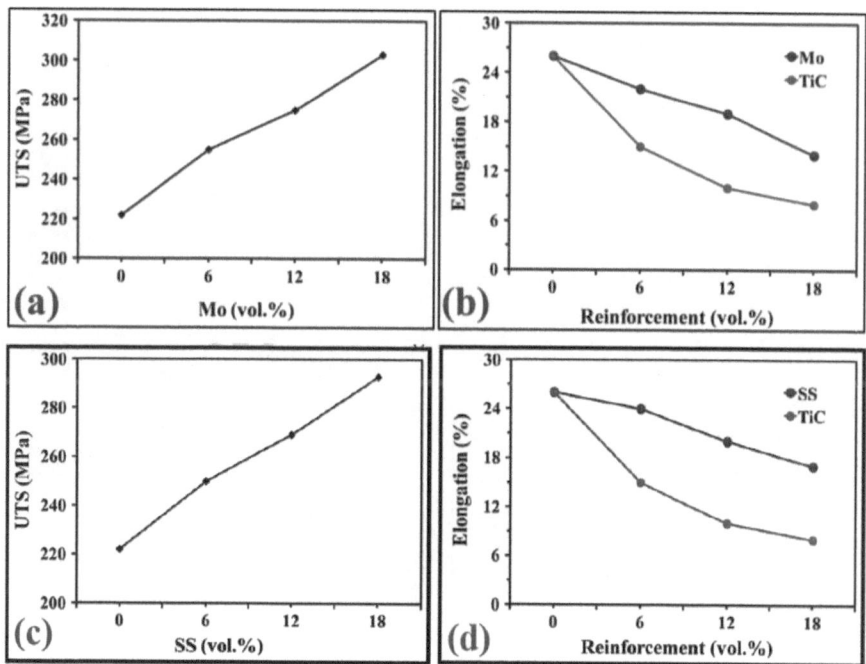

FIGURE 9.9 Effect of Mo addition in AA 6082 on (a) tensile strength and (b) total elongation; and effect of SS addition in AA 6082 on (c) tensile strength and (d) total elongation.

Source: (a, b) Selvakumar *et al.,* 2017a; (c, d) Selvakumar *et al.,* 2017b; all reprinted with permission.

with metallic reinforcement particles (Dixit *et al.*, 2007; Yadav and Bauri, 2010; Yadav and Bauri, 2015; Bauri *et al.*, 2015; Shyam Kumar *et al.*, 2015). It is also inferred from Figure 9.9b and d, that the total elongation is higher in MMCs with metallic reinforcements rather than MMCs with ceramic reinforcement for the same volume fraction. In addition, the surface morphology of the fractured tensile specimens of AA 6082 with metallic reinforcements (Mo and SS) showed a greater number of dimples and micro voids, exhibiting the ductile mode of fracture. Figure 9.10 shows the diverse range of dimples observed on fractured surface of MMCs produced by FSP. The smaller dimples occur owing to the ductile failure of Al matrix, whereas the large dimples are attributed to reinforcements (both Mo and SS). A significant plastic deformation and a flow of AA 6082 matrix before the failure were also reported. Better distribution of metallic reinforcements is clearly visible in the fracture surface morphology (Figure 9.10b–g). Furthermore, during the application of tensile force, a proper load transfer happened between the matrix and the reinforcements as a result of strong bonding. This could be validated by both reinforcement fracture by itself and by pull out (see encircled regions in Figure 9.10), but there is no disintegration of fractured reinforcement particles from the AA 6082 matrix. This suggests that the interfacial bonding of matrix and reinforcement (both Mo and SS) was strong enough.

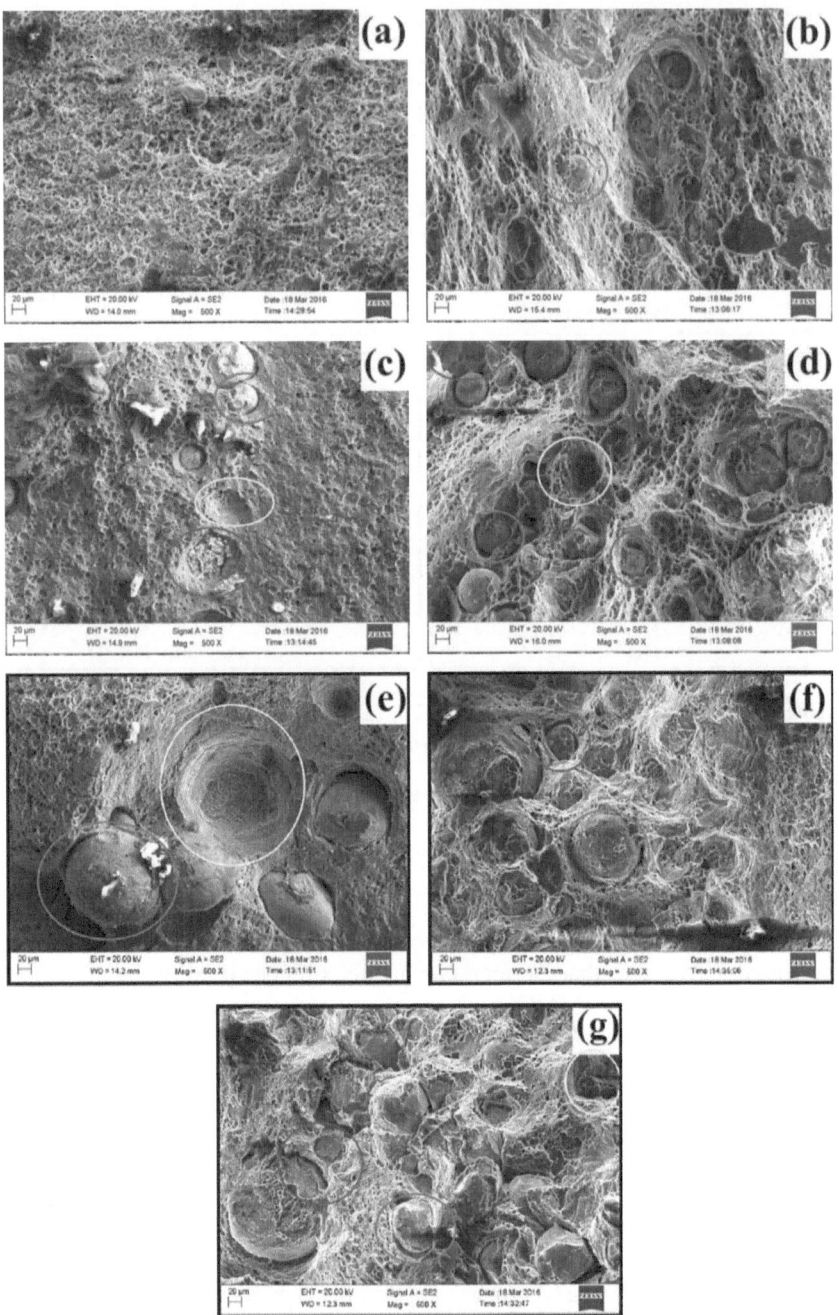

FIGURE 9.10 Fracture surface morphology of fractured tensile specimens of AA 6082 with varying reinforcement processed by FSP: (a) base alloy without any reinforcement, (b) 6 vol.% Mo, (c) 12 vol.% Mo, (d) 18 vol.% Mo, (e) 6 vol.% SS, (f) 12 vol.% SS, (g) 18 vol.% SS.

Sources: (a–d) Selvakumar *et al.,* 2017a; (e–g) Selvakumar *et al.,* 2017b); all reprinted with permission.

9.3.3 EFFECT OF REINFORCEMENT SIZE AND TOOL ROTATIONAL SPEED ON TENSILE STRENGTH

Zohoor *et al.*, 2012, presented some details on the fabrication of AA 5083 MMC with Cu composites. The reinforcement particles are chosen both in nanometric and micrometric levels. Furthermore, the FSP tool rotational speed is varied as 750 and 1900 RPM. The effect of particle size and FSP tool rotational speed on the tensile behaviour is depicted in Figure 9.11. The nanometric Cu particles in MMCs show increased strength compared to the micron-size Cu particles, mostly owing to the fine uniform dispersion of the smaller particles. In addition, the reduction in grain size, which increased the grain boundary area and therefore restricted dislocation movement decreased the thermal mismatch between matrix and particle further supported the strength. The uniform dispersion of the nanometric Cu particles in MMCs also the

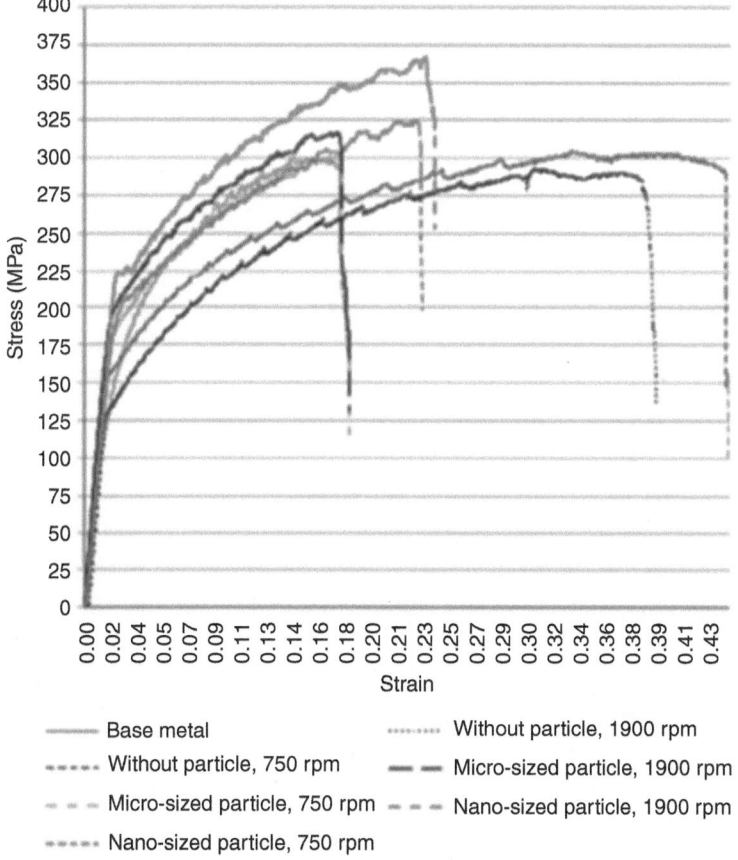

FIGURE 9.11 Stress-strain curves of MMCs fabricated by FSP with nano- and micron-sized Cu reinforcement in AA 5083 processed at two different tool rotational speeds.

Source: Zohoor *et al.,* 2012, reprinted with permission.

ductility. However, the reduced grain size and elevated dislocation density slightly de-
teriorates the elongation irrespective of the fine dispersion of nanometric Cu particles
in the AA 5083 matrix.

The MMCs fabricated at the higher rotational speed (1900 RPM) had greater
tensile strength than those fabricated at the lower speed (750 RPM). At higher tool
rotational speeds a fine intermetallic phase that is smaller than the micron-sized Cu
particles forms. Hence, grain boundary migration is limited, and also the dispersion
of these intermetallic particles causes Orowan strengthening (Yazdipour *et al.*, 2009;
Zhang *et al.*, 2006).

9.3.4   EFFECT OF NUMBER OF FSP PASSES ON TENSILE STRENGTH

Stress-strain curves of Al-W MMC fabricated by FSP with different numbers of
passes are displayed in Figure 9.12. A significant increase in strength and ductility is demonstrated with an increased pass count. The potential reasons for the
improvement in strength are (i) grain size reduction, (ii) increased dislocation density
around reinforcement particles creating geometrically necessary dislocations, (iii)
the W reinforcement acting as nucleation sites and retarding grain boundary migration, leading to further grain refinement, and (iv) the W particles pinning the grain
boundaries and causing Orowan strengthening. With an increasing pass count the
matrix-particle bond is enhanced, thereby improving the ability for load transfer.
Al-W MMCs produced by most other routes produce brittle intermetallic compounds, whereas FSP creates an Al-W solid solution. This solid solution is capable

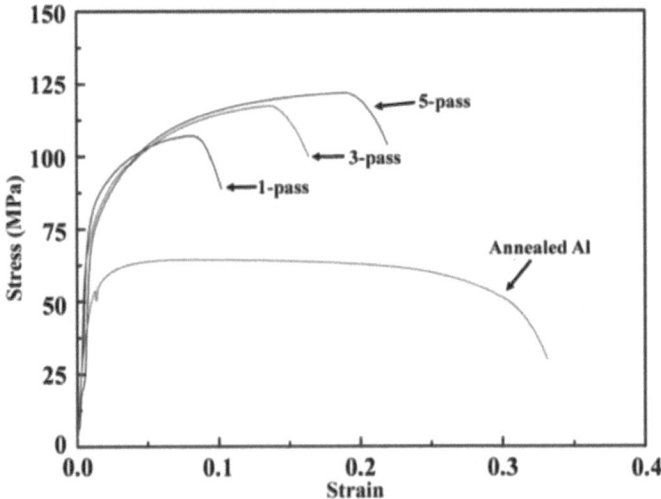

FIGURE 9.12 Variation in stress-strain behaviour of Al-W MMC with respect to the number
of passes during FSP.

Source: G. Huang *et al.*, 2016, reprinted with permission.

of undergoing significant plastic deformation without cracking, whereas the Al-W intermetallic compounds are much more brittle (G. Huang *et al.*, 2016). Similarly, Khorrami *et al.*, 2015, reported improvement in both strength and ductility in Al-Fe MMCs produced by FSP.

9.3.5 EFFECT ON TENSILE STRENGTH OF FSP OF AL-BASED MMCS CONDUCTED UNDERWATER

Huang and Shen, 2017, prepared MMCs with an AA 5083 matrix and Ti reinforcement by FSP. The FSP was conducted in the standard atmosphere as well as underwater for different samples. Stress-strain curves of their results are presented in Figure 9.13. The FSP sample that was prepared underwater demonstrated the highest strength compared to the standard atmosphere–prepared sample and base metal. The strength enhancement of both friction-stirred specimens occurs for the same reasons referred to in Section 9.3.4. However, the additional strength attained by the MMC prepared underwater is due to the presence of smaller grains that inhibit dislocation movement. The smaller grains are due to rapid cooling that slows down the grain growth.

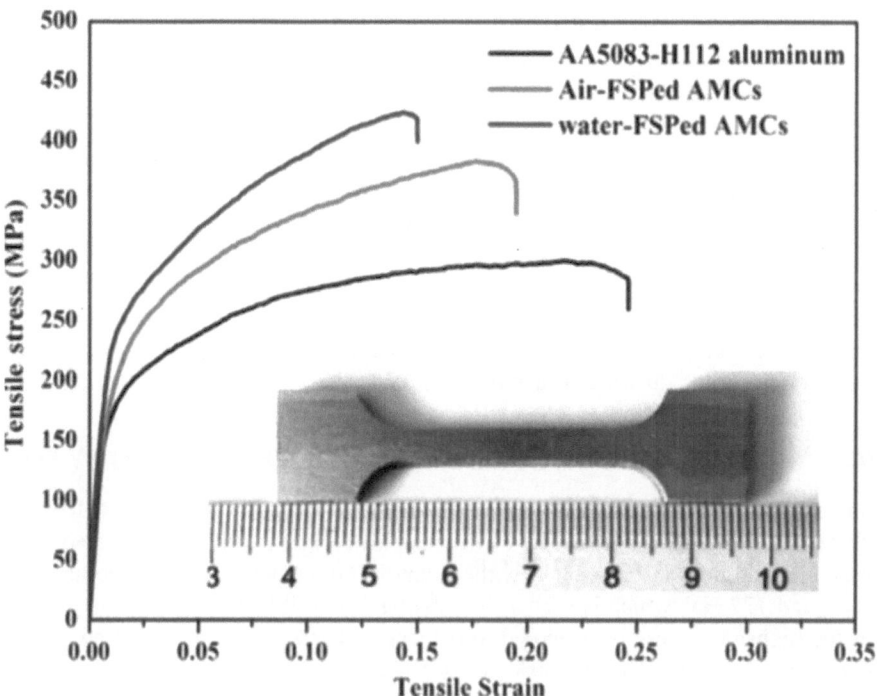

FIGURE 9.13   Stress-strain plot of AA 5083 base material and AA 5083 + Ti AMCs fabricated by FSP in the atmosphere and underwater.

Source: Huang and Shen, 2017, reprinted with permission.

9.3.6 Adaptability of MMCs Fabricated by FSP for Secondary Processing and Their Tensile Behaviour

Dixit *et al.*, 2007, fabricated AA 1100-based MMCs by reinforcing with NiTi and observed the effects of annealing and cold rolling during FSP. The introduction of the reinforcement particles by FSP increases the tensile strength compared to the base metal. This is largely due to a sound bond being realized between the matrix and reinforcing particles. Upon subsequent annealing of the MMCs, tensile strength improved drastically owing to a phase transformation. Cold rolling of the MMCs also resulted in an enhanced strength attributed to work hardening. This indicates that MMCs fabricated by FSP can produce a sound reinforcing particle-matrix bond with the capability to undergo post processing.

9.3.7 Potential of Realizing MMCs with Enhanced Combinations of Strength and Ductility

Table 9.3 presents a comparison of the tensile properties of various MMCs produced by FSP. The data indicate that MMCs produced by FSP display a unique combination of strength and ductility for several reasons.

The notable mechanisms of strengthening in MMCs are (i) Orowan strengthening, (ii) grain boundary strengthening, (iii) work hardening (due to misfit strain between matrix and particle), and (iv) formation of geometrically necessary dislocations due to a thermal mismatch between the matrix and particle. As FSP creates uniform dispersion of particles in the matrix, Orowan strengthening is believed to be dominant. The significant retained ductility displayed by MMCs processed by FSP is due to (i) the absence of any intermetallic compound phases, (ii) well-developed bonds, and (iii) uniform distribution due to the stirring action of FSP (Bauri and Yadav, 2018).

MMCs with a homogeneous reinforcement particle distribution will display enhanced strength compared to a non-homogenous distribution (Barmouz *et al.*, 2011; Bahrami *et al.*, 2014). Agglomeration or grouping of reinforcement particles in the matrix promotes porosity and affects the matrix-particle bonding. This leads to reduced strength and ductility (Liu *et al.*, 2010).

9.4 WEAR BEHAVIOUR OF AL-BASED MMCS FABRICATED BY FSP

Shyam Kumar *et al.*, 2016, recorded higher hardness for AA 5083–W MMCs produced by FSP (Figure 9.14a). The MMCs display a higher hardness compared to the base material and FSP AA 5083 without any reinforcement. The average hardness is around 127 HV, which is higher than the base and FSP material. The realization of the higher hardness is attributed to dispersion hardening. The dispersed W particles in the matrix retard the dislocation motion and therefore induce an increased hardness. Furthermore, a difference in the coefficient of thermal expansion and the associated deformation pattern of AA 5083–W creates additional dislocations. This enhanced dislocation density in the MMCs causes an elevated hardness. For these reasons, the MMCs demonstrate a reduced wear rate under three different loads (Figure 9.14b). As the load increases, the wear rate of the base and unreinforced

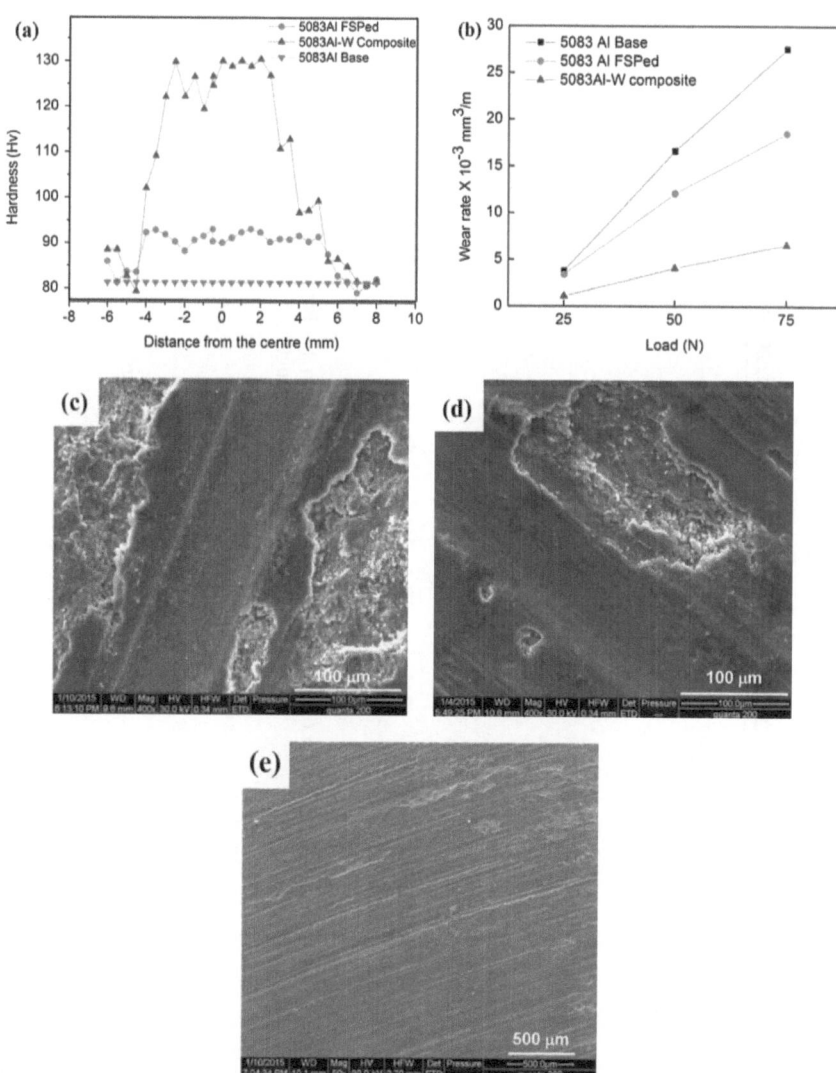

FIGURE 9.14 (a) Hardness profiles, (b) wear rates, and eroded surface morphology of (c) base material, (d) FSP base material, and (e) FSP AA 5083–W MMCs.

Source: Shyam Kumar *et al.,* 2016, reprinted with permission.

FSP material reveal a sharp increase compared to MMCs. Along with the increased hardness, the W particle reinforcement reduced the wear rate of the MMCs even at higher loads. At reduced loads, only mild wear behaviour with shallow grooves is observed. The surface morphology of the eroded regions for the elevated load (75 N) is presented in Figure 9.14c–e for the base material, unreinforced FSP material, and AA 5083–W MMCs, respectively. The surface of the MMCs displays wear from oxidation at the elevated loads. This is due to an increase in frictional heat between

TABLE 9.3
Summary of MMCs Produced by FSP Using Metallic Reinforcement (Process Parameters and Resultant Mechanical Properties)

Reference	Dixit et al., 2007	I. Lee et al., 2008	Yadav and Bauri, 2010	Yadav and Bauri, 2011
MMCs				
Base material	AA 1100	Commercially pure Al	AA 1050	AA 1050
Reinforcement	NiTi	Fe	Ni	Ni
Particle size	2–193 μm	325 mesh size	200 mesh size	200 mesh size
Volume % of reinforcement	—	10	—	7
Composite layer fabrication				
Plate dimension L×W×T (mm)	102 × 76 × 6	12 × 12 × 88	50 W	50 L × 12 T
Method of filling	Hole	Hole	Groove	Groove
Groove/hole depth (mm)	0.9		1	2
Groove width/hole diameter (mm)	1.6		2	1
FSP process parameters				
Tool rotational speed (RPM)	1000	1400	1000	1000
Traverse speed (mm/min)	25	45	60	60
Number of passes	—	4	—	—
Vertical force (kN)	—	—	5	5
Tool				
Tool dimension (mm)	Pin length 2.5	Pin diameter 6; Pin length 1.2	Shoulder diameter 12; Pin diameter4; Pin length 3.5	Shoulder diameter 12; Pin diameter 4; Pin length 3.5
Tensile properties				
UTS (MPa)	175	207 (2nd pass); 217 (4th pass)	127	127
% elongation	-	6.7 (2nd pass); 3.7 (4th pass)	25	25
Reference	Zohoor et al., 2012	Bauri et al., 2014	Yadav and Bauri, 2015	
MMCs				
Base material	AA 5083	AA 5083	Commercially pure Al	
Reinforcement	Cu	W	Cu	
Particle size	Various, nm and μm ranges	—	100 mesh size	
Volume % of reinforcement	—	—	—	
Composite layer fabrication				
Plate dimension L×W×T (mm)	130 × 50 × 5	60 L × 10 T	50 L	
Method of filling	Groove	Groove	Groove	

	Bauri et al., 2015	Shyam Kumar et al., 2015	Khorrami et al., 2015
Groove/hole depth (mm)	1.2	2	2
Groove width/ hole diameter (mm)	1.2	1.5	1
FSP process parameters			
Tool rotational speed (RPM)	750, 1900	1200	1000
Traverse speed (mm/min)	25	24	60
Number of passes	–	–	–
Vertical force (kN)	–	8	–
Tool	H13 steel – threaded and pinless	Threaded pin with spiral grooved shoulders	Hardened steel (M2)
Tool dimension (mm)	Shoulder diameter 16; Pin diameter 6; Pin length 3.2	Shoulder diameter 15; Pin diameter 4; Pin length 3.5	Shoulder diameter 15; Pin diameter 5; Pin length 3.5
Tensile properties			
UTS (MPa)	360 (nm particle @ 1900 RPM) 315 (μm particle @ 1900 RPM)	404 ± 23	114
% Elongation	23 (nm particle @ 1900 RPM) 18 (μm particle @ 1900 RPM)	30 ± 1	22
Reference	Bauri et al., 2015	Shyam Kumar et al., 2015	Khorrami et al., 2015
MMCs			
Base material	AA 5083	AA 5083	Commercially pure Al
Reinforcement	Ni	Ni	Fe
Particle size	20 μm	10 μm	10 μm
Volume % of reinforcement	–	–	–
Composite layer fabrication			
Plate dimension L×W×T (mm)	50 L × 10 T	50 L X 10 T	150 × 70 × 4
Method of filling	Groove	Groove	Groove
Groove/hole depth (mm)	2	2	3
Groove width/ hole diameter (mm)	1	1	1
FSP process parameters			
Tool rotational speed (RPM)	1200	1200	1400
Traverse speed (mm/min)	24	24	100
Number of passes	–	–	1, 2, 3
Vertical force (kN)	–	8	–
Tool	Threaded pin with spiral grooved shoulders	hardened steel (M2)	H13 tool steel

(continued)

TABLE 9.3 (Continued)
Summary of MMCs Produced by FSP Using Metallic Reinforcement (Process Parameters and Resultant Mechanical Properties)

Reference	Dixit et al., 2007	I. Lee et al., 2008	Yadav and Bauri, 2010	Yadav and Bauri, 2011
Tool dimension (mm)	Shoulder diameter 15; Pin diameter 4; Pin length 3.5	Shoulder diameter 15; Pin diameter 4; Pin length 3.5	Shoulder diameter 20; Pin diameter 5; Pin length 3.2	
Tensile properties				
UTS (MPa)	339	362 ± 11	85	
% Elongation	31	24 ± 3	21	
Reference	G. Huang et al., 2016	Selvakumar et al., 2017a	Selvakumar et al., 2017b	
MMCs				
Base material	AA 1060	AA 6082	AA 6082	
Reinforcement	W	Mo	SS (316 L)	
Particle size	1–5 μm	25 μm	25 μm	
Volume % of reinforcement	24	0, 6, 12, 18	0, 6, 12, 18	
Composite layer fabrication				
Plate dimension L×W×T (mm)	120 × 60 × 5	100 × 50 × 10	100 × 50 × 10	
Method of filling	Groove	Groove	Groove	
Groove/hole depth (mm)	3	5.5	5.5	
Groove width/ hole diameter (mm)	2	0.4, 0.8, 1.2	0.4, 0.8, 1.2	
FSP process parameters				
Tool rotational speed (RPM)	1200	1600	1600	
Traverse speed (mm/min)	40	60	60	
Number of passes	1,3,5	1	1	
Vertical force (kN)	-	-	-	
Tool	H13 tempered steel	HCHCr – St. cylinder + without pin	HCHCr – St. cylinder + without pin	
Tool dimension (mm)	Shoulder diameter 20; Pin diameter 6 (tip diameter 5); Pin length 4	Shoulder diameter 18; Pin diameter 6; Pin length 5.8	Shoulder diameter 18; Pin diameter 6; Pin length 5.8	
Tensile properties				
UTS (MPa)	120 (for 5th pass)	293 @ 18 vol.%	303 @ 18 vol.%	
% Elongation	12 (for 5th pass)	17% @ 18 vol.%	14% @ 18 vol.%	

the pin and surface that increases the localized temperature, inducing oxidation of the transferring material. However, compared to the base and unreinforced FSP material, the MMCs displayed less overall wear.

In another study, G. Huang *et al.*, 2018b, produced AA 1060–Cu MMCs by FSP with different pass counts and report increased wear resistance with respect to an increase in FSP pass count. The typical wear surface morphologies of MMCs subjected to various pass counts are shown in Figure 9.15a–d. The improved wear resistance is due to the improvement in hardness (Figure 9.15e) and uniform distribution of in-situ Al_2Cu particles. As the pass count increases, the Al_2Cu particles form an improved bond with the matrix, and only a small number of fragments are detached. These fine and evenly distributed (due to multiple passes), high hardness Al_2Cu particles effectively reduce abrasion, ploughing, and steep-sided grooves.

Azizieh *et al.*, 2019, conducted FSP on AA 1100 with Fe_2O_3 reinforcement particles with multiple passes (2, 4, and 6), which resulted in the formation of an in-situ $Al_{13}Fe_4$ composite. Similarly to the studies mentioned previously, this study demonstrates an increase in hardness with an increase in FSP pass count. Furthermore, the distribution of $Al_{13}Fe_4$ is more uniform, and the volume fraction of the particles also increases with pass count leading to an improved hardness. In line with this research, Behnagh *et al.*, 2012, reported that FSP enhanced hardness in the Al-Mg alloy with a proportional decrement in the wear rate.

9.5 METALLURGICAL ASPECTS OF AL-BASED MMCS FABRICATED BY FSP

Yadav and Bauri, 2010, fabricated AA 1050-based MMCs by reinforcing with Ni particles and studied the structure-property correlation in response to the FSP process. Figure 9.16a displays a uniform distribution of Ni particles with no clustering and segregation of particles in the Al matrix. Furthermore, an excellent matrix-particle interface is demonstrated in Figure 9.16b. The cross-section of the stir zone in Figure 9.16c shows that the Ni particles are uniformly distributed within the trace of the stir zone to a depth of approximately 2 mm. The tensile and wear properties of MMCs are affected by the particle distribution. Uniformity with intragranular dispersion of reinforcements is encouraged to obtain improved properties for MMCs. This is mainly made possible with FSP. The uniform dispersion of reinforcement particles observed in the micrographs indicates that the processing parameters selected for the associated FSP were effective.

The refined grain structure is formed as a result of the dynamic recrystallization process that occurs during FSP (Mishra and Ma, 2005). A recrystallized and equiaxed microstructure is observed in Al-Ni MMCs fabricated by Yadav and Bauri, 2011 (Figure 9.17a). The Ni reinforcement particles are effectively embedded in the matrix. The average grain size before FSP was 74 µm. After FSP this reduced to 7 µm (only in the matrix). The refined equiaxed microstructure of unreinforced Al after FSP is displayed in Figure 9.17b, and the corresponding grain size distribution, as well as the misorientation plot, is shown in Figure 9.17c–e), respectively. The low-angle grain boundaries of 5° or less are considered sub-grains formed as a result of dislocation rearrangement. As Al is a high stacking fault energy metal, dynamic

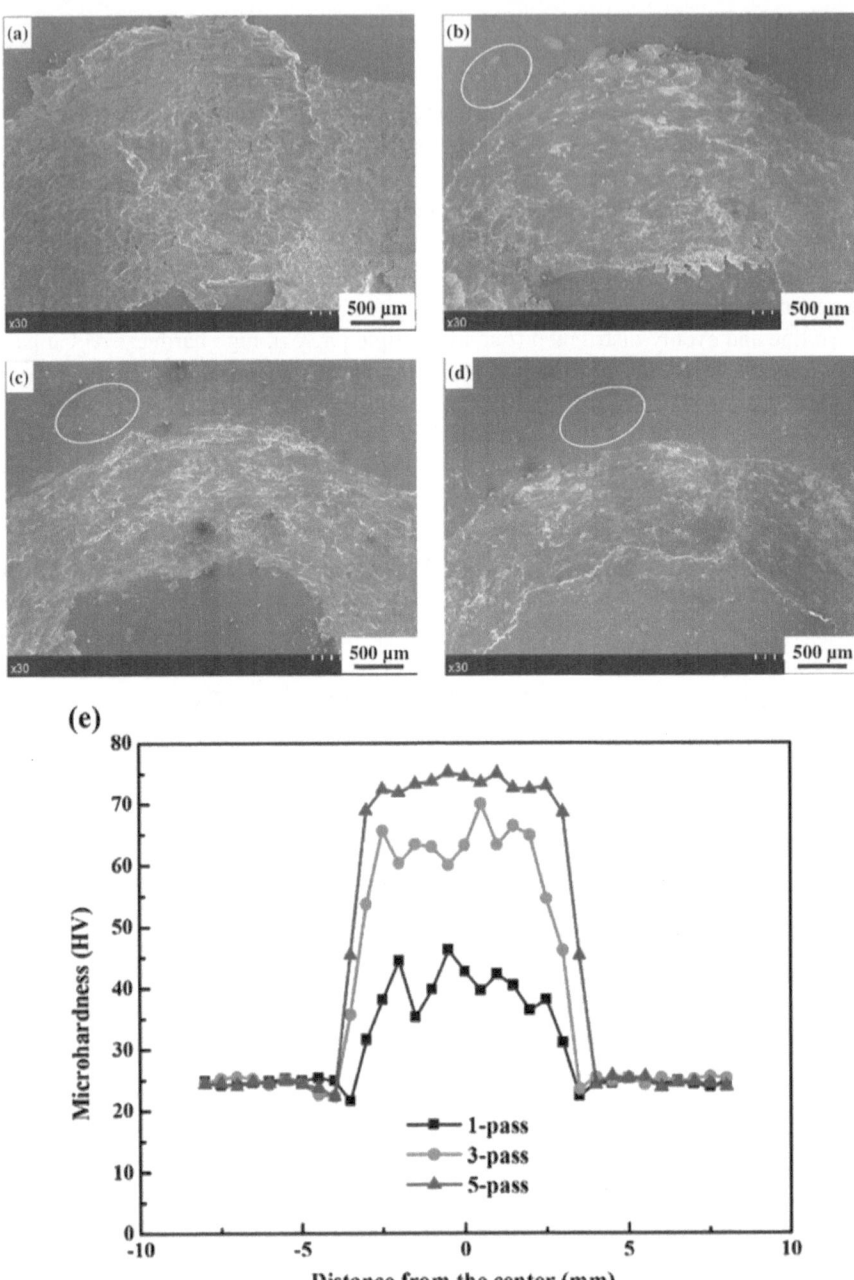

FIGURE 9.15 Wear surface morphology of (a) base material (AA 1060), (b–d) AA 1060–Cu MMCs after first pass, third pass, and fifth pass, respectively. (e) Metallurgical aspects of Al-based MMCs fabricated by FSP.

Source: G. Huang *et al.,* 2018b, reprinted with permission.

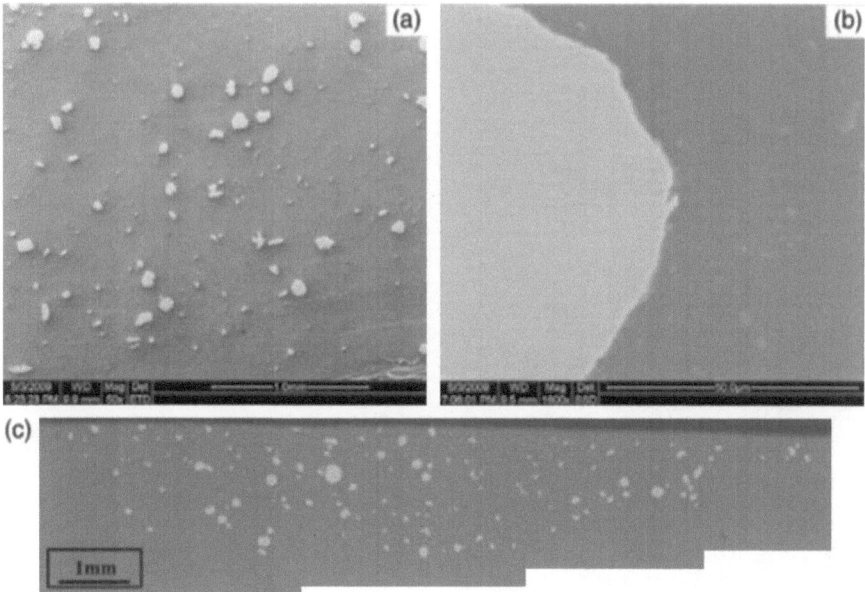

FIGURE 9.16 Al-Ni MMCs fabricated by FSP showing (a) uniform distribution of Ni particles over the surface of the stir zone, (b) the interface of matrix (Al) and particle (Ni), and (c) cross-section of stir zone with Ni particles.

Source: Yadav and Bauri, 2010, reprinted with permission.

recovery is potentially high at the time of FSP and causes the arrangement of disloca-tions into sub-grains.

Furthermore, additional dislocations are formed as a result of the thermal mismatch of Al-Ni that adds to the rearrangement process. The sub-grain boundaries gradually transform into low-angle grain boundaries (LAGBs). The dislocations produced as a result of the deformation process inducted to the sub-grain boundaries increase the misorientation and transform the sub-grains to LAGBs. The dislocation gliding with lattice rotation leads to the formation of high-angle grain boundaries (HAGBs), forming recrystallized fine grains. Hence, continuous dynamic recrystallization oc-curs as a result of the transformation of LAGBs to HAGBs (Jata and Semiatin, 2000; Gourdet and Montheillet, 2003). Therefore, dynamic recrystallization occurs during FSP and dynamic recovery takes place. This microstructural transformation is shown in Figure 9.18, and similar microstructural evolution was recorded by researchers (Bauri *et al.*, 2014; Shyam Kumar *et al.*, 2016) in the fabrication of Al-based MMCs by FSP using tungsten metallic reinforcements.

Selvakumar *et al.*, 2017b, described grain size evolution in AA 6082-SS MMCs produced by FSP. The SEM-EBSD image (Figure 9.19a) of the as-procured AA 6082 shows elongated grains due to the effect of rolling. The MMCs produced with varying volume fractions of SS show refined and mostly equiaxed grains (Figure 9.19b–d). As per Ma, 2008, dynamic recrystallization during FSP induces a fine-grained structure.

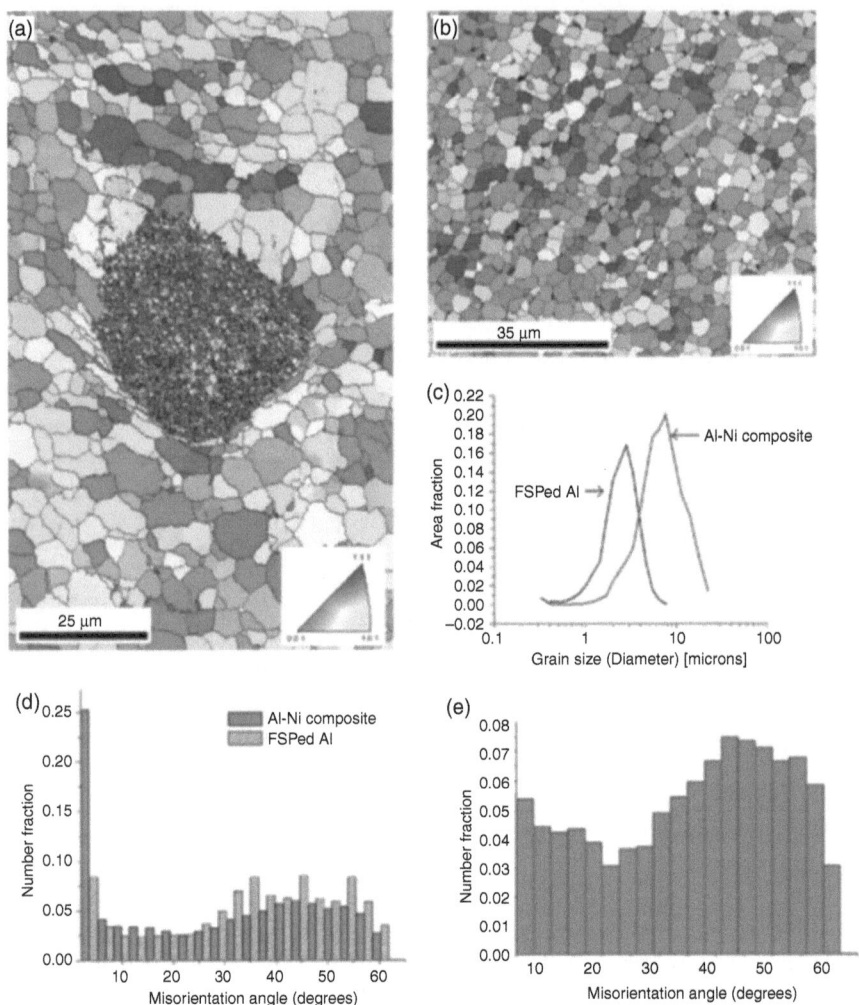

FIGURE 9.17 SEM-EBSD of (a) Al-Ni MMCs, (b) FSP Al without Ni reinforcement, (c) grain size distribution plot, (d) misorientation plots for parts (a and b), (e) misorientation plot for Al-Ni MMCs with grain boundary angles >5°.

Source: Yadav and Bauri, 2011, reprinted with permission.

Recrystallization takes place in FSP as a result of frictional heat and strain. The grain size effectively reduces owing to an increase in the SS volume fraction (Figure 9.19e). This reduction in grain size is also visible in the SEM-EBSD images (Figure 9.19b–d). As the average grain size reported is smaller than that of the SS particles owing to FSP, the probability of the SS particles being effectively surrounded by numerous grains is increased. The SS reinforcement therefore may have the ability to be utilized as a grain refiner for AA 6082. The pinned SS particles effectively induce grain

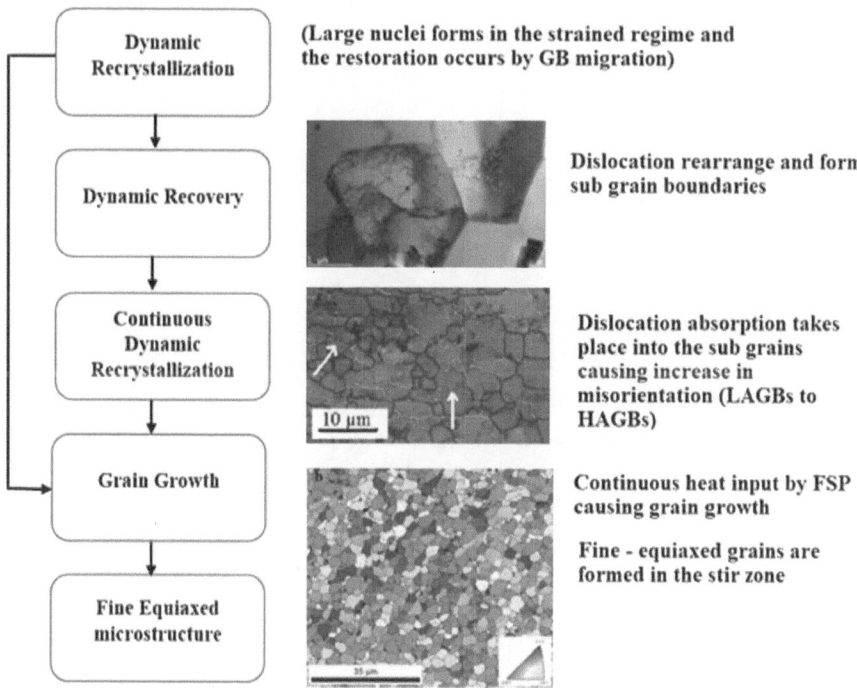

FIGURE 9.18   Microstructural transformation in Al alloys during FSP.

Sources: Flow chart modified from Bauri and Yadav 2018, images reprinted with permission from Yadav and Bauri, 2011.

refinement. Similarly, Selvakumar *et al.*, 2017a, demonstrated that Mo particles may also pin the grain boundaries in an Al matrix, causing a refined grain structure.

9.6   SUMMARY AND FUTURE ORIENTATION

This chapter presents an overview of the fabrication of Al-based MMCs reinforced with metallic particles by using friction stir processing (FSP). The microstructural, tensile, and wear behaviour of different Al-based MMCs reinforced with metallic particles are explored. The effects of process parameters, shoulder and other tool geometry, and processing methods are also considered. The influence of process parameters on the link between the structure and properties are also introduced. It is shown that an increase in tool rotational speed and an increase in FSP pass count enhances the uniformity of the particle distribution. FSP is an effective method to fabricate MMCs with uniform distribution of reinforcement particles by selecting appropriate process parameters and tool design. Utilizing FSP, microstructurally sound MMCs may be fabricated that display homogeneous reinforcement particle dispersion with fewer defects than with the liquid and powder metallurgy routes. FSP typically produces an effective interface between the matrix and reinforcement without any formation

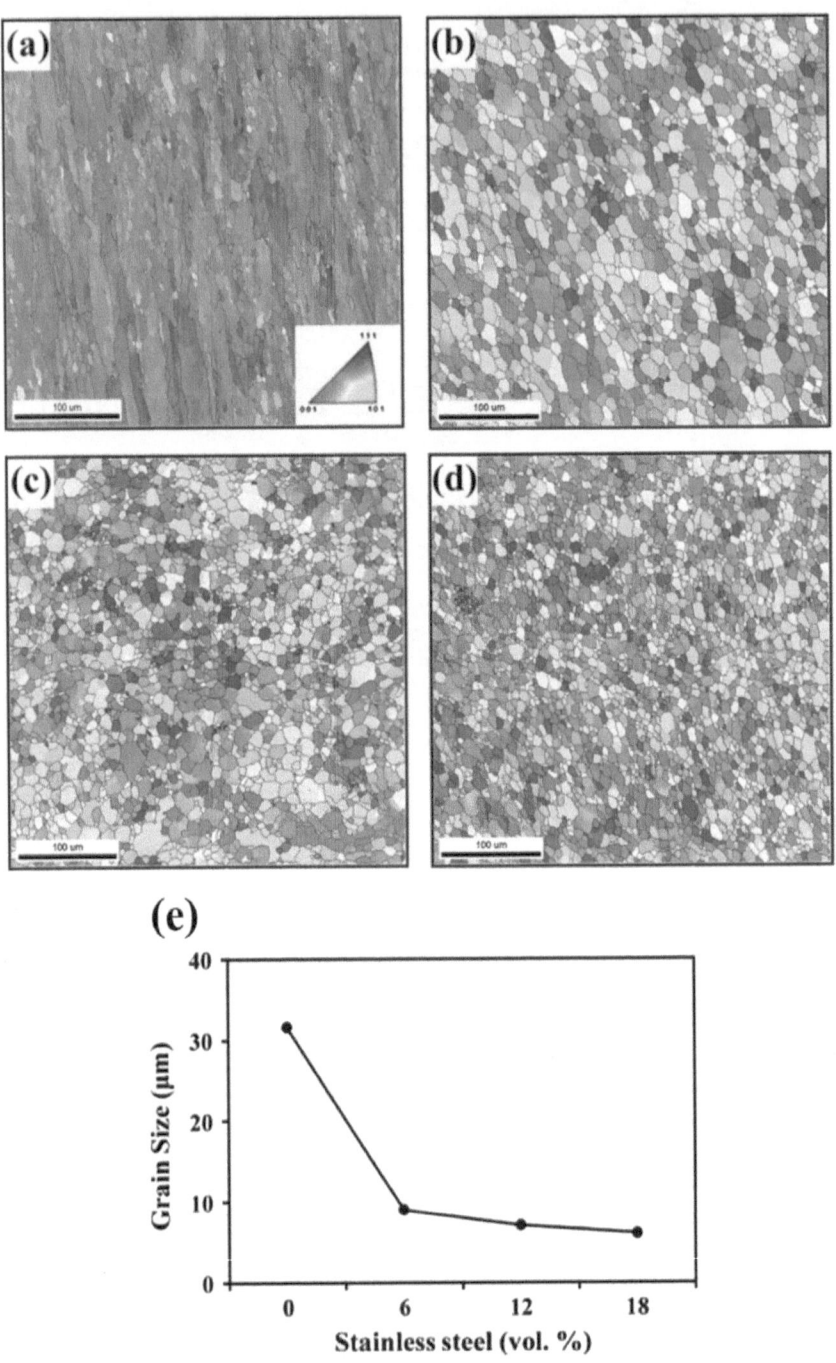

FIGURE 9.19 SEM-EBSD images of (a) AA 6082 base material, (b) AA 6082 + 6 vol.% SS, (c) AA 6082 + 12 vol.% SS, (d) AA 6082 + 18 vol.% SS, (e) effect on grain size of SS addition to AA 6082 matrix.

Source: Selvakumar et al., 2017b, reprinted with permission.

of deteriorating intermetallic phases. As the bond between matrix and reinforcement is typically excellent, the load-bearing and load-transfer ability are high. The fine-grained microstructure evolved through FSP contributes to these elevated properties. Groove filling and hole filling or any multi-track techniques can be adapted for making MMCs. MMCs fabricated by FSP typically display improved mechanical strength, mostly owing to grain refinement and Orowan strengthening (particle pinning effect). Along with the improved strength, sufficient ductility is retained because of the excellent matrix-particle interface. Hence, MMCs demonstrate a combination of both good strength and ductility. Improvement in hardness and therefore the corresponding wear resistance is observed in the MMCs compared to the base material.

Most research on FSP has concentrated on its application to lightweight materials such as Al and Mg. Limited work exists on high strength materials including Ti and steel. The general use of FSP may be significantly expanded if more research is conducted with higher strength materials and composites. One of the major challenges that will have to be overcome relates to the tool life, especially for the processing of high strength materials. Tools made of cubic boron nitride and tungsten-based materials are available for FSP on high strength materials. However, there is no detailed study on tool wear on such materials when FSP is applied to high strength materials, which needs further attention. Polymer-based composites are also attracting more interest, especially when fabricated by FSP. Surface modification and/or alloying is another possible use for the application of MMC fabrication by FSP. This may lead to new developments in regard to surface behaviour, including tribology. A few solid-state processes other than FSP, including high pressure torsion (Milhorato *et al.,* 2020) and accumulative roll bonding (Ramkumar and Dinaharan, 2020), can produce surface composites, which gain attractive mechanical and tribological properties.

REFERENCES

Arunachalam, R., P.K. Krishnan and R. Muraliraja. 2019. A review on the production of metal matrix composites through stir casting – Furnace design, properties, challenges, and research opportunities. *Journal of Manufacturing Processes* 42 (June): 213–245. https://doi.org/10.1016/j.jmapro.2019.04.017.

Azizieh, M., R. Pourmodheji, A.N. Larki, M.A.G. Dezfuli, A. Rezaei, and H.S. Kim. Effect of multi-pass friction stir processing on the microstructure and hardness of AA1100/Al13Fe4 in situ composites. *Materials Research Express* 6, no. 4 (2019): 046558.

Bahrami, M., M.K.B. Givi, K. Dehghani and N. Parvin. 2014. On the role of pin geometry in microstructure and mechanical properties of AA7075/SiC nano-composite fabricated by friction stir welding technique. *Materials & Design* 53, (January): 519–527. https://doi.org/10.1016/j.matdes.2013.07.049.

Barmouz, M., M.K.B. Givi and J. Seyfi. 2011. On the role of processing parameters in producing Cu/SiC metal matrix composites via friction stir processing: Investigating microstructure, microhardness, wear and tensile behavior. *Materials Characterization* 62, no. 1 (January): 108–117. https://doi.org/10.1016/j.matchar.2010.11.005.

Bauri, R., and D. Yadav. 2018. *Metal Matrix Composites by Friction Stir Processing.* Butterworth-Heinemann.

Bauri, R., D. Yadav and G. Suhas. 2011. Effect of friction stir processing (FSP) on microstructure and properties of Al–TiC in situ composite. *Materials Science and Engineering A* 528, no. 13–14 (May): 4732–4739. https://doi:10.1016/j.msea.2011.02.085.

Bauri, R., D. Yadav, C.N. Shyam Kumar and B. Balaji. 2014. Tungsten particle reinforced Al5083 composite with high strength and ductility. *Materials Science and Engineering A* 620, no. 3 (January): 67–75. https://doi.org/10.1016/j.msea.2014.09.108.

Bauri, R., G.D. Janaki Ram, D. Yadav and C.N. Shyam Kumar. 2015. Effect of process parameters and tool geometry on fabrication of Ni particles reinforced 5083 Al composite by friction stir processing. *Materials Today: Proceedings* 2, no. 4–5 (September): 3203–3211. https://doi.org/10.1016/j.matpr.2015.07.115.

Behnagh, R.A., M.K.B. Givi and M. Akbari. 2012. Mechanical properties, corrosion resistance, and microstructural changes during friction stir processing of 5083 aluminum rolled plates. *Materials and Manufacturing Processes* 27, no. 6 (May): 636–640. https://doi.org/10.1080/10426914.2011.593243.

Bhanu Prasad, V.V., K.S. Prasad, A.K. Kuruvilla, A.B. Pandey, B.V.R. Bhat and Y.R. Mahajan. 1991. Composite strengthening in 6061 and Al-4 Mg alloys. *Journal of Materials Science* 26 (January): 460–466. https://doi.org/10.1007/BF00576543.

Blucher, J.T. 1992. Discussion of a liquid metal pressure infiltration process to produce metal matrix composites. *Journal of Materials Processing Technology* 30, no. 3 (April): 381–390. https://doi.org/10.1016/0924-0136(92)90227-J.

Clyne, T.W., and F.R. Jones. 2006. Metal matrix composites: Matrices and processing. In *Concise Encyclopedia of Composite Materials,* ed. A. Mortensen, Amsterdam: Elsevier Science.

Dash, K., D. Chaira and B.C. Ray. 2013. Synthesis and characterization of aluminium–alumina micro- and nano-composites by spark plasma sintering. *Materials Research Bulletin* 48, no. 7 (July): 2535–2542. https://doi.org/10.1016/j.materresbull.2013.03.014.

David Raja Selvam, J., I. Dinaharan, S. Vibin Philip and P.M. Mashinini. 2018. Microstructure and mechanical characterization of in situ synthesized AA6061/(TiB$_2$ + Al$_2$O$_3$) hybrid aluminum matrix composites. *Journal of Alloys and Compounds* 740 (April): 529–535. https://doi.org/10.1016/j.jallcom.2018.01.016.

Dinaharan, I., N. Murugan and E.T. Akinlabi. 2020. Friction stir processing route for metallic matrix composite production. In *Encyclopedia of materials: composites*, ed. D. Brabazon, Amsterdam: Elsevier.

Dixit, M., J.W. Newkirk and R.S. Mishra. 2007. Properties of friction stir-processed Al 1100–NiTi composite. *Scripta Materialia* 56, no. 6 (March): 541–544. https://doi.org/10.1016/j.scriptamat.2006.11.006.

Dolatkhah, A., P. Golbabaei, M.K.B. Givi and F. Molaiekiya. 2012. Investigating effects of process parameters on microstructural and mechanical properties of Al5052/SiC metal matrix composite fabricated via friction stir processing. *Materials & Design* 37 (May): 458–464. https://doi.org/10.1016/j.matdes.2011.09.035.

Faraji, G., and P. Asadi. 2011. Characterization of AZ91/alumina nanocomposite produced by FSP. *Materials Science and Engineering A* 528, no. 6 (March): 2431–2440. https://doi.org/10.1016/j.msea.2010.11.065.

Gecu, R., Ş.H. Atapek and A. Karaaslan. 2017. Influence of preform preheating on dry sliding wear behavior of 304 stainless steel reinforced A356 aluminum matrix composite produced by melt infiltration casting. *Tribology International* 115, 608–618. https://doi.org/10.1016/j.triboint.2017.06.040.

Girish, G., and V. Anandakrishnan. 2019. Investigations on microstructural and texture evolution during recursive friction stir processing of aluminium 7075 alloy. *Materials Research Express* 6, 126574. https://doi.org/10.1088/2053-1591/ab58ed.

Gourdet, S., and F. Montheillet. 2003. A model of continuous dynamic recrystallization. *Acta Materialia* 51, no. 9 (May): 2685–2699. https://doi.org/10.1016/S1359-6454(03)00078-8.

Guo, J.F., J. Liu, C.N. Sun, S. Maleksaeedi, G. Bi, M.J. Tan and J. Wei. 2014. Effects of nano-Al$_2$O$_3$ particle addition on grain structure evolution and mechanical behaviour of friction-stir-processed Al. *Materials Science and Engineering A* 602 (April): 143–149. https://doi.org/10.1016/j.msea.2014.02.022.

Huang, G., and Y. Shen. 2017. The effects of processing environments on the microstructure and mechanical properties of the Ti/5083 Al composites produced by friction stir processing. *Journal of Manufacturing Processes* 30 (December): 361–373. https://doi.org/10.1016/j.jmapro.2017.10.007.

Huang, G., W. Hou and Y. Shen. 2016. Fabrication of tungsten particles reinforced aluminum matrix using multi-pass friction stir processing: Evaluation of microstructural, mechanical and electrical behavior. *Materials Science and Engineering A* 674 (September): 504–513. https://doi.org/10.1016/j.msea.2016.07.124.

Huang, G., W. Hou and Y. Shen. 2018a. Evaluation of the microstructure and mechanical properties of WC particle reinforced aluminum matrix composites fabricated by friction stir processing. *Materials Characterization* 138 (April): 26–37. https://doi.org/10.1016/j.matchar.2018.01.053.

Huang, G., W. Hou, J. Li, and Y. Shen. 2018b. Development of surface composite based on Al-Cu system by friction stir processing: Evaluation of microstructure, formation mechanism and wear behavior. *Surface & Coatings Technology* 344 (June): 30–42. https://doi.org/10.1016/j.surfcoat.2018.03.005.

Huang, Y., T. Wang, W. Guo, L. Wan and S. Lv. 2014. Microstructure and surface mechanical property of AZ31 Mg/SiC surface composite fabricated by direct friction stir processing. *Materials & Design* 59 (July): 274–278. https://doi.org/10.1016/j.matdes.2014.02.067.

Jata, K.V., and S.L. Semiatin. 2000. Continuous dynamic recrystallization during friction stir welding of high strength aluminum alloys. *Scripta Materialia* 43, no. 8 (September): 743–749. https://doi.org/10.1016/S1359-6462(00)00480-2.

Jayaraman, M., R. Sivasubramanian, V. Balasubramanian and A.K. Lakshminarayanan. 2009. Optimization of process parameters for friction stir welding of cast aluminium alloy A319 by Taguchi method. *Journal of Scientific and Industrial Research* 68 (January): 36–43.

Kaczmar, J.W., K. Pietrzak, W. Wøosin Âski. 2000. The production and application of metal matrix composite materials. *Journal of Materials Processing Technology* 106, no. 1–3 (October): 58–67. https://doi.org/10.1016/S0924-0136(00)00639-7.

Kalra, C., S. Tiwari, A. Sapra, S. Mahajan and P. Gupta. 2018. Processing and characterization of hybrid metal matrix composites. *Journal of Materials and Environmental Sciences* 9, no. 7 (April): 1979–1986. www.jmaterenvironsci.com/Document/vol9/vol9_N7/218-JMES-3296-Kalra.pdf.

Kaur, K., and O.P. Pandey. 2010. Microstructural characteristics of spray formed zircon sand reinforced LM13 composite. *Journal of Alloys and Compounds* 503, no. 2 (August): 410–415. https://doi.org/10.1016/j.jallcom.2010.04.249.

Khorrami, M.S., S. Samadi, Z. Janghorban and M. Movahedi. 2015. In-situ aluminum matrix composite produced by friction stir processing using FE particles. *Materials Science and Engineering A* 641 (August): 380–390. https://doi.org/10.1016/j.msea.2015.06.071.

Kingsley, T., and O.M. Suárez. 2011. Study of boride-reinforced aluminum matrix composites produced via centrifugal casting. *Material and Manufacturing Process* 26, no. 2: 338–345. https://doi.org/10.1080/10426910903124829.

Kong, L.B., J. Ma and H. Huang. 2002. $MgAl_2O_4$ spinel phase derived from oxide mixture activated by a high-energy ball milling process. *Materials Letters* 56, no. 3 (October): 238–243. https://doi.org/10.1016/S0167-577X(02)00447-0.

Kumar, Anil K., S. Natarajan, M. Duraiselvam and S. Ramachandra. 2019. Synthesis, characterization and mechanical behavior of Al 3003–TiO_2 surface composites through friction stir processing. *Materials and Manufacturing Processes* 34, no. 2 (January): 183–191. https://doi.org/10.1080/10426914.2018.1544711.

Kurt, A., I. Uygur and E. Cete. 2011. Surface modification of aluminium by friction stir processing. *Journal of Material Processing Technology* 211, no. 3 (March): 313–317. https://doi.org/10.1016/j.jmatprotec.2010.09.020.

Lee, C.J., J.C. Huang and P.J. Hsieh. 2006. Mg based nano-composites fabricated by friction stir processing. *Scripta Materialia* 54, no.7 (April): 1415–1420. https://doi.org/10.1016/j.scriptamat.2005.11.056.

Lee, I.S., P.W. Kao and N.J. Ho. 2008. Microstructure and mechanical properties of Al–Fe in situ nanocomposite produced by friction stir processing. *Intermetallics* 16, no. 9 (September): 1104–1108. https://doi.org/10.1016/j.intermet.2008.06.017.

Liu, X., Y. Liu, D. Huang, Q. Han and X. Wang. 2017. Tailoring in-situ TiB2 particulates in aluminum matrix composites. *Materials Science and Engineering A* 705 (September): 55–61. https://doi.org/10.1016/j.msea.2017.08.047.

Liu, Z.Y., Q.Z. Wang, B.L. Xiao, Z.Y. Ma. 2010. Clustering model on the tensile strength of PM processed SiCp/Al composites. *Composites Part A: Applied Science and Manufacturing* 41, no. 11 (November): 1686–1692. https://doi.org/10.1016/j.compositesa.2010.08.007.

Ma, Z.Y. 2008. Friction stir processing technology: A review. *Metallurgical and Materials Transactions A* 39 (February): 642–658. https://doi.org/10.1007/s11661-007-9459-0.

Mertens, A., A. Simar, H.M. Montrieux, J. Halleux, F. Delannay and J. Lecomte Beckers. 2012. Friction stir processing of magnesium matrix composites reinforced with carbon fibres: Influence of the matrix characteristics and of the processing parameters on microstructural developments. In *Proceedings of the 9th International Conference on Magnesium Alloys and Their Applications,* ed. W. J. Poole and K. U. Kainer, 845–850. Vancouver: Conference Proceedings.

Milhorato, F.R., R.B. Figueiredo, T.G. Langdon and E.M. Mazzer. 2020. Development of an Al 7050-10 vol.% alumina nanocomposite through cold consolidation of particles by high-pressure torsion. *Journal of Materials Research and Technology* 9, no. 6 (November): 12626–12633. https://doi.org/10.1016/j.jmrt.2020.09.014.

Miracle, D.B. 2005. Metal matrix composites – From science to technological significance. *Composite Science Technology* 65, no. 15–16 (December): 2526–2540. https://doi.org/10.1016/j.compscitech.2005.05.027.

Mishra, R.S., and Z.Y. Ma. 2005. Friction stir welding and processing. *Materials Science and Engineering R* 50, no. 1–2 (August): 1–78. https://doi.org/10.1016/j.mser.2005.07.001.

Mishra, R.S., Z.Y. Ma and I. Charit. 2003. Friction stir processing: A novel technique for fabrication of surface composite. *Materials Science and Engineering A* 341, no. 1–2 (January): 307–310. https://doi.org/10.1016/S0921-5093(02)00199-5.

Muralidharan, N., K. Chockalingam, I. Dinaharan and K. Kalaiselvan. 2018. Microstructure and mechanical behavior of AA2024 aluminum matrix composites reinforced with in situ synthesized ZrB2 particles. *Journal of Alloys and Compounds* 735 (February): 2167–2174. https://doi.org/10.1016/j.jallcom.2017.11.371.

Ni, D.R., J.J. Wang, Z.N. Zhou and Z.Y. Ma. 2014. Fabrication and mechanical properties of bulk NiTip/Al composites prepared by friction stir processing. *Journal of Alloys and Compounds* 586 (February): 368–374. https://doi.org/10.1016/j.jallcom.2013.10.013.

Oosterkamp, A. A. N. A., L. Djapic Oosterkamp, and A. Nordeide. Kissing bond phenomena in solid-state welds of aluminum alloys. *Welding Journal-New York* 83, no. 8 (2004): 225–S.

Padmanaban, G., and V. Balasubramanian. 2009. Selection of FSW tool pin profile, shoulder diameter and material for joining AZ31B magnesium alloy – An experimental approach. *Materials & Design* 30, no. 7 (August): 2647–2656. https://doi.org/10.1016/j.matdes.2008.10.021.

Palanivel, R., R.F. Laubscher, S. Vigneshwaran and I. Dinaharan. 2018. Prediction and optimization of the mechanical properties of dissimilar friction stir welding of aluminum alloys using design of experiments. *Journal of Engineering Manufacture* 232, no. 8 (June): 1384–1394. https://doi.org/10.1177/0954405416667404.

Rajan, H.B.M., I. Dinaharan, S. Ramabalan and E.T. Akinlabi. 2016. Influence of friction stir processing on microstructure and properties of AA7075/TiB$_2$ in situ composite. *Journal of Alloys and Compounds* 657 (February): 250–260. https://doi.org/10.1016/j.jallcom.2015.10.108.

Ramkumar, K.R., and I. Dinaharan. 2020. Accumulative roll bonding route for composite materials production. In *Encyclopedia of Materials: Composites*, ed. D. Brabazon, Amsterdam: Elsevier.

Rathee, S., S. Maheshwari, A.N. Siddiquee and M. Srivastava. 2018. A review of recent progress in solid state fabrication of composites and functionally graded systems via friction stir processing. *Critical Reviews in Solid State and Materials Sciences* 43 (September): 334–366. https://doi.org/10.1080/10408436.2017.1358146.

Ratna Sunil, B. 2016. Different strategies of secondary phase incorporation into metallic sheets by friction stir processing in developing surface composites. *International Journal of Mechanical and Materials Engineering* 11, no. 12 (November): 1–8. https://doi.org/10.1186/s40712-016-0066-y.

Razavi, M., A.R. Farajipour, M. Zakeri, M.R. Rahimipour and A.R. Firouzbakht. 2017. Production of Al$_2$O$_3$–SiC nano-composites by spark plasma sintering. *Boletín de la Sociedad Española de Cerámica y Vidrio* 56, no. 4 (July): 186–194. https://doi.org/10.1016/j.bsecv.2017.01.002.

Reddy, M.P., R.A. Shakoor, G. Parande, V. Manakari, F. Ubaid, A.M.A. Mohamed and M. Gupta. 2017. Enhanced performance of nano-sized SiC reinforced Al metal matrix nanocomposites synthesized through microwave sintering and hot extrusion techniques. *Progress in Natural Science: Materials International* 27, no. 5 (October): 606–614. https://doi.org/10.1016/j.pnsc.2017.08.015.

Rejil, C. Maxwell, I. Dinaharan, S.J. Vijay and N. Murugan. 2012. Microstructure and sliding wear behavior of AA6360/(TiC+ B4C) hybrid surface composite layer synthesized by friction stir processing on aluminum substrate. *Materials Science and Engineering A* 552: 336–344. https://doi.org/10.1016/j.msea.2012.05.049.

Rozak, G.A., J.J. Lewandowski, J.F. Wallace, S. Altmy and A. Oglu. 1992. Effects of casting conditions and deformation processing on A356 aluminum and A356-20 vol% SiC composites. *Journal of Composite Materials* 26 (December): 2079–2106. https://doi.org/10.1177/002199839202601405.

Sahraeinejad, S., H. Izadi, M. Haghshenas and A.P. Gerlich. 2015. Fabrication of metal matrix composites by friction stir processing with different particles and processing parameters. *Materials Science and Engineering A* 626, (February): 505–513. https://doi.org/10.1016/j.msea.2014.12.077.

Seenuvasaperumal, P., K.D. Dudekula, A. Basha, A. Singh, A. Elayaperumal and K. Tsuchiya. 2018. Wear behavior of HPT processed UFG AZ31B magnesium alloy. *Materials Letters* 227 (September): 194–198. https://doi.org/10.1016/j.matlet.2018.05.076.

Selvakumar, S., I. Dinaharan, R. Palanivel and B. Ganesh Babu. 2017a. Characterization of molybdenum particles reinforced Al6082 aluminum matrix composites with improved ductility produced using friction stir processing. *Materials Characterization* 125 (March): 13–22. https://doi.org/10.1016/j.matchar.2017.01.016.

Selvakumar, S., I. Dinaharan, R. Palanivel and B. Ganesh Babu. 2017b. Development of stainless steel particulate reinforced AA6082 aluminum matrix composites with enhanced ductility using friction stir processing. *Materials Science and Engineering A* 685 (February): 317–326. https://doi.org/10.1016/j.msea.2017.01.022.

Shaarbaf, M., and M.R. Toroghinejad. 2008. Nano-grained copper strip produced by accumulative roll bonding process. *Materials Science and Engineering A* 473, no. 1–2 (January): 28–33. https://doi.org/10.1016/j.msea.2007.03.065.

Sharma, V., U. Prakash and B.V.M. Kumar. 2015. Surface composites by friction stir processing: A review. *Journal of Materials Processing Technology* 224 (October): 117–134. https://doi.org/10.1016/j.jmatprotec.2015.04.019.

Shirvanimoghaddam, K., S.U. Hamim, M.K. Akbari, S.M. Fakhrhoseini, H. Khayyam, A.H. Pakseresht, E. Ghasali, M. Zabet, K.S. Munir, S. Jia, J.P. Davim and M. Naebe. 2017. Carbon fiber reinforced metal matrix composites: Fabrication processes and properties. *Composites Part A: Applied Science and Manufacturing* 92: 70–96. https://doi.org/10.1016/j.compositesa.2016.10.032.

Shyam Kumar, C.N., D. Yadav, R. Bauri, G.D. Janaki Ram. 2015. Effects of ball milling and particle size on microstructure and properties 5083 Al-Ni composites fabricated by friction stir processing. *Materials Science and Engineering A* 645 (October): 205–212. https://doi.org/10.1016/j.msea.2015.08.026.

Shyam Kumar, C.N., R. Bauri and Yadav D. 2016. Wear properties of 5083 Al–W surface composite fabricated by friction stir processing. *Tribology International* 101 (September): 284–290. https://doi.org/10.1016/j.triboint.2016.04.033.

Singh, R., D. Podder and S. Singh. 2015. Effect of single, double and triple particle size SiC and Al_2O_3 reinforcement on wear properties of AMC prepared by stir casting in vacuum mould. *Transactions of the Indian Institute of Metals* 68, (February): 791–797. https://doi.org/10.1007/s12666-015-0512-1.

Surappa, M.K. 2003. Aluminium matrix composites: Challenges and opportunities. *Sadhana* 28: 319–334. https://doi.org/10.1007/BF02717141.

Taha, M.A. 2001. Practicalization of cast metal matrix composites (MMCCs). *Materials & Design* 22, no. 6 (September): 431–441. https://doi.org/10.1016/S0261-3069(00)00077-7.

Thangarasu, A., N. Murugan, I. Dinaharan and S.J. Vijay. 2014. Synthesis and characterization of titanium carbide particulate reinforced AA6082 aluminum alloy composites via friction stir processing. *Archives of Civil and Mechanical Engineering* 15, no. 2 (February): 324–334. http://dx.doi.org/10.1016/j.acme.2014.05.010.

Venkatesan, S., and M. Anthony Xavior. 2018. Tensile behavior of aluminum alloy (AA7050) metal matrix composite reinforced with graphene fabricated by stir and squeeze cast processes. *Science and Technology of Materials* 30, no. 2 (May): 74–85. https://doi.org/10.1016/j.stmat.2018.02.005.

Yadav, D., and R. Bauri. 2010. Nickel particle embedded aluminium matrix composite with high ductility. *Materials Letters* 64, no. 6 (March): 664–667. https://doi.org/10.1016/j.matlet.2009.12.030.

Yadav, D., and R. Bauri. 2011. Processing, microstructure and mechanical properties of nickel particles embedded aluminium matrix composite. *Materials Science and Engineering A* 528, no. 3 (January): 1326–1333. https://doi.org/10.1016/j.msea.2010.10.035.

Yadav, D., and R. Bauri. 2015. Development of Cu particles and Cu core-shell particles reinforced Al composite. *Materials Science and Technology* 31, no. 4 (August): 494–500. https://doi.org/10.1179/1743284714Y.0000000644.

Yang, M., C. Xu, C. Wu, K. Lin, Y.J. Chao and L. An. 2010. Fabrication of AA6061/ Al_2O_3 nano ceramic particle reinforced composite coating by using friction stir processing. *Journal of Materials Science* 45 (April): 4431–4438. http://dx.doi.org/10.1007/s10853-010-4525-1.

Yang, W., Q. Zhao, L. Xin, J. Qiao, J. Zou, P. Shao, Z. Yu, Q. Zhang and G. Wu. 2018. Microstructure and mechanical properties of graphene nanoplates reinforced pure Al matrix composites prepared by pressure infiltration method. *Journal of Alloys and Compounds* 732 (January): 748–758. https://doi.org/10.1016/j.jallcom.2017.10.283.

Yazdipour, A., M.A. Shafiei and K. Dehghani. 2009. Modeling the microstructural evolution and effect of cooling rate on the nanograins formed during the friction stir processing of Al5083. *Materials Science and Engineering A* 527, no. 1–2 (December): 192–197. https://doi.org/10.1016/j.msea.2009.08.040.

Zhang, Z., and Chen, D.L. 2006. Consideration of Orowan strengthening effect in particulate reinforced metal matrix nanocomposites: A model for predicting their yield strength. *Scripta Materialia* 54, no. 7 (April): 1321–1326. https://doi.org/10.1016/j.scripta mat.2005.12.017.

Zohoor, M., M.K.B. Givi and P. Salami. 2012. Effect of processing parameters on fabrication of Al–Mg/Cu composites via friction stir processing. *Materials & Design* 39: 358–365. https://doi.org/10.1016/j.matdes.2012.02.042.

10 Composites for Corrosive Wear Applications

Titus Thankachan,[1]* V. Kavimani,[2a] P. M. Gopal[2b]

[1]Karpagam College of Engineering, Coimbatore, 641 032, India.

[2]Department of Mechanical Engineering, Karpagam Academy of Higher Education, Coimbatore 641 021, India.

*Corresponding author: titusmech007@gmail.com

Other email IDs: [a]manikavi03@gmail.com, [b]gopal33mech@gmail.com

CONTENTS

10.1 INTRODUCTION

Corrosive wear, a combined effect of mechanical and corrosion wear on materials leading to degradation of material, is an important factor that has to be considered while selecting a material for various industries including mining, chemical and mineral processing industries, and oil and gas industries (Mao, Cai, and Wang 2018). As these industries include the usage of advanced machines working in harsh conditions at high speed and load in corrosive environments, wear along with corrosion has to be monitored to keep production running smoothly. In the marine and mining industries, use of conventional steel and cast iron has been promoted to meet specific needs of applications which, however, subject these metals to corrosive conditions (alkaline in nature) along with the movement of sea water and mine water containing minerals and abrasive particles (Hernandez-Rodriguez et al. 2007; Zavareh et al. 2014). This leads to corrosion of ferrous metal, and the erosive nature of water movement will speed up wear on the metal. These problems pushed researchers to exploit the field

DOI: 10.1201/9781003109723-10

of composite materials, in which the required property is achieved through the proper selection and combination of materials.

Proper selection of reinforcement particles is obligatory, as addition of improper reinforcement particles reduces the corrosion resistance of the base metal, thereby increasing the risk of greater corrosive wear. This reduction in corrosion resistance can be attributed to the electrode potential difference between the base metal and reinforcement materials, as well as to the passive oxide layer being affected by the addition of certain reinforcement particles (Yadav and Dixit 2019a). Hence appropriate particles are a major consideration in selecting a material for a specified application in which corrosive wear is a major drawback.

The next set of composites mainly preferred in high corrosive wear situations including mining and marine applications are polymer matrix composites, in which a set of reinforcements including fibers and filler materials are added into polymers like epoxy resin to develop a new set of materials that better withstand the corrosive wear and load requirement of a given application. Polymer-based composites are lightweight materials that exhibit high strength, resistance to corrosion and friction, and good electrical and thermal insulating properties (Dagdag et al. 2020). With the proper addition of hard filler material into the polymer matrix along with the fiber reinforcements, the fiber-based polymer matrix composites can also resist corrosive wear, making them suitable for pipeline applications in oil and gas industries as well as mining industries (Oliveira, Rocha, and Galdino 2019).

Ceramic matrix composites are another set of composites that has been widely used in highly corrosive situations because these materials are highly resistant to chemical reactions and air oxidations (Chen et al. 2018). This set of composites mainly consists of ceramics based on oxides, carbides, silicates, or nitrides as matrix material and has continuous fibers, whiskers, particles or platelets dispersed as reinforcement materials. These types of composites can be mainly observed in cutting tool applications and are under further investigations for commercial production (Ouyang, Zhu, and Li 2014). This chapter discusses the various ceramics that have found applications or are under research for high corrosive wear applications.

10.2 METAL MATRIX COMPOSITES FOR CORROSIVE WEAR APPLICATIONS

The conventional candidate metals used in high corrosive wear applications have been replaced with metal matrix composites designed specifically for a particular application. These applications include marine structures, impellers and agitators in the marine field, where the material will be exposed to sea water and at the same time to minerals and other particulates. Metal matrix composites like aluminum composites reinforced with SiC particles have also found their appropriate place in the fields of dredging and drilling operations where slurry is being transported elsewhere by a piping system (Khan and Dixit 2017). Apart from this, metal matrix composites for use in pumps, propellers, impellers, valves and other applications in oil, gas and chemical industries are also gaining notice (Das, Saraswathi, and Mondal 2006). Some metal matrix composites put forward by researchers to replace a conventional material contained reinforcement particles that had an adverse effect on the environment or a problem with structural properties. However, some of the composites proved to be promising candidates for the applications mentioned. Silicon carbide (SiC) has been a prominently used reinforcement in aluminum-based metal matrix composites, and these composites have found a good market in many applications.

Studies of corrosive wear on aluminum-based composites have proved that the addition of silicon carbide into aluminum matrix improves the corrosion-erosion resistance in marine as well as in acidic conditions, making it suitable for application in marine, mineral processing, mining and chemical industries (Yadav and Dixit 2019a). Variation in corrosion-erosion of the LM13 with respect to increase in SiC content at various slurry conditions is provided in Figure 10.1a–b. In a marine environment, a decrease in wear rate with respect to increase in addition of reinforcement particles was observed at different sand concentrations, which can be attributed to the dispersed SiC particles. In an acidic environment, the presence of reinforcement particles had a negative impact on wear resistance properties. This reduction in composites could be due to the discrepancy created by SiC particles on the passive layer, making it weaker and subject to corrosion in the acidic environment.

The effect of NaCl solution on different specimens is shown in Figure 10.2. In contained reinforcement particles that this case of LM13 alloy composites, corrosion attack preferentially took place at the Al-Si interface, and at the same time abrasion and erosion also took place leading to the corrosive wear. But with increased SiC particle dispersion into aluminum alloy, a reduction in abrasion and erosion is noted. This reduction in abrasion and erosion can be credited to the protruding silicon carbide particulates as seen in Figure 10.2. These particles carry the load acting on the material, thereby reducing the corrosive wear rate. The resulting surfaces can be observed from Figure 10.2b and c showing the aftereffects of corrosion and erosion. The corroded part gets eroded away by the abrasive and erosive movements of the sand particles, revealing a fresh surface of the material.

Addition of SiC into aluminum matrix has found a positive report with respect to wear resistance; as with addition of SiC reinforcement, corrosive wear resistance tends to increase. This enhancement in property reveals the passive layer formation which inhibits corrosive wear of surface and performs as a protective layer (Ramachandra and Radhakrishna 2006). Yttria reinforcement into aluminum matrix showed positive behavior in sodium chloride, nitric acid and sulfuric acid environments, making it a likely candidate for marine applications and chemical industries (Zhang and Li 2001; Bouaeshi and Li 2007). Aluminum metal matrix composites reinforced with SiC can also be considered an eminent material for slurry transportation in mineral processing and mining industries as the developed MMC has found to be resistant to wear even at high concentration of sand, owing to the passive layer formation and work hardening of surface (Das et al. 1999). Titanium carbide (TiC) along with zirconium sand in aluminum 5052 also had a positive result in corrosive wear properties, as the presence of hard TiC particles enhances hardness of developed composites while zirconium sand enhances the bonding strength (Kumar Patel et al. 2018). An optimized percentage of 2% of SiC along with fly ash in aluminum matrix also enhanced corrosive wear resistance (Ramachandra and Radhakrishna 2007). Aluminum-based metal matrix composites with titanium boride (Yadav and Dixit 2019b) and boron nitride (Sanman and Sreenivas Rao 2018) have been tested with positive results, and it was further confirmed that titanium boride is a better candidate than silicon carbide in acidic and aqueous condition. Studies have put forward many aluminum-based composites

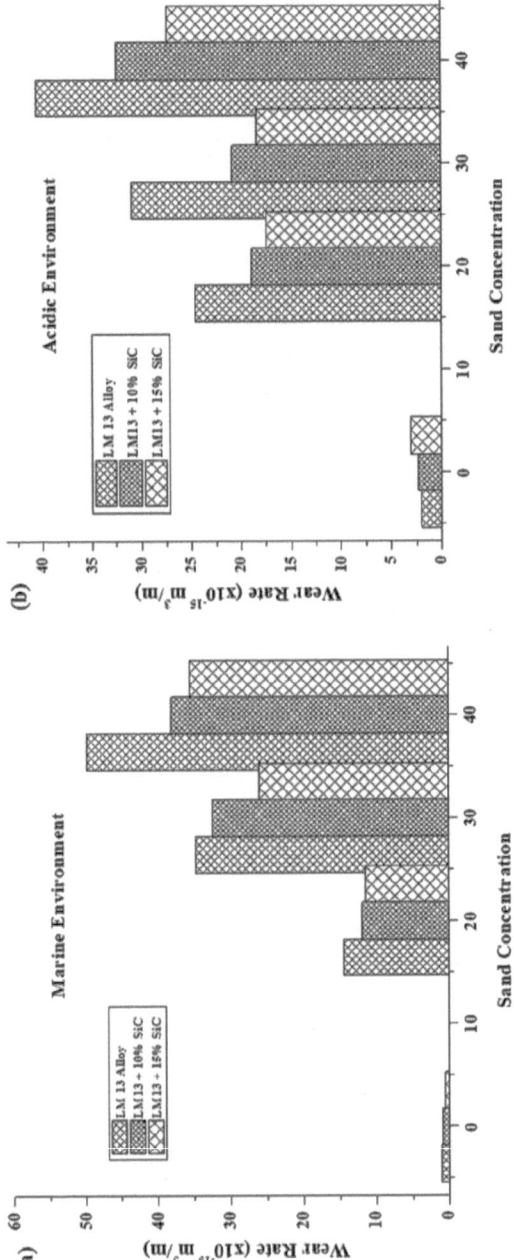

FIGURE 10.1   Effect of sand concentration on corrosion-erosion of LM13 and its composites.

Source: Das, Saraswathi, and Mondal (2006).

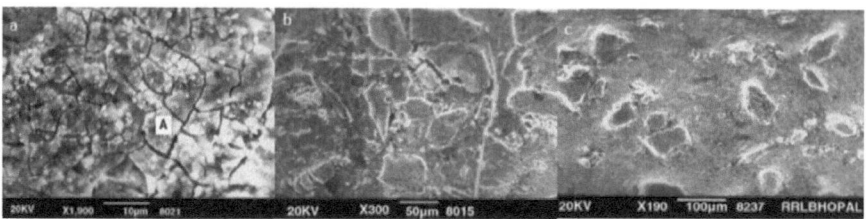

FIGURE 10.2 Eroded–corroded surface tested in NaCl solution for (a) alloy LM13, (b) LM13–10% SiC composite, (c) LM13–15% SiC composite.

Source: Das, Saraswathi, and Mondal (2006).

for corrosive wear applications, but certain reinforcement additions such as graphite (Sharma et al. 2006) and albite (Mosleh-Shirazi, Akhlaghi, and Li 2016) had a negative impact on the corrosive wear resistance.

Aluminum metal, because of its low weight and high strength-to-weight ratio, has been a preferred material, but for certain applications the hardness and toughness has to be high enough to withstand the impact created in production applications, for example, in rock and sand industries. In scenarios such as hydro transport processes, tungsten carbide–based composite has been found to be the best choice owing to its ability to withstand the corrosive wear created by slurry and chlorinated water (Neville et al. 2006). Tungsten carbide reinforced with magnesium oxide is a prominent material for tool applications in mining and mineral processing plants. Again, its hardness and ability to withstand corrosive wear has made it a promising material for valves, seal rings, fluid mixers, conveyor belt scrapers and other uses (Ouyang, Zhu, and Li 2014). Certain bronze-based composites are also used in the high corrosive wear applications owing to their ability to withstand corrosion at varying working conditions (Ragab et al. 2010; 2013). Studies are still being carried out by many research centers to develop possible replacements for conventional materials so as to enhance industrial production rates.

10.3 POLYMER MATRIX COMPOSITES FOR CORROSIVE WEAR APPLICATIONS

Superior properties of polymer matrix composites including their light weight, ease to design, high specific strength and modulus, along with superior fracture resistance, have earned them a top spot in material election for applications requiring high corrosion resistance including mining and mineral processing industries. Fiber-reinforced polymer composites have replaced many conventional metals in coal mining industries for uses from pipeline to fastener applications, where they enhance the strength of components. Rubber polymer–based composites also have a consistent place in the coal mining industries including for use in water transportation hose and conveyor belts, where corrosive wear has a certain influence Coal mine shafts have to undergo high load and at the same time withstand corrosive wear of both acidic and alkaline natures. Researchers put forward fiber-reinforced polymer composites as a replacement for traditional steel owing to high corrosion–resistance properties (Y. Q. Wang, Zhang, and Li 2013).

Certain general properties that favor glass fiber–reinforced polymers over other conventional materials include its high specific strength, which is about 4 to 6 times that of steel and 10 times that of cast iron and concrete–based pipes (Ainsworth 1981). The ability to withstand corrosion with high flexibility, toughness and ease of joining makes these composites potential candidates for sewage pipes as well as pipes in mining and mineral processing industries where corrosive wear is found to be a major concern (Ainsworth 1981). Glass fiber polymer composites have been extensively used in high corrosive wear applications such as pipeline for carrying chemical effluents, sewage effluents, and coal powder slurry in coal mining industries. Its properties have also made it an apt material for undersea lines in oil and gas industries where corrosive wear has a major impact from the saline conditions along with the erosive sediments that come into contact with the pipes.

Steel-based telescopic hydraulic cylinders on trucks used for soil transportation have been effectively replaced with fiber-reinforced epoxy polymer composites in trucks, which resulted in a 50% whole weight reduction in the hydraulic systems (Solazzi and Buffoli 2019). Turbine blades operating at high temperature conditions have been developed with carbon fiber reinforced with silicon carbide particles dispersed in phenolic resin solution. The resulting blades exhibit a high bending strength and fracture toughness at an optimized percentage of carbon fiber dispersion (Lu et al. 2014).

Coming to the marine environment, carbon fiber dispersed in polytetrafluoroethylene has been found to be a potential candidate for hydraulic drive systems in sea water and can be a good replacement for GCr15 steel and Ni-Cr-WC alloy (J. Z. Wang, Yan, and Xue 2009). The conventional marine composite materials are mainly glass fiber composites reinforced in polyester matrix or vinyl matrix, which are widely used in fabricating small marine crafts and offshore drilling platforms. However, studies have proved that the carbon fiber–based polymer composites are excellent choices for these applications (Kootsookos and Mouritz 2004). Certain studies have pointed out that the conventional materials such as bronze, nickel, manganese bronze, aluminum and stainless-steel alloy can be replaced with the carbon fiber–reinforced polymer composite for propeller applications. Studies further state that the problems raised in propellers such as corrosion, cavitation, corrosive wear and galvanic cell formation can be reduced to an extent with the efficient use of carbon fiber–based composites, as they contribute to the high strength and corrosion resistance required for the application (A. Kumar, Krishna, and Subramanian 2019).

Certain matrix materials used with varying reinforcement include (i) polyphenylene sulfide, which has a high resistance to chemical and temperature; (ii) polysulfone, known for low moisture absorption, low creep and high strength; (iii) polyethylene, well suited for corrosion-resistant applications; (iv) polypropylene and epoxy resin, known for their chemical resistance and strength, respectively. For these materials, research is ongoing to achieve composites with high strength and corrosion resistance by adding proper reinforcement fibers and filler materials, and usage of these products has yet to be carried out in real-life applications.

10.4   CERAMIC MATRIX COMPOSITES FOR CORROSIVE WEAR APPLICATIONS

Ceramic composites have been widely used in applications where the capability to withstand high temperature is a must, such as applications in aerospace applications and energy turbine parts. Ceramic matrix reinforced with fibers provides efficient materials compared with other superalloys owing to its capability to perform at high temperature with high fracture toughness. An overview of where ceramic matrix composite ranks compared with other, monolithic or composite materials are provided in Figure 10.3.

Owing to their properties, ceramic matrix composites are mainly seen as cutting tool inserts where corrosive wear has little influence. However, fiber-reinforced ceramic composites used for tubes for heat exchangers are prone to corrosive wear. Ceramic matrix composites reinforced with aluminum oxide–silicon carbide have a wide variety of applications in high corrosive wear environments including the mining industry. Applications such as sludge pump housing that come in contact with slurry, pump shaft and liners where high corrosive wear resistance is required, and seals in high temperature environments employ the ceramic composite materials. Ceramic matrix composites are also used in manufacturing cyclone apexes, liners, scraper blades, and other components where resistance to high corrosive wear is mandatory (Markgraaff 1996).

Ceramic composite with silicon nitride reinforced with hexagonal boron nitride was found to be an efficient material in terms of corrosive wear and can be used as part of a sliding pair, so it can replace many conventional materials including polymers (Chen et al. 2018). Ceramic composite reinforced with aluminum

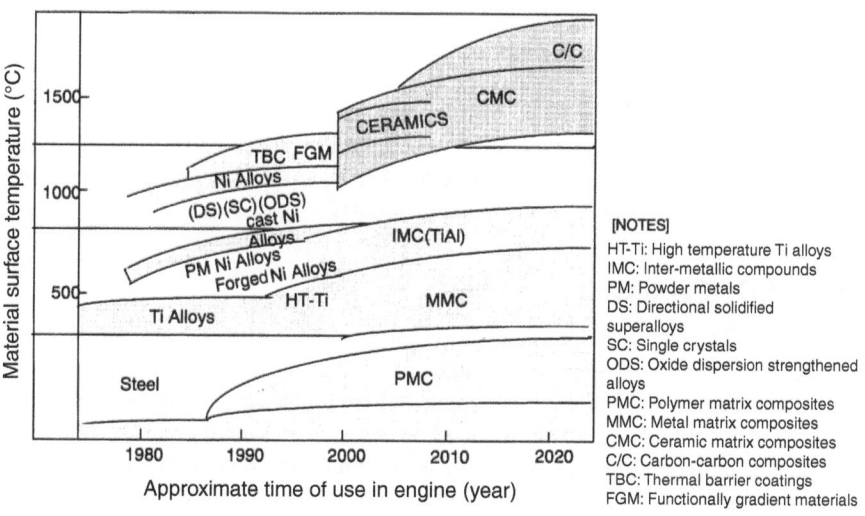

FIGURE 10.3   Trends in selection of high temperature materials.

Source: Ohnabe et al. (1999).

oxide–titanium oxide has been adapted as a material for submarine parts owing to its high corrosion resistance, wear resistance and bond strength (Maitra and Roy 2018). This reduces the number of preventive maintenances per annum, thereby reducing maintenance cost. Boron carbide composite with calcium hexaboride developed through hot pressing technology has proved to be an efficient candidate in marine applications owing to its capability to withstand corrosion reaction and at the same time offer high toughness (Qiao et al. 2015). Other than the usage of ceramic matrix composites for corrosive wear applications, ceramic composite–based coating has also been found efficient in overcoming the corrosive wear in oil and gas industries. Ceramic composite coatings like Al_2O_3–40 wt.% TiO_2 and Cr_3C_2–20NiCr have proved to significantly protect the overlying and underlying carbon steel petroleum tubes against corrosive wear (Zavareh et al. 2014; 2016). Compared with metal matrix and polymer matrix composites, application of ceramic matrix composites is minimal in the field of high corrosive wear, but investigations has are in the pipeline from various research institutes for developing capable ceramic-based composites for this purpose.

10.5 FAILURES DUE TO CORROSIVE WEAR IN COMPOSITES

Some of the main components where corrosive wear can be observed are in slurry handling equipment, as the transportation of slurry leads to substantial corrosive wear. Slurry particles' momentum is observed to be sufficient to cause abrasive wear, and the sliding of slurry through pipelines also causes damage to the components. In mining industries, components get exposed to corrosive environments along with rubbing action and abrasive action, leading to materials wearing out at a faster pace. In industries that use grinding, such as sugarcane, pulp and paper industries, corrosive wear can be increased due to galvanic coupling caused by electrochemical interactions between the grinding medium and minerals. Components working at elevated temperatures are also prone to high corrosive wear as a result of working temperature and particle sliding. Failure due to corrosive wear is a major concern when it comes to industries where high impact, crushing and slurry movement happen, owing to the environmental conditions and the presence of particle movements. This type of wear cannot be stopped completely but can be reduced to an extent by proper replacement of conventional materials with advanced materials.

10.6 CONCLUSION

High corrosive wear is a dominant source of failure which has to be inhibited, and composite materials have shown a high potential for applications in highly corrosive environments and replaced many conventional materials. All three commercialized forms of composites including metal matrix, polymer matrix and ceramic matrix composites employed in high corrosive wear conditions such as mining industries, marine structures, mineral industries and chemical industries are examined and described in this chapter. Of particular importance, certain developed composites are mentioned that are under research but have not been used in real-life applications, which needs clarification. Aluminum metal matrix composites are major research materials because of their exciting properties, with different reinforcements being tried out to find future materials for petroleum pipelines. Polymer composites are also being studied for this purpose. These materials also need to be investigated for methods of developing them at appropriate scales and for how they can meet various applications' requirements with confidence.

REFERENCES

Ainsworth, L. 1981. "Fibre-Reinforced Plastic Pipes and Applications." *Composites* 12 (3): 185–90. https://doi.org/10.1016/0010-4361(81)90501-2.

Bouaeshi, W. B., and D. Y. Li. 2007. "Effects of Y_2O_3 Addition on Microstructure, Mechanical Properties, Electrochemical Behavior, and Resistance to Corrosive Wear of Aluminum." *Tribology International* 40 (2): 188–99. https://doi.org/10.1016/j.triboint.2005.09.030.

Chen, Wei, Kui Wang, Yimin Gao, Nairu He, Hua Xin, and Huaqiang Li. 2018. "Investigation of Tribological Properties of Silicon Nitride Ceramic Composites Sliding against Titanium Alloy under Artificial Seawater Lubricating Condition." *International Journal of Refractory Metals and Hard Materials* 76 (June): 204–13. https://doi.org/10.1016/j.ijrmhm.2018.06.011.

Dagdag, O., R. Hsissou, A. El Harfi, Zaki Safi, Avni Berisha, Chandrabhan Verma, Eno E. Ebenso et al. 2020. "Epoxy resins and their zinc composites as novel anti-corrosive materials for copper in 3% sodium chloride solution: Experimental and computational studies." *Journal of Molecular Liquids* 315: 113757.

Das, S., D. P. Mondal, O. P. Modi, and R. Dasgupta. 1999. "Influence of Experimental Parameters on the Erosive–Corrosive Wear of Al–SiC Particle Composite." *Wear* 231: 195–205. https://doi.org/10.1017/cbo9780511582196.006.

Das, S., Y. L. Saraswathi, and D. P. Mondal. 2006. "Erosive-Corrosive Wear of Aluminum Alloy Composites: Influence of Slurry Composition and Speed." *Wear* 261 (2): 180–90. https://doi.org/10.1016/j.wear.2005.09.013.

Hernandez-Rodriguez, M. A. L., D. Martinez-Delgado, R. Gonzalez, A. Pérez Unzueta, R. D. Mercado-Solís, and J. Rodriguez. 2007. "Corrosive Wear Failure Analysis in a Natural Gas Pipeline." *Wear* 263 (1–6): 567–71.

Khan, M. M., and Gajendra Dixit. 2017. "Erosive Wear Response of SiCp Reinforced Aluminium Based Metal Matrix Composite: Effects of Test Environments." *Journal of Mechanical Engineering and Sciences* 14: 2401–14.

Kootsookos, A., and A. P. Mouritz. 2004. "Seawater Durability of Glass- and Carbon-Polymer Composites." *Composites Science and Technology* 64 (10–11): 1503–11. https://doi.org/10.1016/j.compscitech.2003.10.019.

Kumar, Ashok, G. Lal Krishna, and V. Anantha Subramanian. 2019. "Design and Analysis of a Carbon Composite Propeller for Podded Propulsion." In *Proceedings of the Fourth International Conference in Ocean Engineering (ICOE2018)*. Singapore: Springer. https://doi.org/10.1007/978-981-13-3119-0.

Kumar Patel, Surendra, Raman Nateriya, B. Kuriachen, and Virendra Pratap Singh. 2018. "Slurry Abrasive Wear, Microstructural and Morphological Analysis of Titanium Carbide and Zirconium Sand Aluminium Alloy (A5052) Metal Matrix Composite." *Materials Today: Proceedings* 5 (9): 19790–98. https://doi.org/10.1016/j.matpr.2018.06.342.

Lu, Z. L., F. Lu, J. W. Cao, and D. C. Li. 2014. "Manufacturing Properties of Turbine Blades of Carbon Fiber-Reinforced SiC Composite Based on Stereolithography." *Materials and Manufacturing Processes* 29 (2): 201–9. https://doi.org/10.1080/10426 914.2013.872269.

Maitra, Saikat, and Jagannath Roy. 2018. Nanoceramic Matrix Composites: Types, Processing, and Applications. In *Advances in Ceramic Matrix Composites*, 2nd ed. Elsevier. https://doi.org/10.1016/B978-0-08-102166-8.00003-7.

Mao, Liangjie, Mingjie Cai, and Guorong Wang. 2018. "Effect of Rotation Speed on the Abrasive–Erosive–Corrosive Wear of Steel Pipes against Steel Casings Used in Drilling for Petroleum." *Wear* 410: 1–10. https://doi.org/10.1016/j.wear.2018.06.002.

Markgraaff, J. 1996. "Overview of New Developments in Composite Materials for Industrial and Mining Applications." *Journal of the South African Institute of Mining and Metallurgy* 96 (2): 55–65.

Mosleh-Shirazi, S., F. Akhlaghi, and D. Y. Li. 2016. "Effect of Graphite Content on the Wear Behavior of Al/2SiC/Gr Hybrid Nano-Composites Respectively in the Ambient Environment and an Acidic Solution." *Tribology International* 103: 620–28. https://doi.org/10.1016/j.triboint.2016.08.016.

Neville, A., F. Reza, S. Chiovelli, and T. Revega. 2006. "Assessing Metal Matrix Composites for Corrosion and Erosion-Corrosion Applications in the Oil Sands Industry." *Corrosion* 62 (8): 657–75. https://doi.org/10.5006/1.3278293.

Ohnabe, Hisaichi, Shoju Masaki, Masakazu Onozuka, Kaoru Miyahara, and Tadashi Sasa. 1999. "Potential Application of Ceramic Matrix Composites to Aero-Engine Components." *Composites Part A: Applied Science and Manufacturing* 30 (4): 489–96. https://doi.org/10.1016/S1359-835X(98)00139-0.

Oliveira, Jhonny Dias, Renan Carreiro Rocha, and André Gustavo de Sousa Galdino. 2019. "Effect of Al_2O_3 particles on the adhesion, wear, and corrosion performance of epoxy coatings for protection of umbilical cables accessories for subsea oil and gas production systems." *Journal of Materials Research and Technology* 8 (2): 1729–36.

Ouyang, Chenxin, Shigen Zhu, and D. Y. Li. 2014. "Corrosion and Corrosive Wear Behavior of WC-MgO Composites with and without Grain-Growth Inhibitors." *Journal of Alloys and Compounds* 615: 146–55. https://doi.org/10.1016/j.jallcom.2014.06.137.

Qiao, Yingjie, Lianshi Qu, Xiaohong Zhang, and Hongyan Zhang. 2015. "Boron Carbide Composite Ceramic Preparation and Corrosion Behavior in Simulated Seawater." *Ceramics International* 41 (3): 5026–31. https://doi.org/10.1016/j.ceramint.2014.12.070.

Ragab, Kh. A., R. Abdel-Karim, S. Farag, S. M. El-Raghy, and H. A. Ahmed. 2010. "Influence of SiC, SiO2 and Graphite on Corrosive Wear of Bronze Composites Subjected to Acid Rain." *Tribology International* 43 (3): 594–601. https://doi.org/10.1016/j.triboint.2009.09.008.

Ragab, Kh. A., R. Abdel-Karim, M. Bournane, S. Farag, S. M. El-Raghy, and H. A. Ahmed. 2013. "Corrosive Wear Characteristics of Bronze Friction Resisting Composites." *Tribology – Materials, Surfaces and Interfaces* 7 (1): 52–59. https://doi.org/10.1179/1751584X13Y.0000000033.

Ramachandra, M., and K. Radhakrishna. 2006. "Sliding Wear, Slurry Erosive Wear, and Corrosive Wear of Aluminium/SiC Composite." *Materials Science- Poland* 24 (2/1): 333–49.

Ramachandra, M., and K. Radhakrishna. 2007. "Effect of Reinforcement of Flyash on Sliding Wear, Slurry Erosive Wear and Corrosive Behavior of Aluminium Matrix Composite." *Wear* 262 (11–12): 1450–62. https://doi.org/10.1016/j.wear.2007.01.026.

Sanman, S., and K. V. Sreenivas Rao. 2018. "Effect of Sand Concentration on Erosive–Corrosive Wear Behavior of Chill Cast Aluminum – Boron Carbide Composites." *Materials Today: Proceedings* 5 (1): 2951–54. https://doi.org/10.1016/j.matpr.2018.01.091.

Sharma, S. C., M. Krishna, H. N. Narasimha Murthy, R. Tarachandra, M. Satyamoorthy, and D. Bhattacharyya. 2006. "Study of Corrosive-Erosive Wear Behaviour of Al6061/Albite

Composites." *Materials Science and Engineering A* 425 (1–2): 305–11. https://doi.org/10.1016/j.msea.2006.03.079.

Solazzi, Luigi, and Andrea Buffoli. 2019. "Telescopic Hydraulic Cylinder Made of Composite Material." *Applied Composite Materials* 26 (4): 1189–1206. https://doi.org/10.1007/s10443-019-09772-8.

Wang, Jian Zhang, Feng Yuan Yan, and Qun Ji Xue. 2009. "Tribological Behaviors of Some Polymeric Materials in Sea Water." *Chinese Science Bulletin* 54 (24): 4541–48. https://doi.org/10.1007/s11434-009-0578-4.

Wang, Yu Qing, Peng Fei Zhang, and Jian Zhong Li. 2013. "The Application of Composite Materials in Coal Mine Production." *Advanced Materials Research* 756–759: 49–53. https://doi.org/10.4028/www.scientific.net/AMR.756-759.49.

Yadav, Pradeep Kumar, and Gajendra Dixit. 2019a. "Investigation of Erosion-Corrosion of Aluminium Alloy Composites: Influence of Slurry Composition and Speed in a Different Mediums." *Journal of King Saud University – Science* 31 (4): 674–83. https://doi.org/10.1016/j.jksus.2019.02.003.

Yadav, Pradeep Kumar, and Gajendra Dixit. 2019b. "Erosive-Corrosive Wear of Aluminium-Silicon Matrix (AA336) and SiC_p/TiB_{2p} Ceramic Composites." *Silicon* 11 (3): 1649–1660.

Zavareh, Mitra Akhtari, Ahmed Aly Diaa Mohammed Sarhan, Bushroa Binti Abd Razak, and Wan Jeffrey Basirun. 2014. "Plasma Thermal Spray of Ceramic Oxide Coating on Carbon Steel with Enhanced Wear and Corrosion Resistance for Oil and Gas Applications." *Ceramics International* 40 (9 part A): 14267–77. https://doi.org/10.1016/j.ceramint.2014.06.017.

Zavareh, Mitra Akhtari, Ahmed Aly Diaa Mohammed Sarhan, Parisa Akhtari Zavareh, Bushroa Binti Abd Razak, Wan Jeffrey Basirun, and Mokhtar B. Che Ismail. 2016. "Development and Protection Evaluation of Two New, Advanced Ceramic Composite Thermal Spray Coatings, Al_2O_3-40TiO$_2$ and Cr_3C_2-20NiCr on Carbon Steel Petroleum Oil Piping." *Ceramics International* 42 (4): 5203–10. https://doi.org/10.1016/j.ceramint.2015.12.044.

Zhang, T., and D. Y. Li. 2001. "Improvement in the Resistance of Aluminum with Yttria Particles to Sliding Wear in Air and in a Corrosive Medium." *Wear* 250–251 (part 2): 1250–56. https://doi.org/10.1016/s0043-1648(01)00774-8.

11 Composites for High Temperature Wear Applications

P. M. Gopal,[1]* V. Kavimani,[1a] Titus Thankachan[2b]

[1]Department of Mechanical Engineering, Karpagam Academy of Higher Education, Coimbatore, 641 021, India.

[2]Karpagam College of Engineering, Coimbatore, 641 032, India.

*Corresponding author: gopal33mech@gmail.com

Other email IDs: [a]manikavi03@gmail.com, [b]titusmech007@gmail.com

CONTENTS

11.1 INTRODUCTION

The number of products introduced in the market increases every day, promising a more comfortable life for humanity, and their production also increases rapidly with the aid of innovative technologies to meet requirements. Innovative technology does not only refer to the manufacturing methods or types of equipment; it also includes innovative materials. The material required for advanced applications such as aircraft building is somewhat strange as it requires a material with high strength but low weight. Requirements for materials with exceptional properties lead to breakthroughs

DOI: 10.1201/9781003109723-11

in the material science community as advanced materials like composites with a higher strength-to-weight ratio and better resistance to wear and corrosion are developed.

Composites are classified into different types mainly based on the matrix type such as metal, polymer and ceramics. Different methodologies are also used by various researchers for producing composites, and new approaches are still being designed with different focuses. Production methods include stir casting, powder metallurgy, friction stir processing, diffusion bonding, accumulated roll bonding and so on.

The development and application of composites for a specific application does not come easily, as these advanced materials have some setbacks like reinforcement distribution, machinabililty and formability. Further, the production methods have great impacts on the properties of composite materials, as they affect the reinforcement distribution. In addition to all these difficulties, the major property that has to be focused on is a composite's properties at elevated temperatures, such as wear resistance at high temperature. The high temperature properties of the composite are the focus here because of the greater difference at high temperature in the thermal expansion of matrix and reinforcement materials (Prakash et al. 2020). This greater difference in thermal expansion coefficient values among matrix and reinforcement has a greater impact on a composite's properties at higher temperatures. A number of researchers have analyzed the effect of temperature on the wear characteristics of composites and discussed the wear mechanism at different loading conditions.

Wear characteristics like wear rate and friction during sliding at high temperature and loading conditions were analyzed. The loading conditions applied – load, velocity and sliding distance – are considered along with temperature in order to analyze the performance of composites. This chapter analyzes the composites for high temperature applications, discussing in detail the effect of temperature on the wear performance of composites.

11.2   WEAR MECHANISMS

Wear in material occurs through five basic mechanisms: abrasive wear, adhesive wear, fatigue wear, corrosive wear and erosive wear, which are briefly discussed next.

11.2.1   ABRASIVE WEAR

There are two types of abrasive wear in general, namely, two-body and three-body abrasive wear. Two-body abrasive wear takes place when asperities of sliding hard material plough into the soft material and remove the material. In three-body abrasive wear, a third particle, which may come from the environment, is trapped between the two sliding parts, where it ploughs into soft material and removes the material. The removed particle either sticks into the opposite material or comes free. The presence of grooves is a general indication of abrasive wear.

11.2.2   ADHESIVE WEAR

When higher load is applied between the sliding parts, hard asperities on the opposing bodies weld together and form micro level joints. The movement of these bodies results in rupture of micro welds, and material from one body is transferred to the opposite part. Loss of material as a result of adhesion that takes place between the sliding parts is known as adhesive wear.

11.2.3   Fatigue Wear

Fatigue wear is nothing but the loss of material as a result of cyclic loading while sliding, and it occurs when the load applied goes beyond the material's fatigue strength. This type of wear usually starts as a crack at the upper surface of the material, and the crack penetrates into the connected subsurface, resulting in delamination of material from the surface.

11.2.4   Corrosive Wear

When two parts slide over one another, heat is generated and thus a protective oxide layer is produced at the top of the surface. But this layer will be removed during subsequent sliding of the parts and again a fresh oxide layer will be produced at the surface which will also be removed. This kind of continuous formation of a hard oxide layer and its removal results in material loss and is called corrosive wear. The removed oxides get trapped between the sliding parts and further increase the wear by abrasion.

11.2.5   Erosive Wear

Removal of material fragments from the material surface as a result of the momentum of impinging particles (solid, liquid or gaseous) results in erosive wear. For example, air bubbles in the lubricating fluid cause this kind of wear in engine bearings.

11.3   HIGH TEMPERATURE WEAR BEHAVIOR OF COMPOSITES

This part of the chapter explains the high temperature wear characteristics of various types of composite materials such as metal matrix composites, including aluminium-based composites, in situ composites and self-lubricating composites, and carbon-carbon composites.

11.3.1   Metal Matrix Composites

Metal matrix composites (MMCs) are a kind of advanced materials that are fabricated with metal as the base matrix and reinforcement added in various forms like particle, fiber and whiskers. Metal matrix composites are applied in many industries, mainly in mechanical parts that serve under critical environmental conditions. So analyzing the metal matrix composites under high temperature environmental conditions is vital, and several authors have analyzed the wear performance of MMCs under different temperatures.

Q. Zhang et al. (2020) compared the dry sliding wear characteristics of Al–12Si–CuNiMg and Al–12Si–CuNiMg/Al_2O_3 fiber composites at room and elevated temperatures of 27 °C, 100 °C, 200 °C and 300 °C, with sliding velocity fixed at 1 m/s while applied load varied from 2.5 to 10 N. Results of the study confirm that the fiber-reinforced composite exhibits better wear resistance than base alloy up to 200 °C. At 300 °C both the alloy and composite exhibit less resistance to wear, and very different delamination is observed in the composite surface at high temperature, while no plasticity flow is noticed. Results of high temperature wear testing conducted for Al–Si alloy reveal that the wear rate of the alloy decreases with increase in temperature because the elevated temperature aids oxide layer development at the interface during sliding. The presence of this protective layer results in enhanced wear resistance (Rajaram et al. 2010). Further, it was reported by Du Jun et al. (2004) that the alumina-reinforced Al alloy also exhibited better resistance to wear at different experimental temperatures. But both studies reported that the resistance to wear decreases when the temperature is increased over 100 °C.

Suresh Kumar and colleagues used silicon carbide and zircon sand reinforcement of LM 13 alloy and tested for wear under severe circumstances. The outcomes of the study showed that the hybrid mixture of silicon carbide and zircon particle reinforcement yielded better wear resistance in Al alloy than the single-particle-reinforced composite. The wear performance of the composite with the SiC and zircon reinforcement mixture is found to be better in all conditions including high load and temperature. Improvements in wear performance of the composites are found to be significant up to 200 °C, beyond which the wear rate increases (Suresh Kumar, Panwar, and Pandey 2013). Scanning electron micrographs (SEMs) of worn surfaces of the zircon-reinforced composites and hybrid-reinforced composite at two different temperatures are shown in Figure 11.1 and Figure 11.2. More delamination and abrasion can be found in the samples tested at 300 °C (Figure 11.1b and Figure 11.2b) than the samples tested at 200 °C (Figure 11.1a and Figure 11.2a).

Harichandran, Selvakumar, and Venkatachalam (2017) compared nano- and micro-sized B_4C reinforcements on the elevated temperature wear behavior of Al MMCs and stated that both composites have better wear resistance than the base alloy. Among the developed composites, Al MMC with nano reinforcement has better wear resistance than composite with micro-sized B_4C reinforcement when tested at temperatures of 30, 100 and 200 °C. This enhanced high temperature wear performance of nano particle–reinforced composites is attributed to their better interfacial bonding with the aluminium matrix.

David Raja Selvam, Dinaharan, and Mashinini (2017) tried fly ash particle as reinforcement in developing aluminium matrix composite for high temperature wear applications. They found that the fly ash particles are stable thermodynamically at elevated temperature without decaying. They also found that the reinforced fly ash particles are capable of reducing plastic deformation at room temperature, and the wear mechanism was found to shift from abrasion to adhesion when the temperature increased. Similarly, Dabral et al. (2017) developed and analyzed the elevated temperature wear performance of Al 6061–based MMC reinforced with red mud up to 473 K and reported that the wear rate is high at elevated temperature. This higher wear loss is attributed to softening of the matrix at elevated temperature. For red mud–reinforced MMCs, the wear mechanism shifted from abrasion to delamination when the temperature increased.

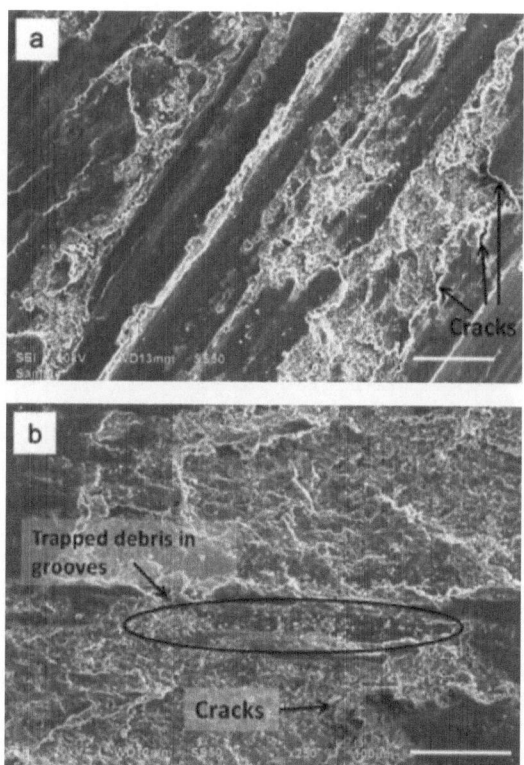

FIGURE 11.1 SEM images of the wear tracks for composites with zircon tested at (a) 200 °C and (b) 300°C.

Source: Suresh Kumar, Panwar, and Pandey (2013).

Yuan et al. (2020) developed a novel NiAl-based composite with Ti_3AlC_2 particles as reinforcement for high temperature wear applications through a thermal explosion method. A composite with low Ti_3AlC_2 content was subject to oxidation wear when tested at elevated temperature, whereas abrasion combined with oxidation is the predominant mechanism for a composite with higher Ti_3AlC_2 content. Selvakumar and Ramkumar (2016) reinforced Ti–6Al–4V matrix with nano boron carbide through powder metallurgy and analyzed its wear behavior at different temperatures. A severe to mild wear transition is observed when the nano reinforcement is used, as the wear resistance offered by the composite is better at higher temperature. It was also recommended based on the results that the Ti–6Al–4V/nano B_4C composites can provide stable properties for elevated temperature wear applications up to 150 °C.

Yongzhong and Guoding (2005) analyzed the feasibility of copper composite with SiC reinforcement of different sizes for high temperature wear usages. SiC particles of 2.5, 14 and 40 μm were reinforced with copper and analyzed for wear capabilities up to 700 K. Wear-resistant capabilities of the developed composites are good up to 550 K, and wear rate starts to increase steeply for temperatures above 550

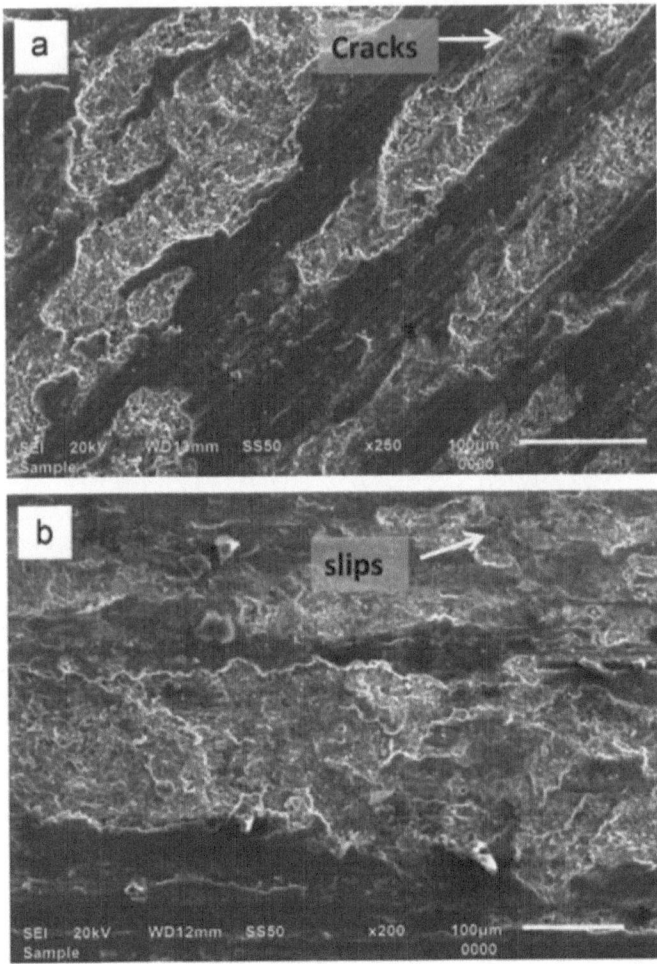

FIGURE 11.2 SEM images of the wear tracks for hybrid composites tested at (a) 200 °C and (b) 300 °C.

Source: Suresh Kumar, Panwar, and Pandey (2013).

K. It was also reported that the fine particle–reinforced composites possess superior wear resistance at higher temperature, while better low temperature wear resistance is offered by larger particles.

The results of wear analysis of CoCr matrix composite with nano TiO_2 reinforcement show that nano reinforcement enhances the hardness and high temperature wear resistance of CoCr during sliding against silicon nitride balls. The optimal reinforcement percentage that yields better wear and friction properties from room temperature to 1000 °C is 4%. This better wear behavior is attributed to the synergetic outcome of extraordinary mechanical characteristics, solid lubricants that develop in situ, and firm oxide films. The dominant wear mechanism at lower temperatures (below 600 °C) is abrasion whereas it is oxidation at 1000 °C (Cui et al. 2020a).

Gui et al. (2016) developed an intermetallic $NiMo/Mo_2Ni_3Si$ composite and compared its high temperature wear behavior above 500 °C with 1Cr18Ni9Ti stainless steel. The resistance to wear of developed intermetallic composite is seven times better than that of the rival material.

11.3.2 Metal Matrix Composites with In-Situ Reinforcements

Most commonly the in situ reinforcements are used in fabricating aluminium-based metal matrix composites. In situ composites are a kind of composite in which the reinforcements are developed with the chemical reaction in the molten metal during composite fabrication. Among the several in situ reinforcements, broadly employed or developed materials are alumina, TiC, ZrC, TiB_2 and ZrB_2 (G. N. Kumar et al. 2010; S. Zhang et al. 2007; H. Zhu et al. 2008; Zhao et al. 2007; Yilmaz, Ozenbas, and Yaz 2009). Jerome et al. (2010) added K_2TiF_6 and graphite (Gr) powder into the aluminium molten metal in an attempt to develop TiC reinforcement in situ and succeeded. The developed composite is analyzed for its wear resistance capability under high temperature and high loading conditions. The results indicated that the developed composite showed better resistance to thermal softening by forming an oxidation transfer layer, and the increase in TiC content also improved the wear resistance.

Similarly, S. Kumar, Sarma, and Murty (2009) developed in situ TiB_2 reinforcement in the Al–7Si alloy matrix through chemical reaction and analyzed the wear behavior in severe conditions up to 80 N load and 573 K temperature. The results also confirmed that the addition of in situ reinforcement enhances the wear resistance of Al–7Si alloy. Mild to severe wear transformation at 373 K temperature takes place at 80 N for the base alloy, whereas it occurs at a load of 120 N at 473 K for the 5% reinforced in situ composite. For the 5% and 10% reinforced composites, the transformation load at 573 K temperature is 80 N and 100 N, respectively, which shows the high temperature wear resistance of the TiB_2 reinforcement. The effect of in situ TiB_2 particles on the high temperature wear characteristics of Al–4Cu alloy composite is also commendable. The severe to mild wear transition at a constant load of 80 N occurs for Al–4Cu alloy and its composites with 5 and 10 wt% of added TiB_2 at 373 K, 473 and 573 K, respectively. This shows that there is an increase of 100 K in transformation temperature for every 5% addition of in situ TiB_2. When tested at a constant temperature of 373 K the transformation load is increased from 80 N (for base alloy) to 100 N and 120 N, respectively, for 5 and 10 wt% of TiB_2 addition. The predominant mechanisms of wear at high temperature for base alloy are adhesion and plastic flow, whereas oxidation, delamination and metal flow are the predominant mechanisms for the composites with TiB_2 reinforcement (S. Kumar, Sarma, and Murty 2010).

Aluminium composite with α-Al_2O_3+Al_3Zr reinforcement also showed better resistance to wear in a high temperature environment. At both of the tested temperatures (373 K and 473 K) the Al composite with 30% reinforcement suffered less wear loss than the composite with 20% reinforcement. Also it was found that the wear resistance of the developed composites is higher than the base material at the higher sliding velocity whereas the wear resistance is lower than the base material at lower sliding velocity. Abrasion, adhesion and oxidation are the three major wear mechanisms for the composites when tested under higher load and temperature (H. Zhu et al. 2012).

11.3.3 METAL MATRIX COMPOSITES WITH SOLID LUBRICANTS

Solid lubricants are generally added with base matrix unaccompanied or with other hard ceramic reinforcements to fabricate self-lubricating composites and hybrid composites, respectively. Owing to their soft nature, solid lubricants are most commonly added as secondary reinforcement along with hard ceramics as primary reinforcement. Generally graphite, molybdenum disulfide, hexagonal boron nitride (h-BN) and mica are used as solid lubricants; these materials lessen the friction between sliding parts and protect them from severe wear at normal working conditions. But their effectiveness as a solid lubricant at higher temperature is uncertain because of their unique molecular structure and chances of resection with base material (Pan et al. 2017). Some results reported by different authors on the performance of solid lubricants at high temperature are contradictory. A comparative study by Rajaram et al. (2010) on the high temperature wear behavior of Al–Si matrix alloy and Al–Si–Gr MMC reported that the Al-Si-Gr hybrid MMC with 3 wt% Gr is stable up to 250 °C. They also reported that the thick oxide and glazing layer formation on the sliding surface is the reason for the stability of the composite. However, the results reported in this research do not agree with those of some other researchers because of the inability of graphite to form a layer at temperatures over 100 °C. Raj and Radhika (2019) developed aluminium composite with LM13 as the base alloy and silicon nitride and graphite as reinforcements, and they analyzed the wear characteristics at different temperatures. They reported that the hybrid composite with 3% Gr showed a good self-lubricating effect at normal temperature, and when the temperature reached 100 °C, the self-lubricating nature of Gr disappears. On the other hand, a report on high temperature wear characteristics of SiC- and Gr-incorporated Cu MMC reveals that the friction coefficient of Cu/SiC/Gr is more stable than Cu/SiC composite when tested at high temperatures ranging from 373 to 723K. It was reported that the Gr addition decreases the wear rate of Cu MMC and the counterpart surface, and most importantly, it hinders the occurrence of severe wear up to 723 K (Zhan and Zhang 2006).

Another study compared the self-lubricating capability of the two primary solid lubricants, namely, graphite and molybdenum disulfide, at elevated temperatures up to 250 °C. The results also confirm the inability of the graphite to hold its self-lubricating nature at higher temperature whereas a better lubrication effect is exhibited by MoS_2 even at a temperature above 200 °C (Monikandan, Joseph, and Rajendrakumar 2018). Daniel et al. (2018) developed an aluminium hybrid composite with SiC and MoS_2 reinforcements and reported that the hybrid composite yielded better wear resistance owing to the development of a tough and firm tribolayer. X. Zhu et al. (2019) reported that the coefficient of friction during high temperature wear testing of Ni–Cr composite with h-BN reinforcement declined at 600 °C whereas better wear characteristics are offered by the composite reinforced with 10 wt% of h-BN. Conversely, S. Ayyanar et al. (2020) stated that the wear resistance of 10% h-BN reinforced composite is higher only up to 225 °C.

When steel matrix composite reinforced with either WS_2 or h-BN and produced through spark plasma sintering were tested for wear properties at 450°C, the WS_2-reinforced composites performed better. On the basis of their observations, the researchers suggested that WS_2 is a better option as a solid lubricant to develop a solid tribolayer that can conquer shearing stresses (Orozco Gomez et al. 2011). To test their self-lubricating capabilities, Mo, Ag, and CuO were incorporated through powder metallurgy into FeCr composite and tested at temperatures up to 800 °C for their wear properties. The best possible combination among the developed composites is Fe(Cr)-14% Mo-10.5% Ag-10% CuO, which exhibits better wear properties at elevated temperature. The wear mechanisms that cause material loss during testing were found to be plastic deformation and fatigue at normal temperature, whereas oxidation was predominant at higher temperature (Cui, Liu, Gao et al. 2020b).

11.3.4 CARBON-CARBON COMPOSITES

The top contestant for applications in aerospace vehicles is carbon-carbon (C/C) composite because of its light weight and superior wear resistance at elevated temperature. Another unique advantage of C/C composite is that it exhibits better strength at elevated temperatures whereas decreased strength is observed for metals and ceramics with increases in temperature. Even though C/C composites exhibit better wear resistance, oxidation at high temperature is the major problem that leads to oxidative wear. The layer formed by oxidation reduces the friction between the parts, and the detached oxide layer is trapped between the sliding parts and causes abrasive wear. To overcome this setback of the C/C composites, a number of researchers tried several techniques such as the application of nano additives, in situ grown carbon nanofibers (CNFs), SiC coating, multilayer coatings and other methods.

Oxidation of carbon fiber reinforcement and carbon matrix is the major reason for wear in C/C composites, resulting in reduced friction between the parts and increased abrasive wear (Policandriotes and Filip 2011). During braking, a huge quantity of kinetic energy is absorbed by the C/C composite friction material in the form of heat, and as a result the temperature at the surface may reach 1000 °C under severe braking situations (Stimson and Fisher 1980). The time interval of this energy absorption in heat form is as short as 15–18 seconds, owing to which temperature at the interface increases steeply (Mohanty 2013). Even though C/C composites are capable of withstanding this much heat, oxidation at this higher temperature limits the composite's lifetime.

Policandriotes and Filip (2011) stated that the properties of C/C composites have to be modified to use in aircraft brakes. The expected properties are lower oxidation and abrasion wear and steady friction demonstration at different energy levels. For that, nano additives of Si, SiC, and single-walled carbon nanotube (SWCNT) were added to C/C composites and analyzed for wear and friction performance. Results confirmed the positive effect of nano additives and it was reported that the added particles stabilize the friction performance of C/C composites.

CNF was grown in situ in C/C composites and its oxidation behavior was compared with CNF-free C/C composites at various temperatures. It was found that the wear loss up to 600 °C is nearly zero for both materials, but C/C composites suffer more loss than CNF-C/C composites over 600 °C. It was also reported that the oxidation

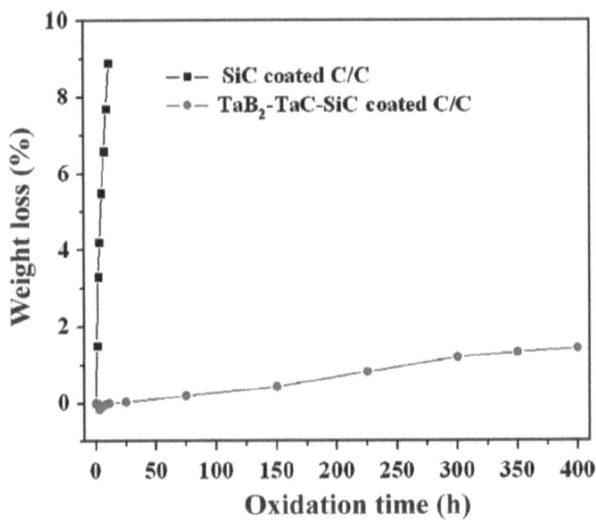

FIGURE 11.3 Oxidation time versus weight loss for coated C/C composites in air at 1773 K.

Source: Ren et al. (2014).

protection capability of the CNFs grown in situ is high at lower temperature whereas the influence of CNFs in slowing down oxidation at high temperature is low (Lu and Xiao 2014). Another prominent methodology used for preventing oxidation of C/C composites at high temperature is ceramic coatings (Jian-Feng et al. 2004; Zheng et al. 2008; Westwood et al. 1996).

Ren et al. (2014) compared the oxidative wear loss at elevated temperature (1773 K) of C/C composites coated with SiC and with ultra-high temperature TaB$_2$–TaC–SiC . The results revealed that the performance of TaB$_2$–TaC–SiC coating is better than the SiC coating in protecting against oxidative wear loss, which shows the efficacy of ultra-high temperature coatings. However, owing to its excellent compatibility with C/C composites, SiC is the most commonly used coating material (Liu et al. 2009; Chen et al. 2014). The time variation between the SiC and ultra-high temperature coatings for oxidation wear is shown in Figure 11.3.

Infiltration of nano particles into carbon–ceramic composites was also tried in order to fabricate brake discs with the aim of dropping the diffusion of moisture in open porosity (Güther, Rosenlöcher, and Bauer 2008). Integration of nano particles into the friction surface can also be done by soaking nano particle–mixed resin into the bulk fiber support structure (Lam, Chen, and Maruo 2008).

11.4 SUMMARY

The chapter gives a detailed review of the different composites developed for high temperature wear applications. High temperature wear behavior of typical MMC, MMC with in situ reinforcements and C/C composites are discussed, and the effect of solid lubricants in high temperature wear applications is also covered. Among the different composites discussed, C/C composites with nano additives or infiltration are capable of use in very high temperature applications such as aircraft brakes. Aluminium, copper and titanium composites with various reinforcements cannot be used for very high temperature applications, whereas Fe, CoCr and NiMo matrix composites can withstand wear at higher temperatures. In terms of self-lubricating composites, graphite-based composites are limited to 100 °C as the Gr loses its lubricating nature above 100 °C. Among self-lubricants such as h-BN, WS_2 and MoS_2, WS_2 can be used as a solid lubricant to develop composites for elevated temperature applications.

REFERENCES

Ayyanar, S, A Gnanavelbabu, K Rajkumar, and P Loganathan. 2020. "Studies on High Temperature Wear and Friction Behaviour of AA6061/B 4 C/HBN Hybrid Composites." *Metals and Materials International.* Springer: 1–18.

Chen, Zishan, Hejun Li, Kezhi Li, Qingliang Shen, and Qiangang Fu. 2014. "Influence of Grain Size on Wear Behavior of SiC Coating for Carbon/Carbon Composites at Elevated Temperatures." *Materials & Design* 53. Elsevier: 412–18.

Cui, Gongjun, Yanping Liu, Sai Li, Huiqiang Liu, Guijun Gao, and Ziming Kou. 2020a. "Nano-TiO_2 Reinforced CoCr Matrix Wear Resistant Composites and High-Temperature Tribological Behaviors under Unlubricated Condition." *Scientific Reports* 10 (1). Nature Publishing Group: 1–12.

Cui, Gongjun, Yanping Liu, Guijun Gao, Huiqiang Liu, and Ziming Kou. 2020b. "Microstructure and High-Temperature Wear Performance of FeCr Matrix Self-Lubricating Composites from Room Temperature to 800 C." *Materials* 13 (1). Multidisciplinary Digital Publishing Institute: 51.

Dabral, R, N Panwar, R Dang, R P Poonia, and A Chauhan. 2017. "Wear Response of Aluminium 6061 Composite Reinforced with Red Mud at Elevated Temperature." *Tribology in Industry* 39 (3).

Daniel, S Ajith Arul, M Sakthivel, P M Gopal, and S Sudhagar. 2018. "Study on Tribological Behaviour of Al/SiC/MoS 2 Hybrid Metal Matrix Composites in High Temperature Environmental Condition." *Silicon* 10 (5). Springer: 2129–39.

David Raja Selvam, J, I Dinaharan, and P M Mashinini. 2017. "High Temperature Sliding Wear Behavior of AA6061/Fly Ash Aluminum Matrix Composites Prepared Using Compocasting Process." *Tribology-Materials, Surfaces & Interfaces* 11 (1). Taylor & Francis: 39–46.

Gui, Yongliang, Chunyan Song, Shuhuan Wang, and Dingguo Zhao. 2016. "Elevated-Temperature Wear Behaviors of NiMo/Mo₂ Ni₃ Si Intermetallic 'in Situ' Composites." *Journal of Materials Research* 31 (1). Cambridge University Press: 66–75.

Güther, Hans-Michael, Jens Rosenlöcher, and Moritz Bauer. 2008. "Nanoparticle-Modified Carbon-Ceramic Brake Discs." Google Patents.

Harichandran, R, N Selvakumar, and G Venkatachalam. 2017. "High Temperature Wear Behaviour of Nano/Micro B_4C Reinforced Aluminium Matrix Composites Fabricated by an Ultrasonic Cavitation-Assisted Solidification Process." *Transactions of the Indian Institute of Metals* 70 (1). Springer: 17–29.

Jerome, S, B Ravisankar, Pranab Kumar Mahato, and S Natarajan. 2010. "Synthesis and Evaluation of Mechanical and High Temperature Tribological Properties of In-Situ Al–TiC Composites." *Tribology International* 43 (11). Elsevier: 2029–36.

Jian-Feng, Huang, Zeng Xie-Rong, Li He-Jun, Xiong Xin-Bo, and Fu Ye-wei. 2004. "Influence of the Preparation Temperature on the Phase, Microstructure and Anti-Oxidation Property of a SiC Coating for C/C Composites." *Carbon* 42 (8–9). Elsevier: 1517–21.

Jun, Du, Liu Yao-hui, Yu Si-rong, and Li Wen-fang. 2004. "Dry Sliding Friction and Wear Properties of Al_2O_3 and Carbon Short Fibres Reinforced Al–12Si Alloy Hybrid Composites." *Wear* 257 (9–10). Elsevier: 930–40.

Kumar, G Naveen, R Narayanasamy, S Natarajan, S P Kumaresh Babu, K Sivaprasad, and S Sivasankaran. 2010. "Dry Sliding Wear Behaviour of AA 6351-ZrB_2 in Situ Composite at Room Temperature." *Materials & Design* 31 (3). Elsevier: 1526–32.

Kumar, S, V Subramanya Sarma, and B S Murty. 2009. "Effect of Temperature on the Wear Behavior of Al-7Si-TiB 2 in-Situ Composites." *Metallurgical and Materials Transactions A* 40 (1). Springer: 223–31.

Kumar, S, V Subramanya Sarma, and B S Murty. 2010. "High Temperature Wear Behavior of Al–4Cu–TiB_2 in Situ Composites." *Wear* 268 (11–12). Elsevier: 1266–74.

Kumar, Suresh, Ranvir Singh Panwar, and O P Pandey. 2013. "Effect of Dual Reinforced Ceramic Particles on High Temperature Tribological Properties of Aluminum Composites." *Ceramics International* 39 (6). Elsevier: 6333–42.

Lam, R C, Y F Chen, and K Maruo. 2008. "Patent US7429418B2 USA." September.

Liu, Chidong, Laifei Cheng, Xingang Luan, Weihua Zhang, and Chuanqing Wang. 2009. "Real-Time Damage Evaluation of a SiC Coated Carbon/Carbon Composite under Cyclic Fatigue at High Temperature in an Oxidizing Atmosphere." *Materials Science and Engineering: A* 524 (1–2). Elsevier: 98–101.

Lu, Xue-feng, and Peng Xiao. 2014. "Short Time Oxidation Behavior and Residual Mechanical Properties of C/C Composites Modified by in Situ Grown Carbon Nanofibers." *Ceramics International* 40 (7). Elsevier: 10705–9.

Mohanty, R M. 2013. "Climate Based Performance of Carbon-Carbon Disc Brake for High Speed Aircraft Braking System." *Defence Science Journal* 63 (5).

Monikandan, V V, M A Joseph, and P K Rajendrakumar. 2018. "Influence of Temperature Variation on Tribological Behavior of Aluminum Matrix Hybrid Composites: A Statistical Analysis." *Metallography, Microstructure, and Analysis* 7 (6). Springer: 735–45.

Orozco Gomez, Solisabel, Karl Delbé, Alberto Benitez, Jean Yves Paris, and Jean Denape. 2011. "High Temperature Tribological Behaviour of Metal Matrix Composites Produced by SPS." *Key Engineering Materials* 482. Trans Tech Publications: 89–100.

Pan, Like, Jianmin Han, Zhiyong Yang, Jialin Wang, Xiang Li, Zhiqiang Li, and Weijing Li. 2017. "Temperature Effects on the Friction and Wear Behaviors of Sicp/A356 Composite against Semimetallic Materials." *Advances in Materials Science and Engineering* 2017. Hindawi.

Policandriotes, T, and P Filip. 2011. "Effects of Selected Nanoadditives on the Friction and Wear Performance of Carbon–Carbon Aircraft Brake Composites." *Wear* 271 (9–10). Elsevier: 2280–89.

Prakash, K Soorya, P M Gopal, M Purusothaman, and M Sasikumar. 2020. "Fabrication and Characterization of Metal-High Entropy Alloy Composites." *International Journal of Metalcasting* 14 (2). Springer: 547–55.

Raj, Nithin, and N Radhika. 2019. "Tribological Characteristics of LM13/Si 3 N 4/Gr Hybrid Composite at Elevated Temperature." *Silicon* 11 (2). Springer: 947–60.

Rajaram, G, S Kumaran, T Srinivasa Rao, and M Kamaraj. 2010. "Studies on High Temperature Wear and Its Mechanism of Al–Si/Graphite Composite under Dry Sliding Conditions." *Tribology International* 43 (11). Elsevier: 2152–58.

Ren, Xuanru, Hejun Li, Qiangang Fu, and Kezhi Li. 2014. "Ultra-High Temperature Ceramic TaB$_2$–TaC–SiC Coating for Oxidation Protection of SiC-Coated Carbon/Carbon Composites." *Ceramics International* 40 (7). Elsevier: 9419–25.

Selvakumar, N, and T Ramkumar. 2016. "Effects of High Temperature Wear Behaviour of Sintered Ti–6Al–4V Reinforced with Nano B$_4$C Particle." *Transactions of the Indian Institute of Metals* 69 (6). Springer: 1267–76.

Stimson, I L, and R Fisher. 1980. "Design and Engineering of Carbon Brakes." *Philosophical Transactions of the Royal Society of London. Series A, Mathematical and Physical Sciences* 294 (1411). Royal Society London: 583–90.

Westwood, M E, J D Webster, R J Day, F H Hayes, and R Taylor. 1996. "Oxidation Protection for Carbon Fibre Composites." *Journal of Materials Science* 31 (6). Springer: 1389–97.

Yilmaz, S O, M Ozenbas, and Mehmet Yaz. 2009. "Synthesis of TiB$_2$-Reinforced Iron-Based Composite Coating." *Tribology International* 42 (8). Elsevier: 1220–29.

Yongzhong, Zhan, and Zhang Guoding. 2005. "Particle Size Effect on the Elevated Temperature Wear Behavior of SiC$_p$/Cu Composites." *Journal of Materials Science* 40 (1). Springer Nature BV: 223–25.

Yuan, Jianjun, Qingzhao Wang, Xinying Liu, Shumei Lou, and Qun Li. 2020. "Preparation and High-Temperature Wear Behavior of NiAl-Ti$_3$AlC$_2$ Fabricated by Thermal Explosion." *Materials Research Express* 7 (2). IOP Publishing: 26539.

Zhan, Yongzhong, and Guoding Zhang. 2006. "The Role of Graphite Particles in the High-Temperature Wear of Copper Hybrid Composites against Steel." *Materials & Design* 27 (1). Elsevier: 79–84.

Zhang, Qing, Shuo Wei, Jie Gu, and Ming Qi. 2020. "High-Temperature Dry Sliding Wear Behavior of Al–12Si–CuNiMg Alloy and Its Al$_2$O$_3$ Fiber-Reinforced Composite." *Metals and Materials International*. Springer, 1–11.

Zhang, Songli, Yutao Zhao, Gang Chen, and Xiaonong Cheng. 2007. "Microstructures and Dry Sliding Wear Properties of in Situ (Al$_3$Zr + ZrB$_2$)/Al Composites." *Journal of Materials Processing Technology* 184 (1–3). Elsevier: 201–8.

Zhao, D G, X F Liu, Y C Pan, X F Bian, and X J Liu. 2007. "Microstructure and Mechanical Properties of in Situ Synthesized (TiB$_2$ + Al$_2$O$_3$)/Al–Cu Composites." *Journal of Materials Processing Technology* 189 (1–3). Elsevier: 237–41.

Zheng, Guo-Bin, Hironori Mizuki, Hideaki Sano, and Yasuo Uchiyama. 2008. "CNT–PyC–SiC/SiC Double-Layer Oxidation-Protection Coating on C/C Composite." *Carbon* 46 (13). Elsevier: 1808–11.

Zhu, Heguo, Hengzhi Wang, Liangqi Ge, Wenjuan Xu, and Yunzhan Yuan. 2008. "Study of the Microstructure and Mechanical Properties of Composites Fabricated by the Reaction Method in an Al–TiO$_2$–B$_2$O$_3$ System." *Materials Science and Engineering: A* 478 (1–2). Elsevier: 87–92.

Zhu, Heguo, Cuicui Jar, Jinzhu Song, Jun Zhao, Jianliang Li, and Zonghan Xie. 2012. "High Temperature Dry Sliding Friction and Wear Behavior of Aluminum Matrix Composites (Al$_3$Zr+ α-Al$_2$O$_3$)/Al." *Tribology International* 48. Elsevier: 78–86.

Zhu, Xuewei, Xiaofeng Wei, Yuxiang Huang, Fu Wang, and Pengpeng Yan. 2019. "High-Temperature Friction and Wear Properties of NiCr/HBN Self-Lubricating Composites." *Metals* 9 (3). Multidisciplinary Digital Publishing Institute: 356.

12 Influence of Wear Parameters on Friction and Wear Behaviour of Friction Stir Processed Al/CaCO₃ Surface Composite

M. Sivanesh Prabhu,[1]* A. Elayaperumal,[1a]
M. Wasim Khan,[1b] M. S. Jagatheeshwaran[2c]

[1]College of Engineering, Guindy, Anna University,
Chennai-600025, Tamil Nadu, India.

[2]S. A. Engineering College, Chennai-600077, Tamil Nadu, India.

Corresponding author: sivanesh.research@gmail.com

Other email IDs: [a]profelaya@gmail.com, [b]wasimmech2@gmail.com, [c]jegee82@gmail.com

CONTENTS

12.1 INTRODUCTION

The most commonly used material in the field of engineering is aluminium and its alloys, because of their high strength-to-weight ratio. Though the aluminium alloys exhibit high strength, their surface property limitations cause them to fall below requirements for a few engineering applications and cause failure. In order to overcome the limitations, several researchers have focused on developing surface metal matrix composites (SMMCs).

Generally in SMMCs, only the surface properties of the material are modified and the remaining material is unchanged. The SMMCs possess high strength and resistance to fatigue and wear for engineering applications. The most commonly used ceramic particles for surface reinforcement include Al_2O_3, SiC, B_4C, WC, TiC and TiB_2 (Gobalakrishnan et al., 2021; Mehta et al., 2019). A proper reinforcement particle is chosen by testing the composite material's behaviour in response to both friction and wear.

Reinforcement with hard particles may lead to severe abrasion caused by three-body abrasive wear. To meet this concern, a self-lubricating $CaCO_3$ particle could be preferable to improve the wear resistance of the material (Sivanesh Prabhu et al., 2019). $CaCO_3$ is a solid lubricant particle which improves the friction and wear behaviour of the aluminium composite. Uniform dispersion of reinforcement particles on the surface is a tedious challenge to achieve. Plasma spraying, electron beam irradiation and laser melt treatment are some of the existing techniques for liquid phase processing for the development of surface composites. Using these techniques could allow the formation of an interfacial reaction between the reinforcement and matrix, leading to detrimental phases. In order to overcome the problem, the substrate is processed below its melting point to obtain a surface composite.

Friction stir processing (FSP) is an energy-efficient and solid-state technique used to develop surface composite through microstructure modification (S. Zhang et al., 2019). FSP follows the principles of friction stir welding developed at the Welding Institute, UK, in 1991. In friction stir processing, a non-consumable tool which consists of pin and shoulder is plunged into the surface of the material and generates friction heat owing to the stirring action of the tool. The friction heat allows the material to undergo dynamic recrystallization, resulting in fine grains (Aktarer et al., 2019). The temperature that occurs in this process is below the melting point of the material. As a result of recrystallization, the microstructure of the aluminium improves from coarse grains to equiaxed grains. The homogeneous dispersion of the reinforcement particles is achieved using FSP and exhibits better mechanical and tribological properties.

The influence of friction stir processing parameters plays a huge role in the material flow. Sivanesh Prabhu et al. (2021) reported the effect of rotation speed and travel speed on FSP of AA6082 and found that the optimal FSP parameters produced a fine grain microstructure that exhibited better mechanical and wear resistance. The improvement in wear resistance of surface composites was due to the formation of a tribolayer or mechanical mixed layer (Sivanesh Prabhu et al., 2020a). Self- or solid-lubricant reinforcement particles tend to smear over the surface and act as an intermediate between the two mating surfaces.

It can be inferred from the literature that very limited research work has been done on friction and wear behaviour of $Al/CaCO_3$ composite. Therefore, the study reported in this chapter focused on the friction and wear behaviour of Al/CaCO₃ composite fabricated by FSP under various loading and sliding conditions. The chapter also presents a detailed discussion of the grain refinement, mechanical and wear behaviour of FSP Al/CaCO₃ composite.

12.2 EXPERIMENTAL PROCEDURE

In this study, a rolled AA6082 plate with a dimension of 150 ' 75 ' 6 mm³ is selected as the base metal. CaCO₃ powder with an average particle size of 3 μm to 5 μm is selected as reinforcement particle. A three-axis servo-controlled friction stir processing (FSP) machine (Figure 12.1) is used to fabricate Al/CaCO₃ composite. A groove 2 mm wide and 2 mm deep is made to be filled with the CaCO₃ particle. Pin and pinless tools are used for the processing. The pinless tool is used to cover the top surface of the groove to prevent the escape of reinforcement particle. Next, the pin tool is used to mechanically mix the aluminium and CaCO₃ particle to form Al/CaCO₃ surface composite. The friction stir process parameters used are rotational speed of 1250 rpm, travel speed of 40 mm/min and tilt angle of 2.5 degrees.

After FSP, the fabricated samples are machined using wire cut electric discharge machining (WEDM) for the various characterization studies. The microstructure evaluation is carried out using an optical microscope (OM) and scanning electron microscope (SEM). The microhardness test is carried out using a Vickers microhardness tester. The wear test is carried out using pin-on-disc tribometer under a dry sliding condition. The wear parameters used are varying applied loads (5, 10 and

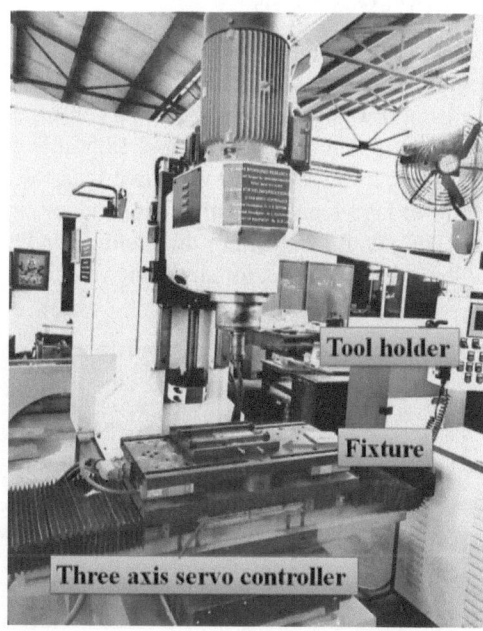

FIGURE 12.1 Experimental setup of friction stir processing machine.

15 N), sliding velocities (0.5, 0.75 and 1.0 m/s) and a constant sliding distance of 1400 m. The worn surface morphology is observed using scanning electron microscopy.

12.3 RESULTS AND DISCUSSION

12.3.1 MICROSTRUCTURAL ANALYSIS

The microstructure of the base metal (AA6082) and friction stir processed Al/CaCO$_3$ composite is shown in Figure 12.2. It is evident that elongated grains are present in the base metal (Figure 12.2a). The measured average grain size of the base metal is ~141 μm. The microstructure of the Al/CaCO$_3$ composite presents fine grains due to the dynamic recrystallization and Zener pinning effect. The average grain size and particle size of the Al/CaCO$_3$ composite are ~3 and ~2 μm, respectively. The dispersion of CaCO$_3$ particle in the Al matrix is homogeneous. This may be due to severe plastic deformation by the stirring action of the tool in the processed region (Bourkhani et al., 2019).

12.3.2 MICROHARDNESS STUDY

The microhardness of base metal (BM) and friction stir processed AA6082/CaCO$_3$ composite is shown in Figure 12.3. It is observed that the average microhardness of FSP AA6082/CaCO$_3$ composite is better than that of BM. This may be due to the grain refinement by FSP (Z. Zhang et al., 2019), and also, the uniform dispersion of reinforcing CaCO$_3$ particles has influenced higher hardness for Al/CaCO$_3$ composite.

12.3.3 FRICTION AND WEAR ANALYSIS

The friction and wear test of base metal and Al/CaCO$_3$ composite is carried out for different loads and sliding velocities. Figure 12.4 and Figure 12.5 show the coefficient of friction (CoF) and wear rate of BM and Al/CaCO$_3$ composite at various loads and sliding velocities. From Figures 12.4 and 12.5, it is observed that the Al/CaCO$_3$ composite exhibited lower CoF and wear rate than the BM. The formation of adhesive wear in BM at high load and high sliding velocity condition resulted in severe adhesive wear (Peng et al., 2019). For all loading conditions, CoF decreases with an increase in sliding velocity. Also, for all sliding velocities, CoF increases with

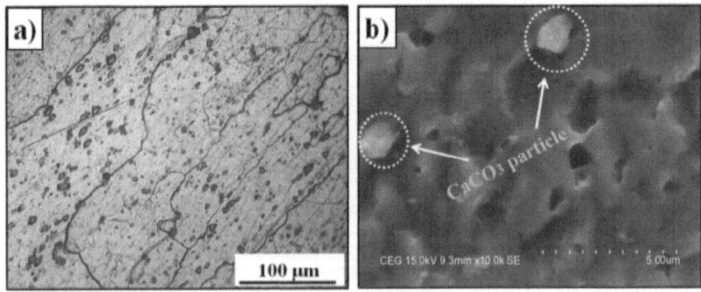

FIGURE 12.2 (a) OM image of base metal and (b) SEM image of Al/CaCO$_3$ composite.

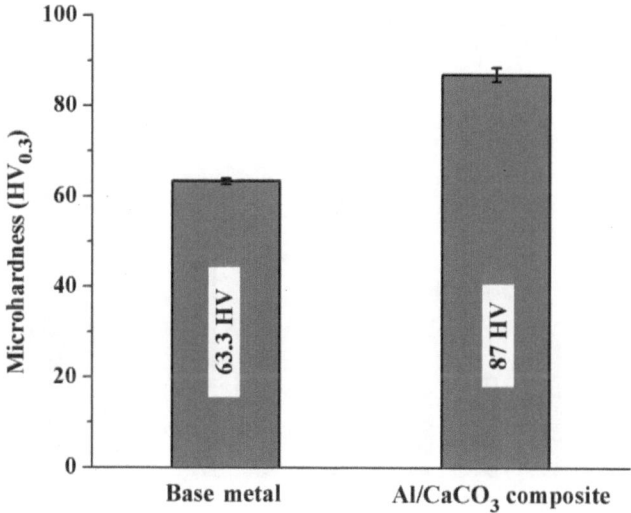

FIGURE 12.3 Vickers microhardness of base metal and Al/CaCO₃ composite.

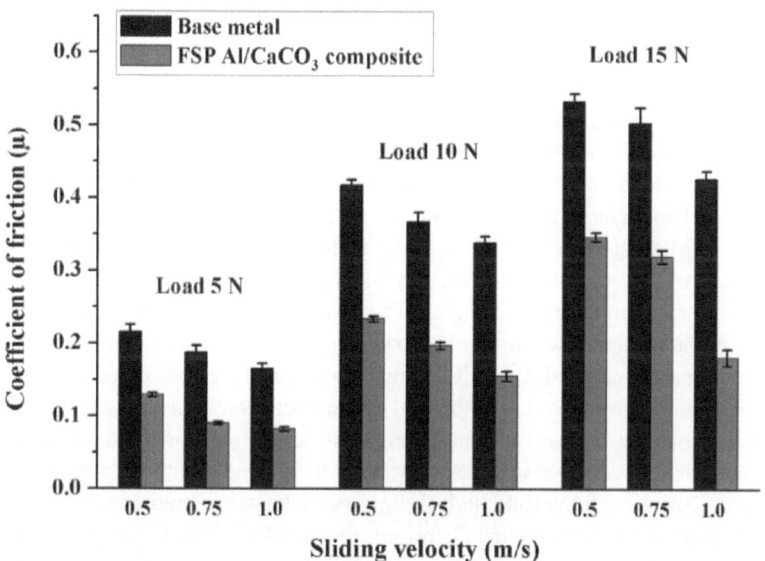

FIGURE 12.4 Coefficient of friction for base metal and Al/CaCO₃ composite under various loads and sliding velocities.

increase in the applied load. This could be due to the real contact between the disc and pin increasing with increase in applied load.

Addition of $CaCO_3$ particle as reinforcement acts as a solid lubricant and reduces the wear to mild adhesion. Also, the homogeneous dispersion of $CaCO_3$ particle in the

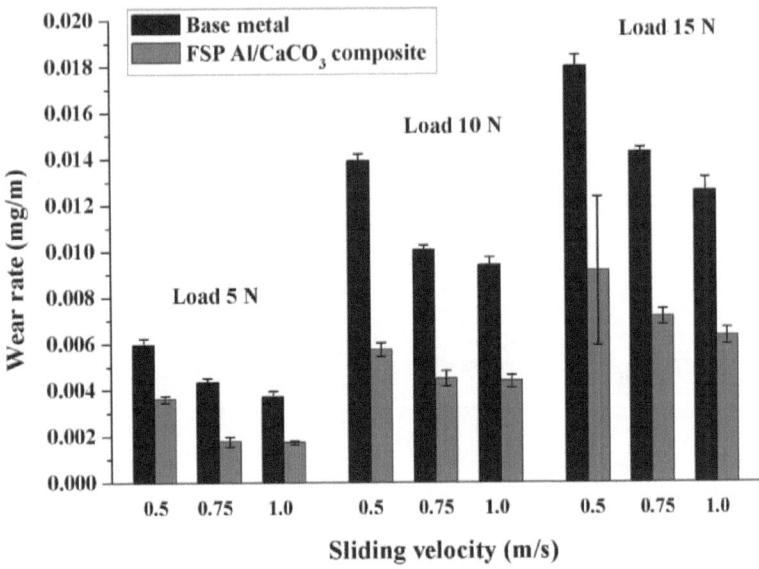

FIGURE 12.5 Wear rate for base metal and Al/CaCO₃ composite under various loads and sliding velocities.

Al matrix by FSP has improved the wear resistance. According to Archard's law, hardness is inversely proportional to wear rate (Sivanesh Prabhu et al., 2020b; Uthayakumar et al., 2013). During dry sliding, reinforced particles pulled from the composite surface act as a barrier and reduce the real contact between the composite pin and disc. The formed tribolayer may also restrict the removal of material and improve the wear behavior (Kishan et al., 2017). By varying sliding velocity, a decrease in CoF and in wear rate is observed with an increase in sliding velocity. This may be due to thermal effect and formation of an oxide layer (Thapliyal et al., 2016) between the contacting surfaces of the pin and disc. By varying load conditions, the increase in CoF and in wear rate of both base metal and Al/CaCO₃ composite with an increase in loading conditions are observed. This may be due to the real contact between the two contacting surfaces increasing with the increase in load. Composite samples at low load and low sliding speed presented a reduction in adhesion wear and replaced it with the presence of abrasion wear. During high load and high sliding speed, there is a presence of mild abrasion with adhesion wear mechanism. Therefore, the AA6082/CaCO₃ composite presented better wear resistance compared to base metal AA6082. This is due to the incorporation of solid lubricant CaCO₃ particle in the Al matrix. Hence, the severe wear of base metal AA6082 is transformed into mild wear of AA6082/CaCO₃ composite.

12.4 CONCLUSIONS

For this study, FSP is developed to fabricate Al/CaCO₃ composite. The effect of CaCO₃ particle reinforcement on friction and wear behaviour are studied in comparison to base metal, and the conclusions are as follows.

- The Al/CaCO$_3$ composite is successfully fabricated using FSP.
- The microstructure of the Al/CaCO$_3$ composite exhibited equiaxed grains rather than the elongated grains of base metal AA6082.
- Significant improvement in the microhardness is exhibited for Al/CaCO$_3$ composite, which is due to the dynamic recrystallization and grain strengthening mechanism.
- The Al/CaCO$_3$ composite exhibited significant improvement in wear resistance compared to base metal. The CaCO$_3$ particle acted as a self-lubricant and reduced the CoF and wear rate.
- By varying the wear parameters, the minimal CoF and wear rate were found to occur at high sliding velocity (1.0 m/s) as a result of thermal softening.

REFERENCES

Aktarer, S. M., T. Küçükömeroğlu, and K. Davut. "Friction stir processing of dual phase steel: Microstructural evolution and mechanical properties." *Materials Characterization* 155 (2019): 109787.

Bourkhani, R. Darzi, A. R. Eivani, and H. R. Nateghi. "Through-thickness inhomogeneity in microstructure and tensile properties and tribological performance of friction stir processed AA1050-Al2O3 nanocomposite." *Composites Part B: Engineering* 174 (2019): 107061.

Gobalakrishnan, B., C. Rajaravi, G. Udhayakumar, P. R. Lakshminarayanan, and M. Sivanesh Prabhu. "Analysis of mechanical properties of cold extruded Al 6061 TiB$_2$ MMCs and validated for finite element analysis." *Materials Today: Proceedings* 43 (2021): 1283–1292.

Kishan, V., Aruri Devaraju, and K. Prasanna Lakshmi. "Influence of volume percentage of NanoTiB$_2$ particles on tribological & mechanical behaviour of 6061-T6 Al alloy nano-surface composite layer prepared via friction stir process." *Defence Technology* 13, no. 1 (2017): 16–21.

Mehta, K. M., and V. J. Badheka. "Wear behavior of boron-carbide reinforced aluminum surface composites fabricated by friction stir processing." *Wear* 426 (2019): 975–980.

Peng, Jinhua, Zhen Zhang, Peng Guo, Yaozu Li, Wei Zhou, and Yucheng Wu. "The effect of the inhomogeneous microstructure and texture on the mechanical properties of AZ31 Mg alloys processed by friction stir processing." *Journal of Alloys and Compounds* 792 (2019): 16–24.

Sivanesh Prabhu, M., A. Elayaperumal, S. Arulvel, and R. Franklin Issac. "Friction and wear measurements of friction stir processed aluminium alloy 6082/CaCO3 composite." *Measurement* 142 (2019): 10–20.

Sivanesh Prabhu, M., A. Elayperumal, S. Arulvel, and M. Wasim Khan. "Significance of tribolayer on the friction and wear resistance of FSPed AA6082/SiCp composite at various load conditions." *Surface Topography: Metrology and Properties* 8, no. 2 (2020a): 025037.

Sivanesh Prabhu, M., A. Elayperumal, and S. Arulvel. "Development of multi-pass processed AA6082SiCp surface composite using friction stir processing and its mechanical and tribology characterization." *Surface and Coatings Technology* 394 (2020b): 125900.

Sivanesh Prabhu, M., A. Elayperumal, S. Arulvel, and M. Wasim Khan. "Assessment on the impact of FSP process parameters on microstructural, mechanical and wear behaviour of FSPed AA6082." *Surface Topography: Metrology and Properties* 8, no. 1 (2021): 015016.

Thapliyal, Shivraman, and Dheerendra Kumar Dwivedi. "Microstructure evolution and tribological behavior of the solid lubricant based surface composite of cast nickel aluminum bronze developed by friction stir processing." *Journal of Materials Processing Technology* 238 (2016): 30–38.

Uthayakumar, M., S. Aravindan, and K. Rajkumar. "Wear performance of Al–SiC–B$_4$C hybrid composites under dry sliding conditions." *Materials & Design* 47 (2013): 456–464.

Zhang, Shuai, Gaoqiang Chen, Jinquan Wei, Yijun Liu, Ruishan Xie, Qu Liu, Shenbo Zeng, Gong Zhang, and Qingyu Shi. "Effects of energy input during friction stir processing on microstructures and mechanical properties of aluminum/carbon nanotubes nanocomposites." *Journal of Alloys and Compounds* 798 (2019): 523–530.

Zhang, Zhen, Yaozu Li, Jinhua Peng, Peng Guo, Pengju Yang, Shan Wang, Chang Chen, Wei Zhou, and Yucheng Wu. "Combining surface mechanical attrition treatment with friction stir processing to optimize the mechanical properties of a magnesium alloy." *Materials Science and Engineering: A* 756 (2019): 184–189.

13 Potential Applications of Nano-Enhanced Phase Change Material Composites

K. Gopi Kannan,[1a] R. Kamatchi,[2a]* D. Dsilva Winfred Rufuss[2b]

[1]SRM Easwari Engineering College, Ramapuram, Chennai – 600 089, India.

[2]Vellore Institute of Technology, Vellore – 632 014, India.

*Corresponding author: rkkamatchi@gmail.com

Other email IDs: [a]gopimetier@gmail.com, [2b]dsilvawinfred.d@vit.ac.in

CONTENTS

13.1 INTRODUCTION

The focus of energy management and auditing is to enhance the storage ability in order to meet the energy demand in industries and commercial applications. Thermal energy storage is a significant technology to diminish the energy demand and to supply heat energy by charging excess heat energy to materials and discharging it when needed (Kannan and Kamatchi, 2020a, Vivekananthan and Amirtham, 2019). In particular, high heat energy potential is stored by phase change material (PCM) when it changes its phase. In general, the thermal conductivity of PCM is low, and integration of various techniques (like fin structures and nanoparticles) has improved the thermal conductivity of PCM. However, the addition of fins and metallic foam increases the weight of PCM storage devices and fabrication cost (Yang et al., 2020).

DOI: 10.1201/9781003109723-13

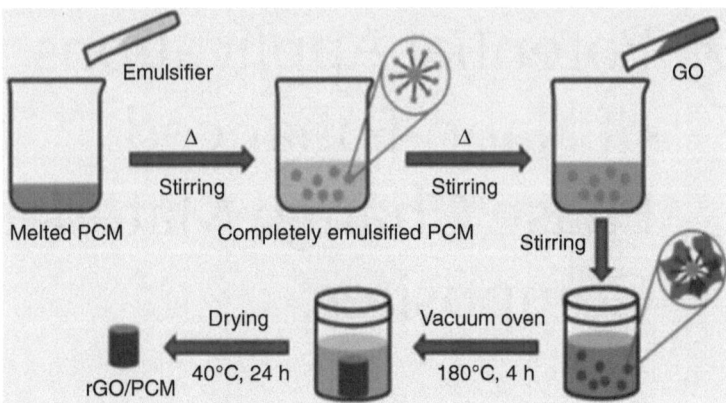

FIGURE 13.1   Preparation methodology of Ne-PCM.

Source: Liu et al. (2018).

This problem is rectified by dispersion of nanoparticles, such as graphene, graphite, metal, metal oxide and carbon nanotubes (CNTs) into base PCM. The preparation methodology of nano-enhanced PCM (Ne-PCM) is depicted in Figure 13.1 (Liu et al., 2018). The inclusion of different nanoparticles in PCM and the properties of those composites are presented by Babapoor et al. (2016). Thermophysical properties and ideal characteristics of Ne-PCM are also investigated for a greater number of cycles.

Yang et al. (2020) reviewed the thermal physical properties of different Ne-PCMs and their applications. The authors inferred that the phase change rate of Ne-PCM increases with the dispersion of nanoparticles, which implies that the amount of heat energy absorbed/liberated during the phase change could be improved. Also, the melting and freezing temperatures are reduced by 6°C and increased by 1.3°C, respectively.

Therefore, applications for Ne-PCM are being widely studied because of the many substantial implementation sectors. Among these, some of the applications such as solar thermal, building materials, cooling and energy recovery from high power density in electronics, and other commercial applications are addressed in this chapter to obtain deeper insights that can help academicians, researchers and industrialists to move forward.

13.2   EVALUATION OF NE-PCM IN SOLAR APPLICATIONS

Solar energy is a momentous renewable energy source which minimizes dependence on non-renewable energy. However, solar energy can be obtained only in the daytime. PCMs are used to mitigate this problem. To augment the latent heat storage capability and thermal conductivity of PCM, Ne-PCM is preferred. Also, only 20% of the incident solar radiation is converted into electrical energy and the remaining 80% leads to an increase in the temperature of cells, which may cause overheating of the photovoltaic (PV) panel (Abdelrazik et al., 2020). This affects the electrical performance and conversion efficiency of the PV system. Irregular temperature scattering throughout the PV panel can be controlled by the high thermal conduction and high storage density of Ne-PCM. Sharma et al. (2017) found that a decrement in the temperature of the concentrated PV system is obtained by adopting Ne-PCM in the finned plate used to disperse excess heat. Moreover, the maximum temperature drop is shown by the micro-finned plate associated with Ne-PCM. For a typical evacuated tube collector (ETC) that incorporates Ne-PCM, its system performance for improving the panel's functionality depends on the heat transfer fluid (HTF) flow, the number of ETC tubes and the concentration of nanoparticles (Algarni et al., 2020). Also, the latent heat stored in the Ne-PCM is used to prolong the period of production of hot water and reduce the fluctuations in the temperature of the hot water. Hexagonal copper nanoparticles of 15–125 nm are added to paraffin wax for improving the efficiency of solar thermal energy storage applications. The results indicated that the inclusion in PCM of nanoparticles at weight percentages of 0.5, 1, 1.5 and 2% improves the thermal conductivity by 14, 23.9, 42.5 and 46.3%, respectively. Also, paraffin wax containing 1 wt% of copper nanoparticles enhances efficiency by 1.7% (S. Lin and Al-Kayiem, 2016).

The thermal performance and economics of a photovoltaic thermal system (PVT) with the design shown in Figure 13.2 were investigated by Al-Waeli et al. (2017). They used silicon carbide with paraffin wax in a tank attached to a PV panel. Nanofluid (water and SiC 0.01%) flows at a constant mass flow rate of 0.175 kg/s through the nano-PCM tank by copper tube. The efficiency increases 7.1–13.7% as compared to the conventional system. The maximum thermal power and outlet water temperature are obtained as 13.8 kW and 39.52°C, respectively. Also, it is reported that the system is economically viable for all kinds of thermal storage applications. The researchers

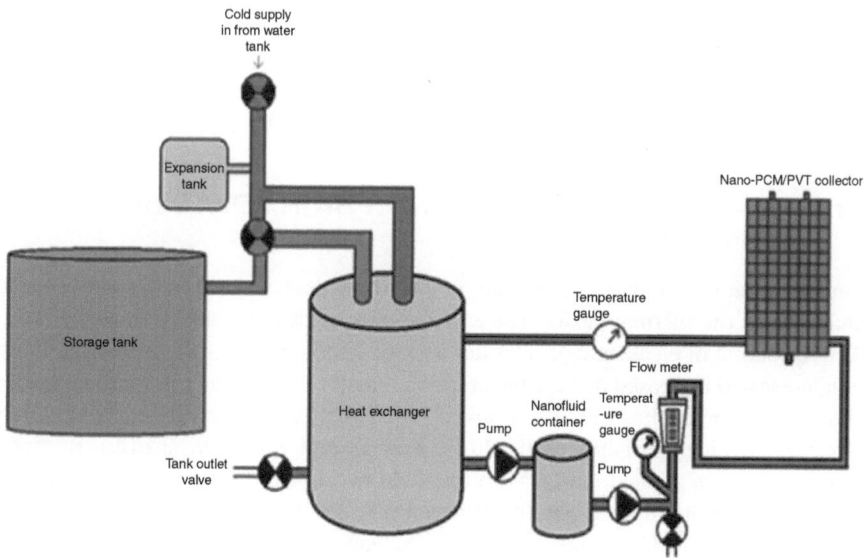

FIGURE 13.2   Nano-PCM/nanofluid in a PVT system.

Source: Al-Waeli et al. (2017).

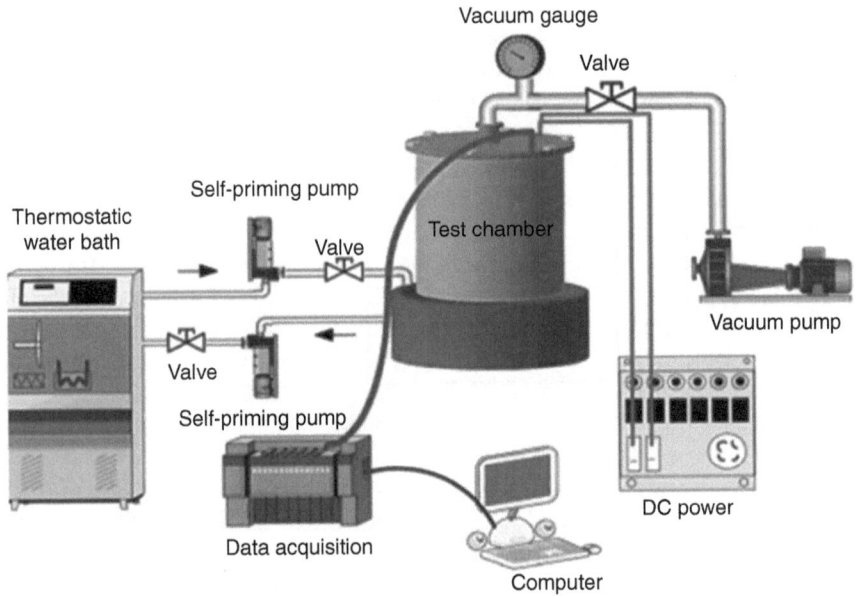

FIGURE 13.3   Experimental setup for solar thermal energy storage with Ne-PCM.

Source: Xiao et al. (2015).

used binary nitrate (50 wt% $NaNO_3$, 50 wt% KNO_3) with a melting temperature of 220°C as the PCM and experimented with various weight fractions of graphite (EG). The result implies that the thermal conductivity of PCM was increased by seven times with the addition of 20 wt% of EG to the base material. Also, the energy storage time and retrieval time decreased by 26.9 and 68.8%, respectively (Xiao et al., 2015). Figure 13.3 represents solar thermal energy storage with Ne-PCM.

13.3   CONTRIBUTION OF NE-PCM IN HIGH POWER DENSITY APPLICATIONS

The integration of high power density in electronic devices generates more heat flux and reduces the thermal performance of the system (Kannan and Kamatchi, 2021). Active cooling of electronic devices has a lot of disadvantages such as internal vibration, noise and increased power consumption. Passive cooling defeats these problems, and PCM-based cooling has also received wide attention for waste heat recovery from electronic devices (Krishna et al., 2017, Kannan and Kamatchi, 2020b). Recently, researchers have applied Ne-PCM to this field because of its elevated thermal management and high storage capability. Praveen et al. (2019) investigated the thermal performance of a finned heat sink using graphene nano-platelets with microencapsulated PCM. It was found that there was an increase in the thermal conductivity of 0.192 to 0.379 watts per meter Kelvin (W/m K), and in addition to this heat sink, the temperature rise rate was delayed. Also, the reduced thermal resistance and nucleation

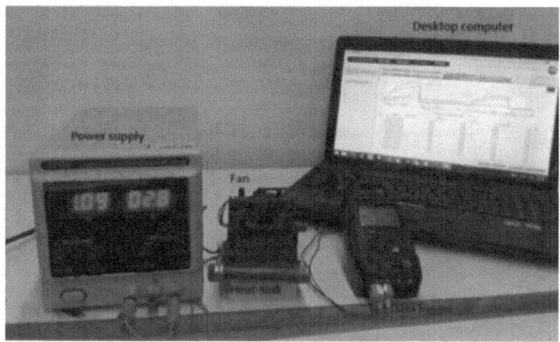

FIGURE 13.4   Electronic chipset cooling using nano-PCM.

Source: Alimohammadi et al. (2017).

effect diminishes the recovery time of the heat sink. The Ne-PCM-based cooling of the chipset and its effect under natural and forced convection is investigated as shown in Figure 13.4 (Alimohammadi et al., 2017). Salt hydrate [Mn(NO$_3$)$_2$] was used as the PCM in which 1 wt% of Fe$_3$O$_4$ nanoparticles was dispersed. The result showed that the constant temperature of the chipset is reduced under both natural and forced convection by 10.5 and 14°C, respectively.

Farzanehnia et al. (2019) investigated the electronic chipsets using pure PCM and Ne-PCM. They reported that the association of nanoparticles with paraffin reduces the cooling time by 6% compared to the virgin PCM. Moreover, the effect of nano-PCM considerably decreased the temperature and increased the operating time of the electronic system. In another study, water and tricosane were used as a PCM and associated with Al$_2$O$_3$ nanoparticles. As shown in Figure 13.5, the nano-PCM was placed near a heat pipe conducting the heat from an evaporator (Krishna et al., 2017). The results indicate that thermal conductivity increased by 32%, the evaporator temperature was reduced to 25.75% and hence fan power consumption was reduced by 53%. In a third study, paraffin/nano-SiO$_2$ was used as a nano-PCM for thermal management in an electronics application. Paraffin wax was added to silicon dioxide nanoparticles at three different ratios, 60:40, 70:30 and 75:25. The researchers reported an increase of thermal protection in electronic devices of 21.8% for the composition of 75% paraffin and 25% nano-SiO$_2$ (Wang et al., 2016). Colla et al. (2015) presented the effects of adding 0.5–1.0 wt% of Al$_2$O$_3$ in RT45 and RT55 paraffin for heat storage and cooling applications. They concluded that nano-PCMs slow down the melting process compared to the reference temperature for paraffin alone.

13.4   CONTRIBUTION OF NE-PCM IN BUILDING APPLICATIONS

Buildings are the most energy consuming (about 45%) among the sectors of global total energy consumption. Adding heat storage in buildings increases the certainty of energy supplies, which enhances the environmental sustainability of a building and diminishes its contribution to global warming. Ne-PCM overcomes the difficulty of

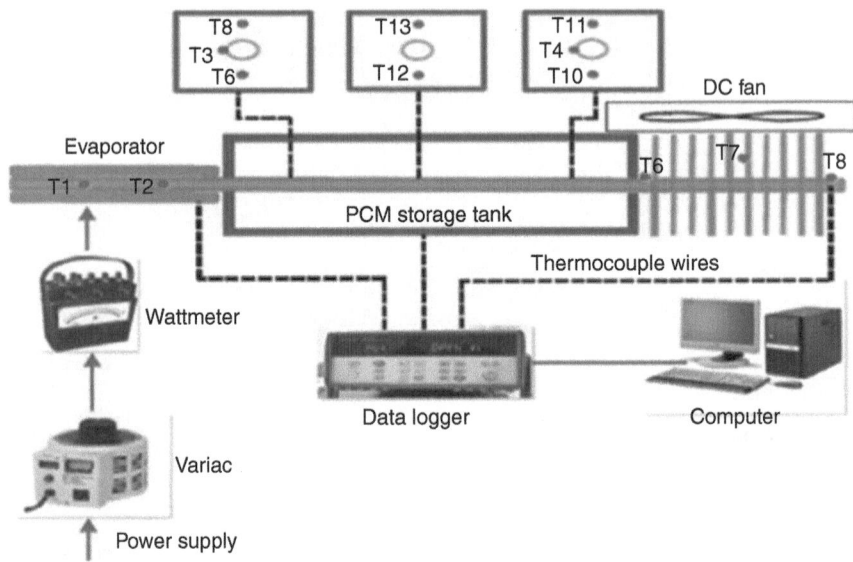

FIGURE 13.5 Schematic arrangement of heat pipe with Ne-PCM for cooling electronics.

Source: Krishna et al. (2017).

low thermal conductivity and also increases the melting and solidification rate compared with pure PCM. In the winter period, 8.3% and 25.1% more heat are absorbed and liberated, respectively, using Ne-PCM instead of pure PCM (Ma et al., 2016). The effect on air conditioning system performance of heat storage using PCM with silver nanoparticles has been estimated (Parameshwaran and Kalaiselvam, 2014). It is reported that 24–51% of daily average energy and 36–58% of on-peak energy can be retained by PCM without the use of nanoparticles. The building thermal mass is enhanced by Ne-PCM adopted in building materials for glass, roof, walls, and even floors. The judicious use of energy in buildings through envelope optimization is promoted by Ne-PCM. A ceiling board using paraffin with alumina nanoparticles was analysed by Guo (2011). It was found that a considerable amount of cooling energy could be saved in summer. A numerical and experimental investigation of Ne-PCM wallboard was done by Biswas et al. (2014). Heat losses and gains through building envelopes were observed. The results show a remarkable reduction in the electrical consumption; however, they apply to a region with long daytime temperatures. Another study was of a home ceiling ventilation system consisting partly of Ne-PCM layers. The system was interconnected with an air-based solar photovoltaic thermal collector for heating and cooling purposes (Lin et al., 2014). In yet another study, nano sheets with graphite and fatty acids were incorporated into gypsum wallboard to improve energy efficiency. Graphite at 8 wt% provided energy savings of 79% and also maintained satisfactory indoor thermal comfort (Sayyar et al., 2014). The reviewed literature indicates that Ne-PCM shows better thermo-physical performance such as peak load reduction, effective thermal comfort and potential energy savings when compared to pure PCM.

13.5   NE-PCM USE IN VARIOUS COMMERCIAL APPLICATIONS

Waste heat recovery is a potential source of energy savings that can be applied in various commercial fields such as automobiles, food preservation, textiles, medical and others. The contribution of Ne-PCM in waste heat recovery plays an increasingly vital role. Venkitaraj et al. (2018) studied the thermal performance of flue gas recovery from internal combustion engine exhaust by solid-state Ne-PCM. They used penta-erythritol as the PCM and mixed it with Al_2O_3 nanoparticles. Results revealed that the charging efficiency and energy savings are increased by 5.48–14.77% and 11.46–18.3%, respectively, for 0.1–0.5% of Al_2O_3. Also, there was considerable improvement in thermal degradation, energy storage capability and dispersion stability of PCM with Al_2O_3 nanoparticles. Johnston et al. (2008) studied the Ne-PCM for food paperboard packages during transport and food short-term storage using calcium silicate (30 wt%) dispersed into alkane PCM. The paperboard package was used to store 2 kg of asparagus. The researchers concluded that the prepared package for perishable food attains the expectation of thermal buffering.

Agrawal et al. (2020) reviewed Ne-PCM use in textiles under various sub-categories such as lamination, fibre technology and coating. They reported that the nano-encapsulated PCM plays a major role by avoiding leakage over the clothes surface during the phase change. The fabrics are prepared by sol-gel method using composite nanoparticles dispersed into thermo-regulating PCM. The thermo-regulating properties could be provided to fibre by adding Ne-PCM to a polymer solution before fibre extrusion. A self-assembled nanostructured PCM from regenerative medicine is emerging towards applications in electronic equipment, and this technology may help in building blocks interaction (Hirst et al., 2008).

13.6   FUTURE RESEARCH POTENTIAL

As the chapter shows, Ne-PCM plays a vital role in various applications as a potential thermal enhancer. The following are a few recommendations for future studies to improve the stability and reliability in thermal energy storage applications.

- Many research studies show that the thermal conductivity of Ne-PCM is higher than PCM, but the sub-cooling effect needs to be minimized to avoid negative performance.
- The maximum dispersion of nanoparticles leads to an increase in mass and volume concentration that causes a negative impact that needs to be studied.
- The addition of nanoparticles into the base PCM causes degradation in latent heat. But in a few studies the presence of optimum nanoparticle concentrations provides a higher latent heat capacity. Researchers should focus on the divergence between these findings to resolve it.
- The major disadvantage is the cost of Ne-PCM. Since the direct purchase of nanoparticles involves a high cost, it is recommended to synthesize the nanomaterial in an eco-friendly approach that minimizes the cost.
- Although the usage of Ne-PCM is found in many applications, no existing work has concentrated on the environmental impact. Hence, it is recommended to evaluate with inherent parameters.

- During continuous melting and solidification, dispersed nanoparticles tend to get damaged, which may affect the crystalline structure of Ne-PCM and reduce its thermo-physical properties after a certain number of repeated cycles. This is a major concern to be assessed with optimized techniques.
- The sedimentation and agglomeration of nanoparticles in PCM during charging and discharging is not mentioned in any existing research work. It must be included as a consideration during future experimentation.
- Future study of various thermal enhancements such as foams, fins and nanoparticle combinations with PCM can further improve the thermal energy storage applications.

13.7 CONCLUSIONS

In this chapter, the addition of nanomaterials into base PCM for improving thermal management in various applications is discussed in detail. The following observations are drawn as conclusions:

i.   Ne-PCM showed remarkable performance in solar thermal applications owing to a reduction in the melting rate. However, this requires a detailed study of the type of nanomaterial and different mass fractions to make a conclusive statement.

ii. Ne-PCM is used as a coolant in high power density applications, where its high heat extraction rate enhances thermal management and performance.

iii. Most studies reported that long-term stability and specific heat capacity at low temperatures can be achieved when nanoparticles are dispersed in PCM. In this case, it is proved that the Ne-PCM has a better performance in building applications.

iv. Ne-PCM is used as an effective thermal enhancer in various commercial applications such as automobiles, food preservation, textiles and medicine owing to the reduced latent heat of Ne-PCM by nanoparticles.

Many experiments and numerical studies have been reported recently on the enhanced performance associated with various applications. However, the challenges of using Ne-PCM, techno-economic aspects, and future directions are also discussed in this chapter. This will help researchers in further investigations of Ne-PCM.

REFERENCES

Abdelrazik, A. S., R. Saidur, and F. A. Al-Sulaiman. "Thermal regulation and performance assessment of a hybrid photovoltaic/thermal system using different combinations of nano-enhanced phase change materials." *Solar Energy Materials and Solar Cells* 215 (2020): 110645.

Agrawal, Rahul, Krishna Deo Prasad Singh, and Mani Kant Paswan. "Review on enhancement of thermal conductivity of phase change materials with nano-particle in engineering applications." *Materials Today: Proceedings* 22 (2020): 1617–1627.

Algarni, Salem, Sofiene Mellouli, Talal Alqahtani, Khalid Almutairi, and Ali Anqi. "Experimental investigation of an evacuated tube solar collector incorporating nano-enhanced PCM as a thermal booster." *Applied Thermal Engineering* 180 (2020): 115831.

Alimohammadi, Mahdieh, Yasaman Aghli, Elaheh Sadat Alavi, Mohammad Sardarabadi, and Mohammad Passandideh-Fard. "Experimental investigation of the effects of using nano/phase change materials (NPCM) as coolant of electronic chipsets, under free and forced convection." *Applied Thermal Engineering* 111 (2017): 271–279.

Al-Waeli, Ali H. A., Kamaruzzaman Sopian, Miqdam T. Chaichan, Hussein A. Kazem, Adnan Ibrahim, Sohif Mat, and Mohd Hafidz Ruslan. "Evaluation of the nanofluid and nano-PCM based photovoltaic thermal (PVT) system: An experimental study." *Energy Conversion and Management* 151 (2017): 693–708.

Babapoor, Aziz, Gholamreza Karimi, and Samad Sabbaghi. "Thermal characteristic of nanocomposite phase change materials during solidification process." *Journal of Energy Storage* 7 (2016): 74–81.

Biswas, Kaushik, Jue Lu, Parviz Soroushian, and Som Shrestha. "Combined experimental and numerical evaluation of a prototype nano-PCM enhanced wallboard." *Applied Energy* 131 (2014): 517–529.

Colla, L., L. Fedele, S. Mancin, B. Buonomo, D. Ercole, and O. Manca. "Nano-PCMs for passive electronic cooling applications." In *Journal of Physics: Conference Series*, vol. 655, no. 1, p. 012030. IOP Publishing, 2015.

Farzanehnia, Amin, Meysam Khatibi, Mohammad Sardarabadi, and Mohammad Passandideh-Fard. "Experimental investigation of multiwall carbon nanotube/paraffin based heat sink for electronic device thermal management." *Energy Conversion and Management* 179 (2019): 314–325.

Guo, Cha Xiu. "Application study of nanoparticle-enhanced phase change material in ceiling board." In *Advanced Materials Research*, vol. 150, pp. 723–726. Trans Tech Publications, 2011.

Hirst, Andrew R., Beatriu Escuder, Juan F. Miravet, and David K. Smith. "High-tech applications of self-assembling supramolecular nanostructured gel-phase materials: From regenerative medicine to electronic devices." *Angewandte Chemie International Edition* 47, no. 42 (2008): 8002–8018.

Johnston, James H., James E. Grindrod, Margaret Dodds, and Katrin Schimitschek. "Composite nano-structured calcium silicate phase change materials for thermal buffering in food packaging." *Current Applied Physics* 8, no. 3–4 (2008): 508–511.

Kannan, K. Gopi, and R. Kamatchi. "Augmented heat transfer by hybrid thermosyphon assisted thermal energy storage system for electronic cooling." *Journal of Energy Storage* 27 (2020a): 101146.

Kannan, K. Gopi, and R. Kamatchi. "Assessment of power harvesting in electronic modules using phase change material with null electricity: An experimental study." *International Journal of Renewable Energy Research (IJRER)* 10, no. 4 (2020b): 1755–1763.

Kannan, K. Gopi, and R. Kamatchi. "Experimental investigation on thermosyphon aid phase change material heat exchanger for electronic cooling applications." *Journal of Energy Storage* 39 (2021): 102649.

Krishna, Jogi, P. S. Kishore, and A. Brusly Solomon. "Heat pipe with nano enhanced-PCM for electronic cooling application." *Experimental Thermal and Fluid Science* 81 (2017): 84–92.

Lin, Saw C., and Hussain H. Al-Kayiem. "Evaluation of copper nanoparticles–Paraffin wax compositions for solar thermal energy storage." *Solar Energy* 132 (2016): 267–278.

Lin, Wenye, Zhenjun Ma, M. Imroz Sohel, and Paul Cooper. "Development and evaluation of a ceiling ventilation system enhanced by solar photovoltaic thermal collectors and phase change materials." *Energy Conversion and Management* 88 (2014): 218–230.

Liu, Lin, Ke Zheng, Yang Yan, Zihe Cai, Shengxuan Lin, and Xiaobin Hu. "Graphene aerogels enhanced phase change materials prepared by one-pot method with high thermal conductivity and large latent energy storage." *Solar Energy Materials and Solar Cells* 185 (2018): 487–493.

Ma, Zhenjun, Wenye Lin, and M. Imroz Sohel. "Nano-enhanced phase change materials for improved building performance." *Renewable and Sustainable Energy Reviews* 58 (2016): 1256–1268.

Parameshwaran, R., and S. Kalaiselvam. "Energy conservative air conditioning system using silver nano-based PCM thermal storage for modern buildings." *Energy and Buildings* 69 (2014): 202–212.

Praveen, B., S. Suresh, and Vignesh Pethurajan. "Heat transfer performance of graphene nanoplatelets laden micro-encapsulated PCM with polymer shell for thermal energy storage based heat sink." *Applied Thermal Engineering* 156 (2019): 237–249.

Sayyar, Mohammad, Rankothge R. Weerasiri, Parviz Soroushian, and Jue Lu. "Experimental and numerical study of shape-stable phase-change nanocomposite toward energy-efficient building constructions." *Energy and Buildings* 75 (2014): 249–255.

Sharma, S., L. Micheli, W. Chang, A. A. Tahir, K. S. Reddy, and T. K. Mallick. "Nano-enhanced phase change material for thermal management of BICPV." *Applied Energy* 208 (2017): 719–733.

Venkitaraj, K. P., S. Suresh, and Arjun Venugopal. "Experimental study on the thermal performance of nano enhanced pentaerythritol in IC engine exhaust heat recovery application." *Applied Thermal Engineering* 137 (2018): 461–474.

Vivekananthan, Mayilvelnathan, and Valan Arasu Amirtham. "Characterisation and thermophysical properties of graphene nanoparticles dispersed erythritol PCM for medium temperature thermal energy storage applications." *Thermochimica Acta* 676 (2019): 94–103.

Wang, Yaqin, Xuenong Gao, Peng Chen, Zhaowen Huang, Tao Xu, Yutang Fang, and Zhengguo Zhang. "Preparation and thermal performance of paraffin/Nano-SiO_2 nanocomposite for passive thermal protection of electronic devices." *Applied Thermal Engineering* 96 (2016): 699–707.

Xiao, X., P. Zhang, and M. Li. "Experimental and numerical study of heat transfer performance of nitrate/expanded graphite composite PCM for solar energy storage." *Energy Conversion and Management* 105 (2015): 272–284.

Yang, Liu, Jia-nan Huang, and Fengjiao Zhou. "Thermophysical properties and applications of nano-enhanced PCMs: An update review." *Energy Conversion and Management* 214 (2020): 112876.

14 Bioshells and Calcium-Based Composite Coating for Tribology Applications

M. Wasim Khan,[1]* A. Elayaperumal,[1a] M. Sivanesh Prabhu,[1b] A. Tajdeen[2c]

[1]College of Engineering, Guindy, Anna University, Chennai, India.

[2]Bannari Amman Institute of Technology, Sathyamangalam, India.

*Corresponding author: wasimmech2@gmail.com

Other email IDs: [a]profelaya@gmail.com, [b]sivanesh.research@gmail.com, [c]tajdeena@bitsathy.ac.in

CONTENTS

14.1 INTRODUCTION

In recent years, researchers have done comprehensive research on the composite materials in the material science field and found the solution to many intricate problems (Ye et al., 2021). Mechanical and tribological properties of the materials can be modified through bulk and surface modification processes (Elayaperumal et al., 2021; Jagatheeshwaran et al., 2017). The most followed surface modification techniques are physical vapour deposition (PVD), chemical vapour deposition (CVD), plasma arc coating, thermal spray and electroless deposition coating. Generally, PVD and CVD methods are preferred in various industrial applications to provide deposits for higher hardness and wear resistance. However, these techniques are more expensive, and the control of the coating process is highly complex. Hence, in recent years electroless coatings are used in the tribology applications.

DOI: 10.1201/9781003109723-14

243

The electroless deposition process is an autocatalytic process which was first developed by Brenner and Riddell in the year 1950. The electroless process is purely a chemical reduction process which has a wide scope in a variety of industrial applications. Also, the electroless process does not require any external source of electricity, unlike the other coating techniques. Its applications have extended to home appliances because of its extensive advantages such as uniform coating thickness throughout the surface and ability to coat even plastics, rubber material and irregular shapes of material with lower cost than the other surface modification techniques. Generally, there are three forms of the electroless coating process, namely, nickel phosphorus coating (Ni-P), nickel boron coating (Ni-B) and copper phosphorus coating (Cu-P), with Ni-P and Ni-B coatings playing major roles in tribological applications.

In the mechanical engineering field, aluminium (Singh et al., 2021) and magnesium (Tajdeen et al., 2020) are mainly used for structural application because of their high strength-to-weight ratio, good machinability and corrosion resistance. However, owing to poor wear resistance and low hardness, they are restricted in tribology applications. Hence, to improve the mechanical and tribological behaviour of the matrix materials, the surface is coated with ceramic particles such as SiC (Liu et al., 2019), $CaCO_3$ (Zhang et al., 2020), graphene (Prashar et al., 2021) and Al_2O_3 (Chen et al., 2021) to improve wear resistance and mechanical properties for use in the automobile and aerospace industries. Among the various reinforcements, calcium-based coatings for tribology applications have received much attention in recent years. Hence, this chapter discusses the role of calcium-based coatings on tribology performance.

Generally, the selection of coating material is based on the application; for example, materials with good biocompatibility are mostly preferred for biomedical applications and materials with good hardness and solid lubricant are used for enhancing wear resistance. In metal-based material, stainless steel was mostly used in the biomedical applications (artificial knee and hip joints) at the beginning stages of research owing to its good biocompatibility (Bekmurzayeva et al., 2018). Even though stainless steel has good biocompatibility, the depletion of chromium content into the body could cause severe problems for humans; hence, biocompatibility coatings were developed in recent years (Ahmed and Rehman, 2020). However, stainless steel has been replaced by titanium in recent years because of its high density.

Titanium and its alloys are widely used as biomaterials in bone tissue engineering owing to their good mechanical strength, light weight and good biocompatibility (Mohammad et al., 2021). Various surface modification processes are used to increase the metal's life span. Next to titanium, magnesium and its alloys are the metals most used in bone-implant applications, on account of their biocompatibility, good corrosion resistance, low density and adequate strength (Heimann., 2020), and the most used materials in coatings for the biomedical application are hydroxyapatite (calcium phosphate) (Bansal et al., 2021) and calcium (Salama, 2019). It is important to review the various coatings used in biomedical applications to extend the research in bio metallic pairs. Hence, this chapter discusses various studies focused on improving the biocompatibility of the surface through different surface modification processes.

14.2 CALCIUM-BASED COMPOSITE FOR TRIBOLOGY APPLICATIONS

The various coating techniques used for the deposition process are shown in Figure 14.1. The appropriate coating technique is generally selected based on the properties and application of the deposits.

Mild steel is used in various industries because of its low cost and availability. However, the poor resistance to corrosion and wear restricts the use of mild steel in marine and tribology applications. Hence, nickel-phosphorus/precipitated-calcium carbonate (Ni-P/P-CaCO$_3$) composite has been coated on the surface of the mild steel using an electroless coating process. The deposited material acted as a strong productive layer on the mild steel surface and prevented wear. The P-CaCO$_3$ acted as lubrication particles when sliding against EN31 steel. However, a higher weight percentage of P-CaCO$_3$ reduced the hardness of the coatings owing to the soft nature of CaCO$_3$ particles (Premkumar et al., 2019), which also influenced the wear resistance. Hence, better hardness and lubrication effect are important for coatings subject to tribology applications.

Calcium hexaboride (CaB$_6$) is a hard ceramic particle which is used as reinforcement in nickel-phosphorus/calcium hexaboride (Ni-P/CaB$_6$) composite prepared using an electroless coating process (Premkumar et al., 2020a). It is clear that both the wear resistance and hardness are considerably increased with the reinforcement of calcium hexaboride in the composite coatings. The lubrication and hardness of Ni-P/

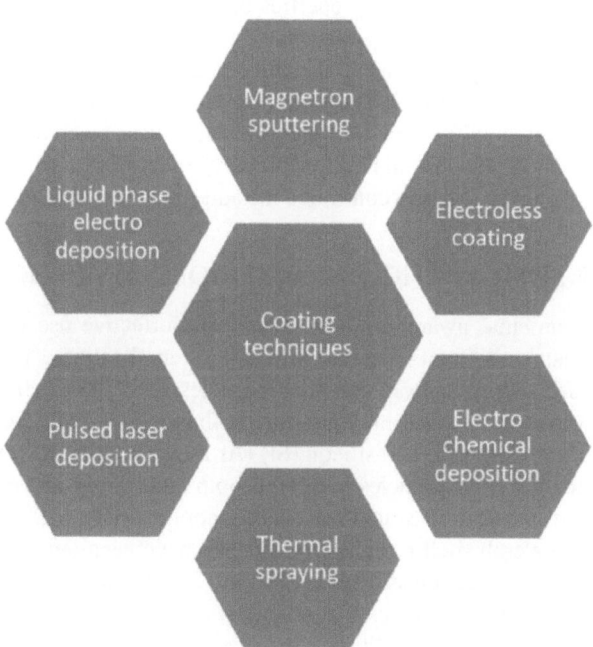

FIGURE 14.1 Coating techniques.

TABLE 14.1

Wear Mechanisms of Calcium-Based Coatings against EN31 Steel and Bone

Disc	Coating techniques	Pin (counterpart)	Sliding condition	Wear mechanism
Ni-P/P-CaCO$_3$–coated mild steel	Electroless coating process	EN31 steel	Dry condition	Adhesive
Ni-P/CaB$_6$–coated mild steel	Electroless coating process	EN31 steel	Dry condition	Abrasion
Ni-P/CaBr$_2$–coated mild steel	Electroless coating process	EN31 steel	Dry condition	Delamination
Calcium titanate–coated AISI steel	Magnetron sputtering	Bone	Both dry and wet conditions	Dry condition: adhesion and abrasion Wet condition: only adhesion

CaB$_6$ composite coating improved tribology behaviour better than the Ni-P/CaCO$_3$ coatings. However, at high load conditions, there is an exhibition of abrasive wear due to the interaction of hard particles (CaB$_6$) against the counterface (Table 14.1). This creates a high frictional force during the sliding, which can be mitigated using additional solid lubricant particles like MoS$_2$ or graphene.

In an electroless coating process, the use of soluble calcium bromide (CaBr$_2$) coating for tribological application is very limited. So, the influence of various weight percentages of CaBr$_2$ on the wear properties of low carbon steel coated with Ni-P/CaBr$_2$ composite has been investigated recently (Premkumar et al., 2020b). For the higher weight percentages of CaBr$_2$, the dissolved calcium bromide increased the surface roughness of the Ni-P/CaBr$_2$ composite coatings. However, the wear rate is considerably decreased with the CaBr$_2$ and the major wear mechanism reported was delamination wear. The semi-lubrication property of Ni-P was certainly enhanced with the CaBr$_2$ particles and also controlled the adhesive wear of the low carbon steel.

14.3   CRAB SHELL PARTICLES FOR TRIBOLOGY APPLICATIONS

Ongoing environmental awareness has induced the effective use of biowaste materials rather than traditional materials for various applications. Hence, extensive research has been carried out on crab shell particles (CSP) in the field of materials science. CSP consists of chitin polymer, magnesium calcite (MgCaCO$_3$) and some impurities of aluminium (Al) and silicon (Si) (Arulvel et al., 2018). An investigation showed that the crab shell particles consist of both amorphous and crystal structure and concluded that it is well suited for coating applications (Arulvel et al., 2016). Nickel-phosphorus/crab shell particles (Ni-P/CSP) have been deposited on the mild steel surface using an electroless coating process. The dry sliding wear behaviour of the prepared specimen has been explored against the EN31 steel using a pin-on-disc tribometer at various loading conditions. The severe adhesive failure exhibited for Ni-P coating at high loading condition has been controlled using the crab shell particles reinforcement. The reinforcement of crab shell particles improved the lubrication as well as the incompatibile pairing with the counterface. Importantly the friction force is low with better wear resistance, which is in contrast to other bio shell reinforcement (Jagatheeshwaran et al., 2015; Jagatheeshwaran et al., 2016).

14.4   COATINGS IN MARINE AND BIOMEDICAL APPLICATIONS

One of the greatest challenges in marine applications is to control the formation of biofouling. Biofouling implies consecutive deposition and growth of unwanted biological structures like biomolecules and microorganisms (bacteria, protozoa and algae). The growth of biofouling agents could possibly impact the density of the marine vehicle and also subject it to corrosion. Hence, biofouling-resistant coatings like 3-D-grafted, hydrophobic polymer; inorganic reinforcements; hierarchical spheres; and ZnO/acrylic polyurethane deposits have been used on various substrates (Huang and Ghasemi, 2020; Xie et al., 2021).

The biomedical devices (biosensors and bioimplants) used in the microbiology lab could also be affected by bacterial agents. To overcome this, a strong hydrophobic coating (antibacterial coating) has been used on the surfaces of biomedical devices in recent years. In recent decades, artificial joints have been fabricated using metals such as stainless steel (Ibrahim et al., 2019) and titanium (Suwanpreecha et al., 2021). Among them, titanium is a prominent biomaterial for bioimplant application because it acts as an antiseptic and biofouling-resistant material. Furthermore, titanium has adequate strength, light weight and good corrosion resistance (Pragathiswaran et al., 2021). Hence, both the femoral head and acetabulum in the hip joint (Figure 14.2) are artificially made using titanium and its alloys to replace the broken bone. But friction and wear rate are gradually increased by the metal on metal contact. This increasing frictional force shortens the life span of the prosthetic joint (Lee et al., 2015). To overcome this problem, biopolymers such as ultra high molecular weight polyethylene (UHMWPE) and chitin are deposited as a film in between the two metals to reduce the friction and wear rate as well as to increase the life span. Use of a biopolymer film gives a better result; however, after a certain duration, wear dominates in the polymer material, which leads to improper mobility of the prosthetic joint (Anstey et al., 2020). For that reason, a strong coating layer of suitable biomaterials is recommended on the surface of the titanium to increase the life span of the prosthetic joint.

The tribological behaviour of bone is analysed against calcium titanate–coated AISI steel under dry and wet conditions using a pin-on-disc tribometer at constant load. Calcium titanate is a biocompatible material and is successfully coated on the AISI steel using magnetron sputtering. For the wear test, the bone is used as a pin

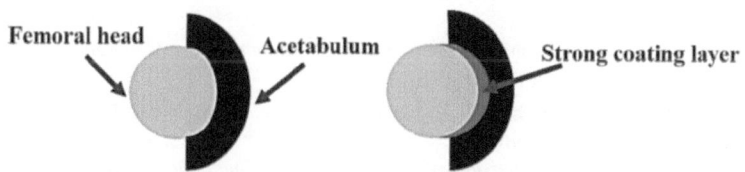

FIGURE 14.2   Schematic representation of hip joint and coating area.

material and a coated specimen is used as a disc. Adhesion and abrasion are the major wear mechanisms obtained under the dry sliding condition. However, only the adhesion wear mechanism occurred at the wet sliding condition. The experimental result revealed that the frictional force and mass loss of the pin material is low for the wet sliding condition compared to the dry sliding condition (Esguerra-Arce et al., 2015); owing to the presence of simulated body fluid in the wet sliding condition.

Hydroxyapatite (HAP) is the one of the most widely used bioceramic coatings in biomedical applications because of its good bioactivity. Furthermore, the crystallographic structure and chemical composition of HAP are similar to bone. HAP is coated on the surface of the titanium using the hydrothermal (HT) method. The hydroxyapatite coating induces the growth of bone in the area surrounding the implant (Arres et al., 2020). However, HAP does not have sufficient mechanical strength to withstand the loading condition. So, hard bioceramics such as strontium titanate (Senthilkumar et al., 2021) and zinc oxide (Bansal et al., 2021) are added with HAP to improve the mechanical and tribological properties. Through the coating techniques, the formation of wear debris during the tribology test can also be controlled, indicating the life span of the prosthetic joint will be increased.

14.5 CONCLUSION

After an extensive literature review on coating techniques and its application, the following conclusions are drawn:

- Depositing solid-state calcium lubricant particles on the surface of base materials reduces the friction and wear rate and decreases the hardness as a result of the particles' soft nature; it also increases the corrosion resistance.
- Depositing hard calcium ceramic particles on the surface of base materials provides excellent wear and corrosion resistance and also increases the hardness of the specimen in the coating zone.
- Biowaste material is also successful in increasing the wear and corrosion resistance owing to strong adhesion between the composite and base metals.

REFERENCES

Ahmed, Yusra, and Muhammad Atiq Ur Rehman. "Improvement in the surface properties of stainless steel via zein/hydroxyapatite composite coatings for biomedical applications." *Surfaces and Interfaces* 20 (2020): 100589.

Anstey, Andrew, Eunse Chang, Eric S. Kim, Ali Rizvi, Adel Ramezani Kakroodi, Chul B. Park, and Patrick C. Lee. "Nanofibrillated polymer systems: Design, application, and current state of the art." *Progress in Polymer Science* (2020): 101346.

Arres, Mar, Mariana Salama, Diogo Rechena, Patrizia Paradiso, Luis Reis, Marta M. Alves, Ana M. Botelho do Rego et al. "Surface and mechanical properties of a nanostructured citrate hydroxyapatite coating on pure titanium." *Journal of the Mechanical Behavior of Biomedical Materials* 108 (2020): 103794.

Arulvel, S., A. Elayaperumal, and M. S. Jagatheeshwaran. "Discussion on the feasibility of using proteinized/deproteinized crab shell particles for coating applications: Synthesis and characterization." *Journal of Environmental Chemical Engineering* 4, no. 4 (2016): 3891–3899.

Arulvel, S., A. Elayaperumal, and M. S. Jagatheeshwaran. "Controlling adhesive wear failure of nickel-phosphorus coating at high load condition using crab shell particle as reinforcement." *Engineering Failure Analysis* 90 (2018): 310–323.

Arulvel, S., A. Elayaperumal, and M. S. Jagatheeshwaran. "Comparative study on the friction-wear property of As-plated, Nd-YAG laser treated, and heat treated electroless nickel-phosphorus/crab shell particle composite coatings on mild steel." *Surface and Coatings Technology* 357 (2019): 543–558.

Bansal, Puneet, Gurpreet Singh, and Hazoor Singh Sidhu. "Improvement of surface properties and corrosion resistance of Ti13Nb13Zr titanium alloy by plasma-sprayed HA/ZnO coatings for biomedical applications." *Materials Chemistry and Physics* 257 (2021): 123738.

Bekmurzayeva, Aliya, Wynter J. Duncanson, Helena S. Azevedo, and Damira Kanayeva. "Surface modification of stainless steel for biomedical applications: Revisiting a century-old material." *Materials Science and Engineering: C* 93 (2018): 1073–1089.

Chen, Long, Hui Liu, Zhanqiang Liu, and Qinghua Song. "Thermal conductivity and anti-corrosion of epoxy resin based composite coatings doped with graphene and graphene oxide." *Composites Part C: Open Access* (2021): 100124.

Elayaperumal, A., S. Arulvel, and M. Wasim Khan. "Assessment on the impact of FSP process parameters on microstructural, mechanical and wear behaviour of FSPed AA6082." *Surface Topography: Metrology and Properties* 9, no. 1 (2021): 015016.

Esguerra-Arce, Johanna, Yesid Aguilar-Castro, William Aperador-Chaparro, Leonid Ipaz-Cuastumal, Gilberto Bolaños-Pantoja, and Carlos Alberto Rincón-López. "Tribological behavior of bone against calcium titanate coating in simulated body fluid." *Ingeniería, Investigación y Tecnología* 16, no. 2 (2015): 279–286.

Heimann, Robert B. "Magnesium alloys for biomedical application: Advanced corrosion control through surface coating." *Surface and Coatings Technology* (2020): 126521.

Huang, Zixu, and Hadi Ghasemi. "Hydrophilic polymer-based anti-biofouling coatings: Preparation, mechanism, and durability." *Advances in Colloid and Interface Science* (2020): 102264.

Ibrahim, Mahmoud Z., Ahmed A. D. Sarhan, T. Y. Kuo, Farazila Yusof, and M. Hamdi. "Characterization and hardness enhancement of amorphous Fe-based metallic glass laser cladded on nickel-free stainless steel for biomedical implant application." *Materials Chemistry and Physics* 235 (2019): 121745.

Jagatheeshwaran, M. S., A. Elayaperumal, and S. Arulvel., "Wear characteristics of electroless NiP/bio-composite coatings on EN8 steel." *Journal of Manufacturing Processes*, 20, Part 1 (2015): 206–214.

Jagatheeshwaran, M. S., A. Elayaperumal A., and S. Arulvel. "The role of calcinated sea shell particles on friction-wear behavior of electroless NiP coating: Fabrication and characterization.," *Surface and Coatings Technology*, 304 (2016): 492–501.

Jagatheeshwaran, M. S., A. Elayaperumal, and S. Arulvel. "Impact of nano zinc oxide on the friction–wear property of electroless nickel-phosphorus sea shell composite coatings." *Materials Science and Engineering: B* 225 (2017): 160–172.

Lee, Yoon-Seok, Mitsuo Niinomi, Masaaki Nakai, Kengo Narita, and Ken Cho. "Predominant factor determining wear properties of β-type and (α+ β)-type titanium alloys in metal-to-metal contact for biomedical applications." *Journal of the Mechanical Behavior of Biomedical Materials* 41 (2015): 208–220.

Liu, Zhenglong, Chengji Deng, Chao Yu, Xing Wang, Jun Ding, and Hongxi Zhu. "Molten salt synthesis and characterization of SiC whiskers containing coating on graphite for application in Al_2O_3-SiC-C castables." *Journal of Alloys and Compounds* 777 (2019): 26–33.

Mohammad, N. F., R. N. Ahmad, N. L. Mohd Rosli, M. S. Abdul Manan, M. Marzuki, and A. Wahi. "Sol gel deposited hydroxyapatite-based coating technique on porous titanium niobium for biomedical applications: A mini review." *Materials Today: Proceedings* (2021).

Pragathiswaran, Chelliah, Govindarajan Thulasi, Mysoon M. Al-Ansari, Latifah A. Al-Humaid, and Muthupandian Saravanan. "Experimental investigation and electrochemical characterization of titanium coated nanocomposite materials for biomedical applications." *Journal of Molecular Structure* 1231 (2021): 129932.

Prashar, Gaurav, and Hitesh Vasudev. "Surface topology analysis of plasma sprayed Inconel625-Al_2O_3 composite coating." *Materials Today: Proceedings* (2021).

Premkumar, A., A. Elayaperumal, S. Arulvel, and M. S. Jagatheeshwaran. "Partial dissolution of precipitated-calcium carbonate (P-$CaCO_3$) in electroless nickel-phosphorus (Ni-P) coating and its surface characterization." *Materials Research Express* 6, no. 6 (2019): 066409.

Premkumar, A., A. Elayaperumal, S. Arulvel, M. S. Jagatheeshwaran, and P. Seenuvasaperumal. "Calcium hexaboride reinforced nickel-phosphorus composite coating for increasing the wear properties of low carbon steel." *Materials Today: Proceedings* (2020a).

Premkumar, A., A. Elayaperumal, S. Arulvel, M. S. Jagatheeshwaran, B. Ramesh, and M. Sivanesh Prabhu. "Optimization of electroless bath process parameter for improving the tribology behavior of Ni-P/$CaBr_2$ composite coating against the hardened EN-31 steel." *Surface Topography: Metrology and Properties* 8, no. 2 (2020b): 025038.

Salama, Ahmed. "Cellulose/calcium phosphate hybrids: New materials for biomedical and environmental applications." *International Journal of Biological Macromolecules* 127 (2019): 606–617.

Senthilkumar, G., Gobi Saravanan Kaliaraj, P. Vignesh, R. Sudharshan Vishwak, T. Nivin Joy, and J. Hemanandh. "Hydroxyapatite–barium/strontium titanate composite coatings for better mechanical, corrosion and biological performance." *Materials Today: Proceedings* 44 (2021): 3618–3621.

Singh, Pratik, Raj Gupta, Saif Izan, Shubham Singh, Rajat Sharma, and Shashi Prakash Dwivedi. "Tribo-mechanical behaviour of aluminium-based metal matrix composite: A review." *Materials Today: Proceedings* (2021).

Suwanpreecha, Chanun, Enrique Alabort, Yuanbo T. Tang, Chinnapat Panwisawas, Roger C. Reed, and Anchalee Manonukul. "A novel low-modulus titanium alloy for biomedical applications: A comparison between selective laser melting and metal injection moulding." *Materials Science and Engineering: A Structural Materials: Properties, Microstructure and Processing* 812 (2021): 141081.

Tajdeen, A., E. Sakthivel Murugan, M. Wasim Khan, S. Praveen Kumar, M. Praveen Kumar, and S. Vidhyaa Prakash. "Optimization of machining parameters in electric discharge machining of magnesium and aluminium by Taguchi technique." In *IOP Conference Series: Materials Science and Engineering* 764, no. 1 (2020): 012052. IOP Publishing.

Xie, Chan, Changquan Li, Yu Xie, Zhenjun Cao, Shiqian Li, Jinsheng Zhao, and Min Wang. "ZnO/acrylic polyurethane nanocomposite superhydrophobic coating on aluminum substrate obtained via spraying and co-curing for the control of marine biofouling." *Surfaces and Interfaces* 22 (2021): 100833.

Ye, Kaixuan, Feng Li, Jin Zhang, Jianfei Chen, Zili Li, Gan Cui, Jianguo Liu, and Xiao Xing. "Effect of SiO_2 on microstructure and mechanical properties of composite ceramic coatings prepared by centrifugal-SHS process." *Ceramics International* 47, no. 9 (2021): 12833–12842.

Zhang, Kegui, Wenzhong Yang, Feng Ge, Bin Xu, Yun Chen, Xiaoshuang Yin, Ying Liu, and Huanzhen Zuo. "A self-curing konjac glucomannan/$CaCO_3$ coating for corrosion protection of AA5052 aluminum alloy in NaCl solution." *International Journal of Biological Macromolecules* 151 (2020): 691–701.

Index